61226213

Understanding Formal Methods

Springer

London
Berlin
Heidelberg
New York
Hong Kong
Milan
Paris
Tokyo

Written and translated by: Jean-François Monin
Translation editor: Michael G. Hinchey

Understanding Formal Methods

This work has been published with the help of the French
Ministère de la Culture - Centre national du livre

 Springer

Jean-François Monin, PhD
France Télécom R&D, Technopole Anticipa, DTL/TAL, 2 avenue Pierre Marzin,
22307 Lannion, France

Translation editor
Michael G. Hinchey, PhD, MSc, BSc
Software Verification Research Centre, University of Queensland, Brisbane,
Queensland 4072, Australia

British Library Cataloguing in Publication Data
Monin, Jean Francois
  Understanding formal methods
  1.Formal methods (Computer science)
  I.Title
  005.1'31
  ISBN 1852332476

Library of Congress Cataloging-in-Publication Data
A catalog record for this book is available from the Library of Congress.

ISBN 1-85233-247-6 Springer-Verlag London Berlin Heidelberg
a member of BertelsmannSpringer Science+Business Media GmbH
http://www.springer.co.uk

Translation from the French language edition of: Introduction aux méthodes formelles. © France Télécom
R&D et HERMES Science Publications, Paris, 2000 (2nd edition) and © Masson, Paris, 1996 (1st edition).

Typesetting: Camera-ready by author
Printed and bound at the Athenæum Press Ltd., Gateshead, Tyne and Wear
34/3830-543210 Printed on acid-free paper SPIN 10749290

Foreword to the First Edition

Slowly, but surely, formal methods are spreading from academic research to industrial applications. The need for certified software for security applications is driven by the increasingly large proportion of software in embedded systems, and by the exponential development of networks, whose reliability and security are essential for the modern economy. In these domains, where zero-defect software is a must, the high cost of these techniques is actually justified by the absolute necessity of certification. In the more traditional domains of software engineering, where zero-defect software is far from being the norm, development methods relying on a rigorous discipline of formal specification are profitable in the long run, thanks to the better structuring of the results, their greater robustness, their better documentation, which entails savings on maintenance and transfer operations, and their greater independence of languages and hardware. The algorithmic solutions are extricated from implementation choices and elaborated with a generality which favors their reuse in other applications. One may then talk about CASE tools, where logical specifications form the conceptual basis of the evolution of a system throughout its lifecycle, from the analysis of customer requirements through to the continuous adaptation to new environments and to new features.

This revolution in the design of software systems has already been successfully undertaken in the domain of hardware design, where formal methods are routinely used on a large scale. The corresponding revolution in software engineering is still to come, because mastering these abstract techniques and the difficulty in using associated tools hampers their penetration of an environment where traditional, or even obsolete programming techniques, die hard. Indeed, it is often tempting to "hack the bug with a patch" in order to urgently satisfy the complaint of a client, even if it means paying dearly, in the long term, for the disorder generated by such practices.

In fact, that part of software which is formally developed is currently tiny, in spite of the considerable amount of research and development which has been devoted to this technology since the 1970s. There is no well-established standard, and tools are still very much at the level of a cottage industry. In fact, the difficulty in learning very abstract methods and a bad estimation of scaling problems gave rise to a number of bitter failures, and even, to some extent, a phenomenon of rejection. The competent programmer feels his or her creativity hindered by the use of bureaucratic shackles that sometimes obscure, using a cryptic set-theoretic jargon, ideas which could be very clear if presented from a more operational perspective. When the notation hampers understanding, one runs the risk of losing the guiding thread of the control flow, and of carrying out symbol pushing to derive meaningless conclusions. Finally, tools are too often used erroneously, because their limitations are insufficiently understood.

There is then a sizeable gap between the specialists in these techniques and real-world engineers, who are pressured by deadlines and cost requirements. It is not easy to keep up with the evolution of tools coming from research laboratories, and in this area, professional offers are sparse and there is a lack of standardization. Comparative studies are rare, as are impartial experts, and the potential user of formal methods often has the impression of making his or her choices as if involved in a game of blind man's buff.

Jean-François Monin's book is therefore of great value, since it sets out a sizeable amount of the knowledge which has to be mastered in order to guide those choices. Far from being an exhaustive hotch-potch, this book proposes an overview of the general techniques for specifying and developing software by stepwise refinement in a modular manner, and elaborating formal proofs, illustrated by concrete examples explained using a number of representative tools. The example of table searching is very well chosen, because it is understandable to everyone, it is small enough to merit a complete treatment, but it is, at the same time, sufficiently complex for illustrating typical issues. The coverage of techniques is satisfactory, and methods are explained without ideological commitment or parochialism. This book relies upon a concrete knowledge of a significant number of tools, and it soberly presents a moderate point of view, without suffering from either the excessive enthusiasm of tool designers nor the exaggerated suspicion of overly pragmatic programmers.

This book is aimed at all those who are rightly puzzled by the complex and controversial panorama of formal methods. It is unique as to its completeness and its compromise between rigorous exposition of underlying mathematical theories and concrete explanations of the implementation of techniques using actual tools. One of its essential merits is to be an up-to-date presentation of the best currently available techniques, in a field where one could easily mistakenly choose an antiquated and rigid technology, or take the risk of a research prototype with an unknown lifespan.

This book is meant to become a reference book for the coming years, and I recommend it to all those who have understood that one should not delay adopting a technology which is unavoidable.

Gérard Huet
May 1996

Preface

This book is intended to help the student or the engineer who wants an introduction to formal techniques, as well as the practitioner who wishes to broaden her or his knowledge of this subject. It mainly aims at providing a synthetic view of the logical foundations of such techniques, with an emphasis on intuitive ideas, so that it can also be considered as a practical complement to classical introductory manuals to logic, which generally focus more detail on specific subjects (e.g. first-order logic), and to books dedicated to particular formal methods.

This book is a translation of the French edition *Introduction aux méthodes formelles*, published by Hermes in 2000. The contents have been updated and somewhat clarified, in particular the discussion of typing which is now at the beginning of Chapter 10.

Many colleagues, researchers, and friends, have had an influence on the form and the content of this text, either through direct comments or enthralling discussions. I would like to cite: Jean-Raymond Abrial, André Arnold, Yves Bertot, Michel Cartier, Paul Caspi, Christine Choppy, Thierry Coquand, Vincent Danos, Pierre Desforges, Gilles Dowek, Jean-Christophe Filliâtre, Laurent Fribourg, Roland Groz, Nicolas Halbwachs, Claude Jard, Gilles Kahn, Claude and Hélène Kirchner, Emmanuel Ledinot, Pierre Lescanne, Fernando Meijia, Max Michel, Kathleen Milsted, Chetan Murthy, Christine Paulin, Simon Pickin, Laurent Régnier, John Rushby, Natarajan Shankar, Joseph Sifakis, Jean-Bernard Stéfani and Daniel Vincent.

I am particularly indebted to Gérard Huet, who wrote the foreword and gave me precious hints. Special thanks to Didier Bégay, Pierre Castéran, Pierre Crégut, Thierry Heuillard, Francis Klay and Jean-Marc Pitié for their careful rereading of the French version, to Mike Hinchey for his considerable work on the translation, and to Catherine Drury, Mekanie Jackson and Rosie Kemp for their kind help.

Finally, I will not forget Caroline, Maxime and Wei, who brought me a band, a book, a small-scale model of a car and, above everything else, constant support.

Acronyms

ASN1	Abstract Syntax Notation 1
BDD	Binary Decision Diagram
BNF	Backus–Naur Form
CASE	Computer Aided Sofware Engineering
CCS	Calculus of Communicating Systems
CSP	Communicating Sequential Processes
CTL	Computation Tree Logic
HOL	Higher Order Logic
ISO	International Standardization Organization
LCF	Logic of Computable Functions
LTL	Linear Temporal Logic
LP	Larch Prover
PLTL	Propositional Linear Temporal Logic
PVS	Prototype Verification System
RRL	Rewrite Rule Laboratory
SADT	Structured Analysis Design Technique
TLA	Temporal Logic of Actions
VDM	Vienna Development Method

Table of Contents

1. Motivation

After a long gestation period, formal methods for software development have reached a maturity level sufficient for use in a range of real applications such as railway or aircraft transportation systems, telecommunications or energy. The fundamental ideas of formal methods have been known for a long time: they emerged with the first computers and have been studied since the 1960s. Independently of any cultural considerations, it transpired that putting them into practice required theoretical improvements as well as complex software support tools, whose principles and architectures became understood over the following decades, resulting in more and more effective prototypes, and, last but not least, machines endowed with powerful computational capabilities.

Various institutions are aware of the progress that has been made in the related technologies. In the domain of security, the European ITSEC (Information Technology Security Evaluation Criteria) has required the use of formal methods in its fourth security level, and above, since the mid 1990s. More recently, the Common Criteria for Information Technology Security, which have been in force as an ISO standard since 1999, recommend the use of formal models from its fifth security level, and above, and require the use of formal verification techniques at the seventh level.[1] By the end of the 1990s, industrial interest in these techniques had been confirmed and significantly widened. This could be observed, for example, on the occasion of the First World Congress on Formal Methods, in September 1999 [WWD99]. As new, and significantly more complex, application areas are emerging (smart cards, highly-secured information systems, robotics, e-commerce, aircraft control, etc.), one can see the increasing importance and relevance of formal methods. New techniques, theories and tools are being used in various applications, and these in turn provide feedback to the theory and evolution of formal methods and their associated proof tools. Nowadays, formal methods are applied in a whole plethora of systems ranging from complicated algorithms, of just a few pages in length, to software systems involving tens of thousands of lines of code. Let us illustrate the evolution of the technology with some industrial applications.

---

[1]The Common Criteria are the result of a joint effort of several countries, in North America, Europe, and Australia/New Zealand. Formal methods have also been mentioned in US security standards as far back as the 1980s.

1.1 Some Industrial Applications

1.1.1 Specification for Re-engineering

One of the oldest large-scale experiments is the CICS project undertaken at IBM (Huxley Park, United Kingdom), in collaboration with Oxford University. Its purpose was to perform a major restructuring of a large existing software system used for transaction management. The overall system was composed of about 800,000 lines of assembly language and of Plas, a high-level proprietary language. 268,000 lines were modified or rewritten, of which 37,000 made use of formal specification with the Z specification notation. Measurement procedures were introduced in order to evaluate the impact of a formal method on productivity and on quality. The quantitative results are detailed in [HK91]. They can be summarized as follows:

- development costs decreased by 9 percent;
- in the first eight months following the installation of the new version of CICS, in 1990, the clients reported 2.5 times fewer errors in the parts developed with Z than in the parts developed with non-formal techniques; moreover, these errors were perceived as being less serious.

This experiment is interesting because of the large amount of code involved. In contrast, its technical goals were rather limited: the issue was to specify software with the Z formal notation, and then to develop the code from the documents resulting from this phase; proof techniques were not taken into account.

1.1.2 Proving Critical Railway Software

When one tackles critical domains, involving human lives or having a potentially great economic or social impact, it becomes important to ensure the correctness of the executable code, or at least to give ourselves the strongest guarantees we can of this correctness. The code should possess no errors or deviations from intended behavior. One means of attaining this goal is to prove that it complies with a carefully written specification, on which competent persons involved in the development agree. Such a requirement entails a large amount of work. It is then important to give the whole system under consideration an appropriate structure, so that the areas where proofs will be performed are suitably delimited. The use of the B method by GEC-Alsthom, and more recently by Matra Transport International-Siemens, in projects such as the Calcutta subway [SDM92] or the Meteor line of the Paris subway [BBFM99] is a good illustration of this approach. The objective is to command and control the speed of a train by means of a device, which can be conceptualized, roughly, in the form of an uninterruptible sequence of instructions which run periodically. This is composed of a phase where pieces of input information are collected, followed by a phase where decisions are made, and finally a phase where commands are sent to physical control devices. It transpires that all of the complexity is concentrated in the second phase. This involves data transformation, which can be reasonably well modeled using the set-theoretic constructs available in Z or in B. However, B, developed more recently than Z, involves a process allowing executable code to be derived in a step-by-step manner; moreover, this code can be proven to conform to the initial specification, thanks to applicable support tools. The result of this procedure was several thousands lines of code written in the C language.

Note that, in the above example, reaction times are relatively long compared to computation times. In other applications, the constraints may be more strict; sometimes several devices have to be handled simultaneously and, generally, this greatly complicates matters. Other formal approaches, based on transition systems or on synchronous languages, for example, are well suited for dealing with such problems.

Finally, more complex applications, such as security components of network services, compilers, or support tools for formal methods themselves, involve both complex data structures and subtle behaviors. Using powerful logics becomes necessary, and we already know of a number of encouraging success stories using tools such as PVS, HOL and Coq.

1.2 What Is a Formal Method?

By "method", one generally means a process aiming at progressively reaching a given objective. For example, the method followed by a high-school student to solve a simple problem of mechanics consists of establishing the balance of forces, modeling them by vectors, then computing the unknowns using linear algebra or vector calculus. We must be aware that today, such a method, in the former sense, is still very underdeveloped in the case of formal methods for software construction. Such methods provide, essentially, a rational framework composed of tools to aid in modeling and reasoning, but they don't bring much from a methodological perspective. We will use the term *formal method*, because it is well established, but *formal technique* would certainly be more appropriate.

The domain of compilation techniques may be an exception. In order to construct a compiler, first the grammar of the source language is defined using suitable formal rules. After a possible transformation of the latter, an efficient parser is automatically derived thanks to general mechanisms determined in the 1960s. We have here all of the ingredients of a formal method. First, we obviously have a formal language for describing the grammar rules in a precise manner — a *BNF*, normal form of Backus–Naur. Furthermore, we have a well-understood *mathematical substratum*, which is the theory of formal languages and automata, and which provides the precise meaning of the grammar rules and justifies the general algorithms to be used. The formal methods we will consider in this book are all based on a formal language, including, for example, set-theoretic or logical notations, or more ad hoc concepts as in the case of the BNF formalism, together with a means of giving a precise mathematical meaning to every statement — its so-called *semantics*.

What can this be useful for? First, to communicate well: a rigorous semantics eliminates ambiguities, and it is an impartial arbiter. This is also an excellent guide for defining support tools. Finally, when a formal model of a system is available, the properties we expect from this system can be stated with precision, then formally verified. This leads us to say a few words on the role of formal methods within software engineering.

1.3 From Software Engineering to Formal Methods

Mastering the complexity and the cost of software proved to be a real technological and economical challenge; this gave birth to a well-established discipline, namely, software engineering. The practical aspects of this discipline are those most well known to developers: languages, compilers, CASE tools and support environments, development methods, programming techniques, methods related to quality management, etc. Design methods appeared: SADT, Jackson, object-oriented techniques, and others. These methods and techniques have non-negligible results to their credit, such as the following:

- a number of key notions have been recognized, for instance the concept of a lifecycle for software (commonly: requirements, specification, general design, detailed design, encoding, unit testing, integration testing, installation and maintenance);
- the introduction of rigorous methods in the production of software;
- the costs of the different stages have been evaluated and compared; for example, one estimates that maintenance takes up at least two-thirds of the overall cost of a software project, and that fixing a specification error requires twenty times more effort if it is detected after the installation stage, and sometimes even much more than that.

1.3.1 Towards More Rigorous Processes

The consequences of a software failure are not limited to recovery issues. In a number of cases (transportation, power plant command and control, medical systems), human lives are concerned. In the domain of telecommunications, major operators have experienced serious failures that entailed heavy losses — for example, the AT&T network in January 1990, following the installation of a new software upgrade to its switching systems. The sad fate of flight number 501 of the satellite launcher Ariane is yet another blasting demonstration that methods in current use are insufficient with regard to the high stakes of today.

We already mentioned that the later a mistake is detected, the more difficult it is to repair. This highlights the concern to devote a large amount of investment to the early stages of the software lifecycle, and the great importance of deriving *reliable specifications*:

1. which actually correspond to what is intuitively expected from the system; and

2. which are consistent.

The techniques considered in this book deal mainly with the second issue. These techniques start with a *formal* specification, and they allow one to develop software in a rigorous way based on this specification.

Regarding the first issue above, note that establishing good specifications necessitates a good knowledge of the users' needs, a knowledge that users themselves do not always possess from the outset. One may remedy this problem by confronting a formal specification with a number of simple properties which we expect. Such properties can be regarded as formal specifications themselves, though partial ones, because their scope is generally limited only to certain aspects.

One may also consider complementary techniques, such as rapid prototyping, in order to quickly develop an easy-to-modify version of the intended system. The most important feature of the technology to be used is then its ability to favor reactivity in the development process; considerations relative to cleanness or efficiency of the software may turn out to be awkward *at this level*. Beyond the stage of prototyping, the order of priorities changes, objectives of quality and rigor come to the forefront. However, it should be noted that, as a side result of formal approaches to computer science, programming languages which are simultaneously powerful, mathematically well defined, efficiently implemented and protected by a strong typing system are now available: functional languages, in particular languages from the ML family [CMP02, Pau91].

1.3.2 Software Development Using Formal Methods

Formal approaches allow one to write rigorous, precise, and complete specifications, and to develop software from them. The main component, as was already mentioned, is a formal specification language. The main benefits of these approaches are the following:

- a formal language comes together with a well-founded and safe semantics, particularly if it is based on well-tried mathematical theories;
- *proving* that the system under consideration satisfies intended properties becomes possible, at the specification level on the one hand, and at the code level on the other — the idea is to prove that a program conforms to a given specification; the latter issue may be tackled using several approaches: Hoare logic, enumeration of reachable states, refinement of specifications, program transformation, program calculation and program extraction;
- a formal language is a good basis for the development of support tools;
- efforts related to testing, maintenance, and sometimes coding may decrease significantly, since one gets a better control over these stages and since documentation becomes more reliable.

1.3.3 Formal Methods for the Customer

Formal methods are also of concern to organizations that contract their software development to others. Indeed, such organizations are mainly involved in the specification stages, and thus have to check that:

1. they are working with correct specifications;
2. the delivered product complies with the specifications.

Regarding the second issue, the customer must at least validate the product. To this end, the product is extensively tested. Designing and debugging test-cases becomes more complicated when the complexity of the desired product increases. Moreover, this is an error-prone and tedious task. There again, formal techniques can serve as a support tool. Automated generation of test cases from formal specifications is an active research topic, and industrial tools are available.

However, validating the product turns out to be insufficient. Tests can only verify that the behavior of the considered system is normal in a finite number of (hopefully) typical situations, but it can only tackle a partial view of the set of all possible behaviors. This can be sufficient for analog systems, which are continuous and regular, but software systems, which are essentially discrete, do not benefit from these properties.[2] In particular, it is illusory to think that software may be specified by the set of test cases to be used in order to validate it. Let us add that it could even be dangerous, because a malicious provider, or simply a provider in a hurry, may well deliver a system which behaves as expected, in the cases corresponding to the specified tests, but badly in other ones.

Clearly, a better perspective is obtained if a product is developed using a formal method: it can be delivered together with the proof that it satisfies the intended properties, for example, in a textual form that the customer may have audited by a contractor, or may check using automated verification software — recall that it is much easier to check a proof than to construct it.

1.4 On Weaknesses of Formal Methods

The previous arguments give some indications of the support which can be provided by formal methods for improving various stages of software development. However, we don't want to pretend that they constitute a miraculous remedy. When we are faced with complex problems, there is no simple way out.

First, we have to keep in mind that there always remains a distance between a formal specification, and the object it is supposed to represent. A similar well-known situation is true of the laws of physics: we cannot prove that they govern

---

[2]Of course, it is not enough to test all "branches" in the code, all *possible combinations of values for data and parameters* have to be taken into account. In general, there are an infinite number of them, or at least a number which is greater than current estimations of the number of atoms in the universe, which is quite a reasonable approximation to the infinite.

the real world, but it is quite reasonable to be confident that this is the case. The certainty of the correctness or appropriateness of a specification can be accepted as relevant only if it has been validated by a process composed of careful reading, reformulation, and confrontation.

When a new formal method is considered, the first obstacle to be overcome is to become fully acquainted with the notation. Beyond this stage, formal methods require an appropriate application, which includes pragmatic aspects — manipulation of tools — and theoretical aspects. Note, in passing, that the mathematical culture developed in traditional scholarly programs often favors analysis to the detriment of discrete mathematics. The situation is improving nowadays, but it is symptomatic that we still feel the need to inform about formal methods for software, whereas in other engineering disciplines, such as electronics or aircraft engineering, mathematical models are *naturally* applied. This acknowledges the rather experimental light in which programming is still commonly perceived.

Finally, let us note that with formal approaches, much more time is devoted to the initial phases of a development (specification, design) than in common processes. However, experiments show that this investment is (partly) compensated in later phases (tests, integration). Indeed, formalization reveals delicate issues very early, whereas, in a conventional lifecycle, these would have to be solved during debugging, or later. Many difficulties that are met when using a formal method are actually a reflection of difficulties that are inherent in the problem at hand. For example, modeling problems will occur just because the situation is intrinsically more complicated than it may appear at first sight. The introduction of complex or abstract concepts — often denoted by mathematical symbols — is then not that surprising. We will see that actual formal techniques offer various degrees of abstraction level and mathematical complexity. But to reassure the reader: basic concepts in logic and set theory, understandable to high-school students, are sufficient for a working knowledge of techniques such as B.

On[3] the issue of formulation, recall that the task of designing a judicious notation requires much care, though it is all too often neglected or overlooked. Both specification and programming languages may suffer from that. As this topic is rarely dealt with explicitly, let us mention here the books [vG90a], [Mey92] and [Set89].

1.5 A Survey of Formal Methods

There are various kinds of formal methods, which we can collect into several families. Most of them can be characterized by:

– an underlying prominent theory (examples: transition systems, set theory, universal algebra, λ-calculus);

---

[3]The meaning of the Möbius band is explained in § 1.7.

- a preferred application field (examples: data processing, real-time systems, protocols);
- a research and user community, themselves sometimes divided into several variants or schools.

We will not go into a detailed taxonomy of the domain, but we can suggest a number of design choices which determine important characteristics of most formal methods.

1.5.1 Specialized and General Approaches

The specification of a system includes various issues, including: architecture, interfaces, visible behaviors and algorithms to be implemented. Some formal methods consider systems which are presented from the outset in a given form, for example, in the form of data transformers, or of data flow, or even of finite state machines; information exchanges are supposed to be performed by data sharing, by synchronous or asynchronous message transmission, by function or procedure calls. Other formal methods stand back from such a view of the world, and limit themselves to a flexible general mathematical framework.

In the first category, one finds specialized formalisms, which may have been designed for protocols, for reactive devices, or for data handlers. This specialization favors the methodological aspects and the development of effective support tools, but it may have an undesirable effect: making irrevocable choices, which are relevant at a given stage of a technology, but may turn out to be a burden in later stages. For example, there are techniques for animating a formal specification: one then uses a so-called *executable specification*. But limiting oneself to the executable fragment of a general language tends to make some descriptions obscure, by forcing the use of ad hoc contortions. Thus, a convincing logical statement may lose much of its original clarity once it is translated to Prolog.

Conversely, methods closer to logic and mathematics offer much more freedom of expression. They have a big theoretical advantage, particularly when one has to model real systems and to reason about them, because reality often reveals an unexpected complexity. But such methods say nothing at the methodological level. The way of using them consists of reconstructing paradigms of specialized methods — with, sometimes, a suitable adaptation or generalization. It is also possible to combine several techniques, in order to work simultaneously on several facets of a given system using a unified framework. But this is still a topic for research.

Example: role of states. We can illustrate the distinction between general methods and specialized methods by means of the importance given to the concept of a *state*. It seems impossible to bypass this notion, since the systems that we want to model, which are a support for software (computers or virtual machines), or for their environment, are essentially memories whose contents change from time to time. On the other hand, this concept is not fundamental in mathematics, which lies in the realm of quantities, shapes, functions, all kinds of spaces — in summary, *immutable values* in a wide sense. It does not mean that, in mathematics, we are unable to talk about states. In general, state changes are represented by a trajectory, that is, a value from a suitable space.

The decision to attach more, or less, importance to states, is significant in practice, because transformations with side effects are rather more complicated to compose than pure (side-effect-free) transformations. When one writes "let $x = 3$" and "let $y = x + 1$", it is absolutely certain that, in the considered scope, the value of x is 3, and that x and y are related by the equation $y = x + 1$. In contrast, if one states "let x be a memory cell which contains 3" and "let y be a memory cell which contains the value of $x + 1$", one can no longer understand the produced effect, without meticulously examining how x and y may be transformed in every state change. This increase in the difficulty is one of the main motivations for introducing simultaneous assignments in imperative languages: it diminishes the number of intermediate states that need to be considered. This idea was proposed by Dijkstra and reused in B (see § 4.3.2 and § 6.3.3). It also explains the interest of functional programming: in its pure and strict version, it consists of describing computations on values; actually, most functional languages include imperative features, because it is sometimes convenient to keep some values in memory and to have side effects. Hence such programs include states, but a good programming discipline limits their impact to a very limited number of areas.

The first formal methods we will consider, Hoare logic or B for example, handle an implicit state. In others, states play an essential role out of necessity: they aim at studying behaviors, and a behavior is nothing but a sequence of states. Some of them will be considered in Chapter 8. Finally, the more abstract formal approaches, such as algebraic specifications, or higher-order logic-based languages, have no predefined concept of a state.

1.5.2 Emphasizing the Specification or the Verification

A formal method is composed of two main ingredients: a specification language and a verification system. The development of these two components is of varying importance depending on the approaches and the associated tools. Thus, the proof assistant of Boyer and Moore puts the emphasis on automating proofs to the detriment of ease of expression. In contrast, the first goal in the design of Z, was to get a very expressive language, but it turned out to be difficult to develop support tools for this language. The first versions of Boyer-Moore and of Z go back to the 1970s.

More recent approaches, such as HOL or Coq or PVS, attempt to provide both advantages: they are based on very powerful logics, together with support tools which aid the user in developing proofs, and some of them are able to check the correctness of the proofs in a very reliable manner.

In this book, we will pay more attention to specification than to automated verification mechanisms. In particular, we will ignore the Boyer–Moore approach, though it can be credited with remarkable successes, such as the partial verification of a complex system, where a hardware processor, an assembly language, a toy Pascal-like language and a basic operating system kernel are stacked.

1.6 Aim of this Book

How does one get one's bearing in the maze of available techniques? Each of them deserves a whole book to describe its foundations and practice. Such books already exist for many of them. On the other hand, it would probably be fruitless to try to tackle all approaches, even if we limit ourselves to a brief presentation. Our aim here is to propose a synthetic view of the subject, by following logic as our main thread. Logic has an influence on all formal methods, and often a direct one. At the same time, logic allows us to understand important and subtle phenomena which occur in practice.

Beyond logic, other mathematical theories play an important role in some formal techniques: notably, algebra and automata. They will be mentioned in order to provide some perspective. At the same time, it should be emphasized that logic has various other application fields in computer science, such as databases, operating systems, and programming languages.[4]

The importance of the different aspects of logic varies a lot, depending on the particular techniques one considers. For example, set theory is essential to formalisms such as Z or B, while intuitionistic logic is a more appropriate basis for the study of typed functional languages and corresponding specification languages. These two approaches share a number of concepts, but they actually belong to different logical traditions, which go back to the beginning of the 20th century.

The reader should find here an overview of logical disciplines which are relevant to computer science, and, more specifically, to formal methods. The aim of covering such a wide domain is moderated by the modesty of the technical contents: most theoretical results are given without demonstration. We hope that the reader will be inspired to gain a deeper knowledge of those topics. We have tried to give appropriate references to the literature, in particular at the end of every chapter.

We tried, whenever possible, to rely on a common simple example: the search for an element in a table. In order to shed light on concepts, without swamping them by irrelevant details, it appeared preferable that the example be as simple as possible. Obviously, the benefits of formalization would be better illustrated on a larger size problem. Indeed, very little will be told about how to tackle a large-sized application in a formal manner. Thus, although we will sometimes give an appreciation of a formalism, it should be clear that we don't have a

---

[4]For example, modern implementations of the functional language ML, which was initially designed from purely logical considerations, can be elegantly and efficiently used in software applications composed of system calls, network modules, and human-machine interfaces. Such examples are the file synchronizer Unison [PJV01] and the Web browser MMM [LR98].

goal of providing a comparative study, which is definitely beyond the scope of this book. The table example is only a support, not a benchmark!

1.7 How to Read this Book

We tried to make this book as self-contained as possible. The three first chapters contain introductory material, including elementary mathematical reminders (in § 3.4). Then, the general idea is to alternate the presentation of (the basics of) concrete formal methods with chapters devoted to their logical foundations. Occasionally, we need to introduce a concept that is not discussed in detail until a later chapter. In such cases, we will provide an intuitive explanation, which should suffice.

Chapter 2 introduces basic concepts related to specification and verification, in an intuitive and semi-formal manner.

The different branches of logic are presented in Chapter 3.

Chapter 4 is devoted to proving the correctness of imperative programs using formal assertions. The ideas contained in this approach, mainly due to Floyd, Hoare and Dijkstra, have an influence on all other techniques.

Chapter 5 presents so-called classical logic, which is a reference for all other logics.

Chapter 6 deals with formal methods based on set manipulations, namely Z, B and VDM.

Chapter 7 is devoted to set theory.

We then propose, in Chapter 8, a synthetic view on formal techniques for specifying complex behaviors, based on transition systems and on temporal logic. More specifically, we consider formalisms such as Unity, TLA and CCS.

Chapter 9 is an introduction to proof theory, which not only provides the essential concepts for understanding computer-aided proof systems, but serves as a foundation for typing systems and computational aspects of logic, to be considered in the last two chapters.

Chapter 10 is essentially a short presentation of the algebraic approach to formal methods, with an emphasis on abstract data types.

The discussion on typing started there is continued in Chapter 11, where we present its relation to λ-calculus and to higher-order constructive logic.

Finally, Chapter 12 is devoted to an implementation of these principles in a very expressive logic, the calculus of inductive constructions, which is supported by proof assistants such as Coq and Lego. This chapter ends with a brief account of other formal techniques based on a higher-order logic, more specifically HOL and PVS, and ends with some research perspectives.

The reading difficulty may vary a lot from one section to the next. The reader already acquainted with basic concepts may skip sections presented in this font; they are also identified by the symbol:

Paragraphs that may be postponed until a second reading, such as somewhat technical asides, are identified by a Möbius band:

Finally, pitfalls are indicated by the following symbol:

We also tried to follow a consistent discipline in our use of fonts. Here are some samples:

- a defined term, for example: **a left gyrating dahu** is a quadruped whose left legs are much shorter than its right legs;
- an arbitrary mathematical object: this is represented by a letter such as d or L;
- a program or a formal specification component, which would be entered on the keyboard: `if l<r then theta:=theta+1;`
- a formal language or a support tool: the method **myth**.

1.8 Notes and Suggestions for Further Reading

A report of the US National Institute of Standard and Technology presents, in its first volume [CGR93a], a set of formal techniques having industrial applications. Its second volume [CGR93b] collects several case studies which were performed prior to 1993. [Rus93] is another useful document on formal methods, written for NASA — and more oriented towards the needs of aerospace systems. It contains many interesting ideas, even if its author claims that it is sometimes biased by his involvment in a particular approach.

The book by Lalement [Lal93] allows one to obtain a deeper knowledge of many topics introduced here. In it, one may find complementary concepts on equational logic, rewriting and resolution. A handbook devoted to mathematical logic for computer scientists has been published [AGM92a, AGM92b]. For a broader introduction, one of the best references is the *Handbook of Theoretical Computer Science* [vL90a, vL90b] which, as indicated by its name, covers all theoretical bases of computer science, far beyond logic. We particularly recommend the second volume [vL90b], devoted to formal models and semantics. Chapters 1 to 14 and 16 to 19 are very readable.

A number of topics are not covered here, even though they could be considered relevant, because tackling them would have carried us too far from our path. This is true in the case of category theory. Developed from the middle of the 20th century, partly for establishing the foundations of mathematics on a more structured basis than set theory,[5] it still plays quite an important role in

---

[5]The initial motivation was actually different: the idea was to transfer results from group theory to topology, in order, for example, to classify geometrical shapes.

theoretical computer science, notably in algebraic specifications [EM85, EM90] and in typing systems [LS86, Hue90, AL91]. The basic reference is [Mac71], intended for mathematicians, but computer science-oriented introductions have been available for several years, amongst them [Hoa89] and [BW90]. The aforementioned manual [AGM92a] also contains a chapter devoted to category theory.

2. Introductory Exercise

A new problem is always tackled, at the outset, via both intuition and empirical methods. The design of software systems is no exception. The first step is to determine the object[1] to be realized. We then have to describe it. Most of the time, one employs the usual means of expression to this effect: our mother tongue, explanatory diagrams. Subsequent steps are devoted to code writing, generally using a high level language. An intuitive understanding of the language constructs is then key. Of course, people involved in this process employ some reasoning: "in that *case*, such an event happens, *then* ... etc."

We will proceed in this manner with an elementary case study. We will introduce — or recall — step-by-step the rudiments of logic and set theory which make up the framework of formal methods, demonstrating how they can enhance specifications and programs: they simply allow one to describe things and to reason in a better way.

In this chapter, concepts are introduced in an intuitive way, with more rigorous definitions coming in later chapters. Our aim is not to solve everything, but to raise a number of questions.

2.1 Exposition

The exercise we propose is quite simple, viz. the search for an element in a table. This is a very banal problem, but we can nevertheless already observe a classic pitfall. This can be illustrated with the following dialog, where S. is in charge of the specification and R. is responsible for the realization (program).

> S.: "Please write a program to search for an element in a table!"
> R.: "What kind of table? A list? An array? A tree? Are the elements sorted? Do they have a key?"
> S.: "I don't want to consider these implementation issues. That is your job."
> R.: "But what should be done if the sought element is not in the table?"
> S.: "Sorry?"

---

[1] In this book the word *object* is to be understood with its usual meaning, without regard to its connotation in computer science.

S. faces the following dilemma:

- either, he plays R.'s game, and then may well end up doing R.'s work;
- or, he sticks to his guns, and R. may well make irrevocable choices — perhaps unconsciously — which could later turn out wrong for S.

2.2 Sketch of a Formal Specification

The formulation given by S., as stated above, is too vague. We need to make it more precise, without going into algorithmic details. Let us see how elementary mathematical concepts could help.

A table is a collection of objects organized in some way. A general mathematical concept for organized collections is that of a *structure*, that is, a set endowed with composition laws. Let us ignore the laws at the moment: they are about organization and we still don't know how to organize the table.

Instead of "collection" we will use the word "set". A set, intuitively, is a collection of objects, termed its elements. What is the point of replacing the word "collection" by another? Actually, there is a whole body of well-established definitions, notations, properties and techniques. This allows us to manipulate sets and reason about them in a secure way. Moreover we will see in Chapter 7 that a collection is not necessarily a set. The statement x *is a member of E* is denoted by $x \in E$.

If we represent the table by a set T, we already know that the element to be found is an x such that $x \in T$. But the previous specification "search for an element in a table" implicitly tells us that we don't want an arbitrary element. In order to characterize it we make use of a property we expect of it. Which property we choose matters little here. In any case, the element has to exist and we must also be able to check whether or not the property holds on given elements.

We formalize this property using the concept of a *predicate*: we introduce a symbol, say P, and make $P(x)$ denote the fact that x satisfies the property P. P is called a predicate symbol.

In summary, we introduce a set T which represents the table, a predicate P defined over T, and we have to search for an element x, which is a member of T and such that $P(x)$. In later chapters we will see how this specification can be expressed in real formal specification languages. For the moment we will content ourselves with a semi-formal presentation, that is, a mixture of formulas (especially in line 4) and informal text.

```
1   T: set   (read: T is a set)
2   P: predicate defined for all elements of T
3   table-search-program
4   x ∈ T and P(x)
```

In line 3 we have the unknown: the expected program. In lines 1 and 2 we have two assertions stating what we know before the execution of the program: they are called the **preconditions**. In line 4 we have another assertion, the **postcondition**, to describe the result. The desired program is then specified by a pair ⟨precondition, postcondition⟩. This is one of the basic principles of formal specification.

What does it meaning? In a real-life (and complete) formal specification, assertions would be *logical formulas*, that can be assigned a mathematical meaning — a *semantics*. For the moment let us content ourselves with their intuitive meaning, as we stated previously. This specification is concerned with the state of the world, or merely that tiny part of it we are interested in here. In concrete terms, it is just computer memory, or at least an abstract version of it. The precondition[2] states here that the state has two components, a set T and a predicate P, whereas the postcondition states that it contains an additional component, the element x; moreover, T, P and x must satisfy the aforementioned conditions. The meaning of a specification expressed in this form ⟨precondition, postcondition⟩ is then:

> *If the program is executed from a state satisfying the precondition,*
> *then, after execution, the state reached satisfies the postcondition.*

Remark. The properties of T and P are actually *invariants* of the program we desire: the latter should return x without changing anything about T and P. Otherwise R. could plainly return a table containing just 0, the predicate "null" and $x = 0$. In order to prevent this, let us rephrase the lines 3 and 4 in the form:

> `program... returns x with postcondition,`

and we agree that everything outside `returns` and `with` is invariant.

```
1    T: set
2    P: predicate defined for all elements of T
3    table-search-program returns
4    x with x ∈ T and P(x)
```

Our new specification indicates what is necessary at this stage and nothing more. No premature design decisions involving a specific representation are made. However, this is more precise than the informal text as a result of the use of mathematical concepts — albeit elementary ones. R. can take advantage of it so long as the implementation data structures faithfully represent sets, elements or predicates. For example, it is easy to convince oneself that a list, an array, or a tree, can represent a set.

 This intuition can be rigorously confirmed by assigning a mathematical meaning to programming statements. This is the topic of

---

[2]The reader having some knowledge of logic may be somewhat reluctant to consider the declarations (e.g. in line 1) as components of logical formulas. This is, however, legitimate in some powerful logics, such as the ones we consider towards the end of this book. For the time-being, it is easier to interpret this as a slight abuse of language.

semantics, particularly denotational semantics. We return to this at the end of the chapter.

In summary, in order to eliminate the original dilemma, the trick was to consider the *correct level of abstraction*. One of the main assets of logic and related mathematics is their provision of a large palette of abstraction mechanisms.

2.3 Is There a Solution?

We still did not answer R.'s last question. Let us reformulate it as follows: what happens if there is no member x of T such that $P(x)$? Several approaches can be considered.

2.3.1 Doing Nothing

Let us first analyze the meaning given to the specification above ⟨precondition, postcondition⟩:

> *If the program is executed from a state satisfying the precondition,*
> *then, after execution, the state reached satisfies the postcondition.*

For this discussion we just need to recognize its logical shape: it is an implication, $A \Rightarrow B$.

A formula such as $A \Rightarrow B$ means "if A then B" and is read A **implies** B. Here A represents the assertion "initially, T is a set, P is a predicate and P is defined for all elements of T"; B represents the assertion "after execution x satisfies $x \in T$ and $P(x)$"; to be more rigorous we should repeat the constraints of A as part of B: see Remark on page 17. This omission has no consequence in what follows.

The use of "after" could suggest that time plays an important role here. On the contrary, we must forget about time because we want to retain the usual framework of plain logic, which is sufficient for our current needs (time will be considered in Chapter 8). We then adopt the viewpoint of an omniscient creature able to consider simultaneously all past, present and future events. Whether this event occurs before that event is no more important than whether this value is smaller than that one.

How can we formalize B, which has two components, "after execution" and "x satisfies $x \in T$ and $P(x)$"? The first term raises a problem because a program may well not terminate its execution — we say that it loops — or may terminate its execution in an abnormal way, for instance as the result of an interrupt. This can happen, for example, if there is an attempt to divide a number by zero. A possible interpretation of "after execution" which takes this into account is: "if the execution of the program terminates, then ...". This is called **partial correctness**.

Let us investigate the consequences of this interpretation. Formally, B can be decomposed into $B_1 \Rightarrow B_2$.

We take here as B_1: "the execution of the program terminates" and as B_2: "x satisfies $x \in T$ and $P(x)$".

It matters little that we don't know whether the postcondition B_2 is false or true: if B_1 is false, B is true whatever the truth value of B_2 — we return to this basic fact in § 3.4.2. As a consequence, R. has the freedom to provide a program which loops or aborts if there is no x in T such that $P(x)$. Actually R. even has the freedom to exaggerate this problem: he could deliver a program which loops in all cases. Of course this is not satisfactory.

2.3.2 Attempting the Impossible

S. could consider that the previous interpretation of B is too wide and then add to his requirements.

"I want your program to terminate[3] normally *and* return an element in the table satisfying P."

This is called **total correctness**. Formally, S. suggests $B_1 \wedge B_2$ (read *A and B*) instead of $B_1 \Rightarrow B_2$. However R. can quite reasonably reply:

"That's impossible: you might as well ask me for the moon on a silver platter!"

Indeed, there are specifications which are unfeasible. Again, division by zero is another example of this kind: "find x such that $ax = 2$" is impossible to realize when $a = 0$ is allowed. In each of the above examples something is required which may not exist. There are more subtle cases of unfeasible specifications. Take a program $\mathcal{P}$, written in the language of your choice and containing a numerical variable. Now ask the question: "will the value of this variable be null during execution?" There is an answer, either yes or no. But in general there is no program for computing it.

It is not sufficient to execute $\mathcal{P}$ and to test the value of the variable at each execution step. The program may well perform many, many computations before finding an assignment to zero. How can we be sure that the next step will not be the last one in this seemingly endless execution?

These somewhat tricky issues are the concern of computability theory, which we tackle in § 3.3.4.

2.3.3 Weakening the Postcondition

Our current specification is unsatisfactory, but we can still try to modify it rather than completely reject it. Total correctness is preferable, so we start with our second interpretation. As the specification is unfeasible, that is, too strong, we will weaken it. The first thing we can do is to weaken the postcondition.

---

[3]Implicitly: "I want *the execution* of your program to terminate." In the following, "program termination" always refers to the termination of executions of that program.

In other words, we will ask that the program returns an x which does not necessarily satisfy $(x \in T) \wedge P(x)$. But, for the program to be useful, we will ask for an additional piece of information that tells us whether x satisfies the required property or not. More precisely, we ask the program to return not only x, but an ordered pair $\langle b, x \rangle$ where b is a Boolean which is true if $(x \in T) \wedge P(x)$ and false otherwise.

It is clear that the postcondition on x is weakened. What about b, which was not even mentioned before? For the sake of comparison, we can suppose that the previous specification asked also for a fake b without any constraint. The last line of the specification would then have been:

$$\langle b, x \rangle \text{ with } (x \in T) \wedge P(x).$$

As the new specification puts a constraint on b, we conclude that the postcondition on b is *stronger*.

The set of Booleans is a set with exactly two elements representing the truth values *true* and *false*. This set is denoted by $\mathbb{B} = \{true, false\}$. More generally one can define a set E by listing its elements in any order. We use the notation $E = \{e_1, e_2, \ldots e_n\}$. This kind of definition is called **by extension**. Only finite sets can be defined in this way. The empty set is often denoted by $\varnothing$ instead of $\{\}$.

A number of programming languages such as Pascal have a built-in boolean datatype. In other languages, such as C, the values *true* and *false* are encoded by the integers 1 and 0, respectively.

Here is the new specification:

```
1    T: set
2    P: predicate defined for all elements of T
3    table-search-program returns
4    ⟨b,x⟩ with b ∈ {true,false}
5           and (x ∈ T) ∧ P(x)        if b=true
6           and (∀ x ∈ T) ¬P(x)       if b=false
```

This possibility, the most satisfactory for S., will be investigated in § 2.4.4 under a somewhat different, but equivalent, form.

The formula at line 6 (literally: **for all** x in T, **not** $P(x)$) means that no x in T satisfies $P(x)$. The set of ordered pairs $\langle a, b \rangle$ where $a \in A$ and $b \in B$ is denoted by $A \times B$, it is the **Cartesian product** of A and B. Be warned that order matters: $\langle a, b \rangle \neq \langle b, a \rangle$. The other important set-theoretic constructs involving two sets A and B are the **intersection** $A \cap B$ and the **union** $A \cup B$; $A \cap B$ is the set of elements which are both members of A and B, while $A \cup B$ is the set of elements which are members of A or B (or both).

2.3.4 Intermezzo: Sum of Sets

Here we have the opportunity to present a simple and key concept, which is ubiquitous in computer science, but often in a hidden form and then, unfortunately, largely underestimated: the sum of two sets,[4] also called their disjoint union.

The ordered pair $\langle b, x \rangle$ is not quite so simple. We could consider it as a member of $\mathbb{B} \times T$. This is not very accurate. When $b = \textit{false}$, nothing is known about x, so we have no reason to suppose that $x \in T$, especially not when $T = \varnothing$!

Let us temporarily forget our previous implementation of the result by the means of an ordered pair. The key idea is that the result is either an element of T *or* the representation of a failure. Let us call R its domain. Can we take $R = T \cup \{\textit{failure}\}$, where *failure* is a value as well as *true, false* and elements of T, rather than $R = \mathbb{B} \times T$? Almost: it works on condition that *failure* is not already a member of T, otherwise nothing could distinguish it. This can be handled at the level of the precondition, but we often prefer to avoid additional constraints. We then introduce a construct combining two sets A and B and providing a way of recognizing where an element comes from. In particular, common elements of A and B will be distinguished. Such a set is called the **sum** of A and B and is denoted by $A + B$. Let us illustrate the idea on $R = T + \{\textit{failure}\}$, which is relevant in our example. Only lines 4 to 6 of the previous specification are modified:

```
1    T: set
2    P: predicate defined for all elements of T
3    table-search-program returns
4    r with r ∈ T + {failure},    such that
5       P(x)                    if r comes from (element x of) T
6       (∀ x ∈ T) ¬P(x) if r comes from {failure}
```

The sum is not a primitive concept in set theory; it is built upon other constructs. The most natural way to proceed is to tag elements of A and B with different tags. Let us call the tagged sets A_T and B_T. Then we take $A + B = A_T \cup B_T$. The tagging operation maps an element x to an ordered pair $c = \langle t, x \rangle$, where t is the tag chosen for x, e.g. *true* if x is taken from A or *false* if x is taken from B.[5] In order to know where c comes from, we just have to check its first component t. Then we again get our specification (page 20).

In summary, $A + B$ is a *subset* of $\mathbb{B} \times (A \cup B)$:

$$A + B = (\{\textit{true}\} \times A) \cup (\{\textit{false}\} \times B) \ .$$

It is easy to generalize this construct to multiple sums and it turns out to be quite useful when one needs to describe data that can take several different formats.

---

[4]Later we consider the sum of two types, but the basic idea is the same.

[5]The choice of *true* and *false* is completely arbitrary, but it happens to be consistent with the specification on page 20.

Note also that, an x of $A \cap B$ yields two distinct elements of $A + B$, $\langle \textit{true}, x \rangle$ and $\langle \textit{false}, x \rangle$.

2.3.5 Strengthening the Precondition

Besides weakening the postcondition, there is another way to weaken a specification: strengthening the precondition. It makes R.'s job easier if he is *a priori* guaranteed that there is an element in the table satisfying the required property. Formally, we use the symbol ∃ (read: there exists). We get the following specification:

```
1    T: set
2    P: predicate defined for all elements of T
3    (∃ x ∈ T) P(x)
4    table-search-program returns
5    x with x ∈ T and P(x)
```

It is up to the engineer in charge of the integration of that piece of software in its environment to ensure that it will be used correctly, that is, that the precondition is satisfied on each occasion that it is used.

Otherwise, he runs the risk of losing control of execution. In particular, the piece of software under consideration can not only abort (which at least can be noticed), or loop, but it could also return a fanciful result without warning. Indeed, recall that the meaning of a specification ⟨precondition, postcondition⟩ is roughly precondition ⇒ postcondition: if the precondition is false, this implication is true even if the postcondition is not satisfied. It is therefore better to avoid strengthening the precondition; this is particularly the case when using assertions which are not easy to verify.

How can we actually use an abstract specification to direct the construction of a correct implementation? This is our next topic. We start with the last specification, which is the easiest version of it to implement.

2.4 Program Development

In order to implement the previous specification, the obvious intuitive idea is to examine every element of T until a suitable x is found. Until now the set T that we used as a model for the table was left undetermined. For a simple program we need to be more specific. We take here $T = \mathbb{N}$.

ℕ is the set of so-called *natural* integers 0, 1, 2 ... Other important sets of numbers are ℤ (positive and negative integers, and zero), ℚ (rationals, i.e. quotients of integers) and ℝ (reals). The latter can be constructed from the natural numbers.
Confusing mathematical integers with the integers of a programming language is slightly improper: generally the latter are bounded. However this issue has no consequence in our example.

The property P will be left abstract. We only assume that there is an expression in the programming language under consideration which computes

$P(x)$ for all x of T.[6] The specified problem then becomes the search for an integer x satisfying $P(x)$. It is at least as general as the search for an element in an array.

2.4.1 Prelude: Correctness of a Loop

The programs we are interested in are made up of a loop allowing a simple operation to be repeated while traversing the table — for us, elements of N. We write it:

while test **do** body **done**

2.4.1.1 Partial Correctness. In order to show that a postcondition Q is true after the execution of a loop, the simplest way is to prove that Q is kept true at each iteration of the loop! More precisely, if Q is true at the starting point of the loop, and if executing the body preserves the truth of Q, it is clear that Q is still true after any number of iterations. Such an assertion is called an **invariant** of the loop. Beware: the invariant can be temporarily violated inside the body; only its status before and after every iteration matters.

This technique is evidently incomplete: if we are interested only in things which do not change, what is the point of executing the body of the loop? Actually the invariant provides only an abstract, partial, view of the state of the program. The state is supposed to change on every iteration; however, this is precisely what we forget with the technique of the invariant.

Surprisingly, a very small addition turns out to be sufficient to derive a proof method which is powerful enough for our needs, at least with partial correctness issues. We just have to take into account the failure of the test which is necessary for exiting the loop. Let C be the assertion corresponding to this test; we decompose the postcondition Q into $I \wedge \neg C$, where I is the invariant of the loop. We can also take advantage of the truth of C at the beginning of an iteration. This yields the following reasoning scheme:

> if I is true at the starting point of the loop
> and, if the body of the loop establishes I from $I \wedge C$,   (2.1)
> then we have $I \wedge \neg C$ at the exit point of the loop.

In order to have total correctness, we still need to ensure that exiting the loop will actually occur. Here again we need to study (an abstract version of) state changes during execution. Somewhat strangely, the key concept is again the concept of invariant.

2.4.1.2 Termination. For the sake of simplicity we exclude abortion or exception mechanisms. We can then informally represent the behavior of our looping program by a sequence

---

[6] In the pseudo-language we employ here we retain the notation $P(x)$. In languages without Booleans, one can use a function f returning 0 or 1, such that $P(x)$ is represented by the test f(x)=1.

$$true, \text{body}, true, \text{body}, \ldots true, \text{body}, false$$
$$\underbrace{\qquad\qquad\qquad\qquad\qquad\qquad\qquad}_{n \text{ iterations}}$$

(n may happen to be zero) if it terminates, or

$$true, \text{body}, true, \text{body}, \ldots true, \text{body}, \ldots$$

if it does not terminate. In order to ensure total correctness of the program, we have to prove that the second case does not occur.

The technique that can generally be used is to identify a value v, called the loop **variant**, which depends only on the state, and which satisfies the following conditions:

v is a natural number (a non-negative integer),	(V_N)
v decreases at every iteration.	$(V_<)$

Indeed, each iteration step results in a distinct value of v; but we have

a strictly decreasing sequence of non-negative integers is necessarily finite.	(2.2)

As a passing remark, (V_N) provides an assertion which must be integrated into the loop invariant. For example, the program

while x$\neq$0 do x:=x-2 done

does not terminate if the initial value of x is odd. This problem becomes apparent if, in an attempt to prove the termination of this loop, we choose the value of x as the variant v: the input condition C in an iteration ensures only $v \neq 0$, which, using the invariant (V_N), yields $v \in \{1, 2, 3, 4 \ldots\}$; after x:=x-2 we would have $v \in \{-1, 0, 1, 2, \ldots\}$, and the allowed value -1 would violate (V_N).

Assuming that the initial value of x is different from 1 would not solve the problem for a similar reason.

This would amount to taking $I \stackrel{\text{def}}{=} v \in \{0, 2, 3, 4 \ldots\}$ as the invariant ($\stackrel{\text{def}}{=}$ means "is defined as"). At the starting point of an iteration we would have $I \wedge v \neq 0$, hence $v \in \{2, 3, 4 \ldots\}$; after x:=x-2 it becomes $v \in \{0, 1, 2 \ldots\}$, which is unfortunately different from the invariant I we expect.

By contrast, if the initial value of x is even, we can take $I \stackrel{\text{def}}{=} v \in \{0, 2, 4 \ldots\}$ as the loop invariant. At the starting point of an iteration we have $I \wedge C$, that is $v \in \{2, 4 \ldots\}$; after x:=x-2 it becomes $v \in \{0, 2 \ldots\}$ which does indeed conform to (V_N).

The behavior of a correct loop can then be roughly summarized as follows:

> while the state is not satisfactory, change it in a way such that the invariant is kept true and the variant decreases.

The concept of a variant can be stated in a much more accurate manner using well-founded relations; we return to this in § 3.5.

2.4.2 Linear Search

We assume here that there is at least one natural integer satisfying P. The search is performed by attemping different integers one by one, hence the term *linear*.

```
1    P: predicate defined for all elements of  N
2    (∃ x ∈ N) P(x)
3    integer-search-program returns
4    x with (x ∈ N) ∧ P(x)
```

The proposed program is of course:

```
1    x:=0 ;
2    while ¬P(x) do x:=x+1 done ;
```

The following reasoning may help to convince ourselves that the above program is correct.

Partial correctness (if the program terminates, then the postcondition is satisfied):

- x, initialized to 0, is incremented by 1 at every step; then we have always $x \in \mathbb{N}$, this invariant is still true at the exit point of the loop;
- $\neg P(x)$ forces the next execution step to be in the loop, then $P(x)$ is necessarily satisfied at the exit point of the loop.

Total correctness (the program terminates).
Let N be an integer such that $P(N)$ is true (the precondition ensures the existence of such an N), and let us take $v = N - x$ as the variant:

(V_N) $N - x$ is an integer because $N \in \mathbb{N}$ and we know (see the above on partial correctness) that $x \in \mathbb{N}$. We still have to show that the *property* $v \geq 0$, which is true after x:=0, is left invariant; let us rephrase this as $x \leq N$ (since $v = N - x$). At the beginning of an iteration step, we necessarily have $\neg P(x)$ which yields $x \neq N$, since N satisfies $P(N)$; hence $x \leq N$ boils down to $x < N$; after the assignment x:=x+1, this yields $x \leq N$ as expected, since N and x are integers.

($V_<$) $N - x$ decreases at every iteration because x increases.

In the above reasoning, N is not necessarily the integer that will be returned by the algorithm: the latter is actually the smaller integer satisfying P. We need an N such that $P(N)$ holds only for purposes guaranteeing termination.

2.4.3 Discussion: Reasoning Figures

The above reasoning is not that long, but that would be the case with more complex specifications and programs. Therefore it is desirable to be able to check a proof in a systematic way. To this effect one reduces this checking to the successive application of primitive reasoning steps, that is, reasoning steps simple enough that we can have no doubt about their validity. Logicians formalize them in a *deduction system*. A great advantage then is that the process can be aided by automated tools. Let us make an inventory of the ingredients needed in the above proof.

2.4.3.1 Logical Laws. A number of steps are purely logical steps: the ones related to connectives such as $\vee$ (or), $\wedge$ (and), $\Rightarrow$ (implies), $\neg$ (not). For example, from $v > 0 \vee v = 0$ (which was written $N - x \geq 0$) and from $v \neq 0$ (coming from $x \neq N$) we deduced $v > 0$. More formally, from $A \vee B$ and from $\neg B$ we deduced A. Such a deduction principle is written in the same way as a fraction, where premises take the place of the numerator while the conclusion takes the place of the denominator:

$$\frac{A \vee B \qquad \neg B}{A} \ . \tag{2.3}$$

The following formula contains a similar idea:

$$(A \vee B) \wedge \neg B \Rightarrow A \ . \tag{2.4}$$

However, the latter must be regarded as an ordinary logical expression, in the same way as $(a + b) \times (-b)/a$ is an arithmetic expression. In contrast (2.3) denotes a deduction step that yields the conclusion A from hypotheses $A \vee B$ and $\neg B$. A complete reasoning consists of a combination of similar steps. This can be viewed as follows:

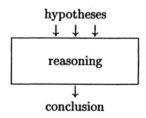

Formulas such as (2.4) allow us to represent the hypotheses, the conclusion, or the fact that the former entails the latter, but not the proof itself. We will see in Chapter 9 how the box "reasoning" can be formalized using rules analogous to (2.3).

Other issues will be tackled, for instance:

– what is the precise link between (2.3) and (2.4)?
– how can we check the validity of a formula like (2.4)?

2.4.3.2 Manipulation of Equalities. Aiming at deducing $x \neq N$ from $\neg P(x)$ and from $P(N)$, let us suppose that $x = N$ and derive a contradiction. We can then replace x with N in $\neg P(x)$, which yields $\neg P(N)$, in a contradiction with the second premise. The general line of reasoning (reduction to the absurd) is a matter for the previous subsection. However, we also used the principle of **substitution of equals by equals**, which is very important in spite of its simplicity.

2.4.3.3 Proper Laws. We also employed laws which are specific to the domain of the model, for example arithmetic rules, allowing us to transform $N - x \neq 0$ into $N \neq x$, or laws about assignments. The behavior of a piece of a program S is described using the notation $\{P\}\, S\, \{Q\}$, which means that starting from the precondition P, executing S establishes the postcondition Q.

$\{P\}\, S\, \{Q\}$ is itself a logical formula, just as are P and Q. The latter two are logical formulas about the state that we get from the variables of the program, whereas $\{P\}\, S\, \{Q\}$ is about its execution. The reasoning scheme (2.1) for verifying the partial correctness of a loop, given on page 23, can also be formalized by means of a premise/conclusion rule:

$$\frac{\{I \wedge C\}\, S\, \{I\}}{\{I\}\, \texttt{while}\ C\ \texttt{do}\ S\ \texttt{done}\ \{I \wedge \neg C\}}\ . \tag{2.5}$$

The formula $\{I\}\,\boxed{\texttt{while}\ C\ \texttt{do}\ S\ \texttt{done}}\,\{I \wedge \neg C\}$ is made up of formulas such as I and C, and of pieces of programs such as S and the part that is framed. In a similar way, an assertion such as $P \wedge (1+1 = 2)$ is made up of another assertion (P) and of integers.

2.4.3.4 Reasoning by Induction. There is a particularly powerful means for proving that a property Q is true for *all* natural integers n. We proceed in two steps:

1. we show that Q is true for $n = 0$;
2. we show that if Q is true of an arbitrary integer, then Q is kept true for the next integer.

This principle, called *induction*, can also be written in the previous format:

$$\frac{Q(0) \qquad \forall n\ n \in \mathbb{N} \wedge Q(n) \Rightarrow Q(n + 1)}{\forall n\ n \in \mathbb{N} \Rightarrow Q(n)}\ . \tag{2.6}$$

Reasoning by induction is ubiquitous, though sometimes in a hidden format. The principle of induction allows us to justify that a loop invariant is true after any number of iterations given that it is initially true and that it is preserved on every iteration. It is also required to prove that a strictly decreasing sequence of natural integers is necessarily finite (which is in turn the key argument for justifying the technique of loop variants, see (2.2) on page 24). All important properties of integers and data structures such as lists or trees require a form of induction. An automated environment for formal methods must support this kind of reasoning; simply handling logical connectors is far from sufficient.

2.4.4 Bounded Linear Search

If no integer satisfies the property P, it is clear that the program on page 25 does not terminate.

If this were the case, we know from partial correctness that, at the exit point of the loop, x would satisfy $P(x)$, in contradiction with the previous hypothesis.

2.4.4.1 Specification. We use the specification given in § 2.3.4 on page 21. With $T = \mathbb{N}$ we can write it as:

```
1   P: predicate defined for all elements of N
2   table-search-program returns
3   r with r ∈ N+{failure},   such that
4     P(x)              if r comes from (elt. x of) N
5     (∀ x ∈ N) ¬P(x) if r comes from {failure}
```

But this is too difficult, mainly because of line 5 where we have a quantification over an infinite number of elements.

If a general program solving this problem could exist, for an arbitrary P, it could in theory be used to solve conjectures or difficult problems of arithmetic. For example, let us consider Fermat's last theorem (recently proved by Wiles): for any n greater than 2 we cannot find three integers a, b and c such that $a^n + b^n = c^n$. We would take, for $P(x)$:

$$\exists n \, \exists a \, \exists b \, \exists c$$
$$(n < x) \wedge (a < x) \wedge (b < x) \wedge (c < x) \wedge a^{n+3} + b^{n+3} = c^{n+3} \ .$$

Here we limit ourselves to finite tables. They are modeled as an interval of integers. We use $[p..q[$ to denote the set of integers greater or equal to p and strictly smaller than q. In particular, if $p = q$, the interval $[p..q[$ is empty.

```
1   (p ∈ N) ∧ (q ∈ N) ∧ p≤q
2   P: predicate defined for all elements of [p..q[
3   table-search-program returns
4   r with r ∈ [p..q[+{failure},   such that
5     P(x)              if r comes from (elt. x of) [p..q[
6     (∀ i ∈ [p..q[) ¬P(i) if r comes from {failure}
```

In the present situation we can take advantage of the structure of the table to avoid the introduction of the Boolean b (see page 20): we simply represent the lack of an element satisfying $P(x)$ in the table by returning a value of x such that $x = q$. In other words, for $T = [p..q[$, we can model $\{failure\}$ by $\{q\}$ and $T + \{failure\}$ by $[p..q[\cup \{q\} = [p..q]$. Thus we get the following specification:

```
1   (p ∈ N) ∧ (q ∈ N) ∧ p≤q
2   P: predicate defined for all elements of [p..q[
3   table-search-program returns
4   x with x ∈ N ∧ p≤x ∧ x≤q
5     and P(x)                        if x<q
6     and (∀ i ∈ N) (p≤i ∧ i<q) ⇒ ¬P(i)  if x=q
```

2.4.4.2 A Naïve Attempt. We could try the following program:

```
1   x:=p ;
2   while x≠q ∧ ¬P(x) do x:=x+1 done ;
```

Aiming at a correctness proof of this program, we consider the loop invariant I that simply tells us that, on the one hand, x is kept confined to the expected domain (I_1) and, on the other hand, values of x investigated so far do not satisfy P (I_2):

$$I \stackrel{\text{def}}{=} I_1 \wedge I_2 \;,$$

$$I_1 \stackrel{\text{def}}{=} \underbrace{x \in \mathbb{N} \wedge p \leq x \wedge x \leq q}_{\text{domain of } x} \;,$$

$$I_2 \stackrel{\text{def}}{=} \underbrace{\forall i \in \mathbb{N}\, (p \leq i \wedge i < x) \Rightarrow \neg P(i)}_{\text{unsuccessful exploration}} \;.$$

This invariant is established before the loop: I_1 comes from the precondition and, with regard to I_2, $p \leq i \wedge i < x$ is necessarily false because $x = p$.

The partial correctness criterion of **while** tells us that the negation of $x \neq q \wedge \neg P(x)$ is verified after line 2 of the program. A logically equivalent formula is

$$x = q \vee P(x) \;. \tag{2.7}$$

In the case where $x = q$, the invariant I_2 can be written $\forall i \in \mathbb{N}\, (p \leq i \wedge i < q) \Rightarrow \neg P(i)$, which agrees with line 6 of the specification. If $x \neq q$, the exit condition (2.7) forces $P(x)$; with I_1 we then get all the ingredients of lines 4 and 5 of the specification.

We still have to examine total correctness. But ...

2.4.4.3 Beware of Limits. There is a well-known snag for the experienced programmer. If there is no element of $[p..q[$ which satisfies P, the exit test of the loop of line 2 is performed for $x = q$, which means that the condition $q \neq q \wedge \neg P(q)$ is computed. The inequality $q \neq q$ is quietly evaluated to *false*; but what about $\neg P(q)$? P is not supposed to be defined at q. The precondition of line 2 has been designed intentionally, because it is a typical programming problem: array overflows.

Let us first remark that usually, in logic, an expression having the form $b \wedge$ *anything* evaluates to *false* if the value of b is *false*. From this point of view we don't hesitate: the assertion $q \neq q \wedge \neg P(q)$ has a value which is *false*.

We will see in Chapter 5 that, in usual logic, all functions are total and predicates are defined everywhere. When we want to model a partial object f (predicate or function), we have to extend it in an arbitrary way over the whole domain under consideration, and to introduce an additional predicate characterizing the elements where f is defined. The expression $f(x)$ is then defined, even if x is outside the expected domain of f (the domain of a function is the set of elements where it is defined). In our case the *assertion* $P(q)$ has a value, but it is arbitrary and unknown: hence $q \neq q \wedge \neg P(q)$ takes the value *false*.

However, the very fact that this assertion has a value does not mean that at the level of the program the corresponding computation succeeds. It is a well-known fact that when executing, a program fragment may starve (hang) in a loop, abort, or raise an exception. This is typically what may happen in the case of an array overflow.[7] Modeling these phenomena requires the introduction of an additional value which represents the indefinite. The mathematical representation of the evaluation of a Boolean expression by a program computation is then more complex than the evaluation of the corresponding logical assertion.

In order to take this into account, a number of programming languages make it explicit that the computation of $A \wedge B$ starts with the computation of A; if $A = \textit{false}$ the result $\textit{false}$ is directly returned without evaluating B. In our example this works quite well. In the general case, if B cannot be evaluated, then $B \wedge A$ cannot either, according to this evaluation strategy. Hence a property as simple as $A \wedge B = B \wedge A$ is lost, and actually many common properties of logical connectors are invalidated at the level of programs. This can make reasoning more complicated.

Another possibility is to ensure that evaluating $P(x)$ is performed only for values of x which are strictly smaller than q. Thus we can content ourselves with the two normal truth values. But, obviously, the previous program needs to be modified. Let us investigate this idea. Rigor would require that we indicate formally that each evaluation of $P(x)$ is performed under good circumstances. To this effect we should insert the assertion $x < q$ before all instructions containing $P(x)$, and prove that those assertions are true in the indicated places. This leads us to mix specifications and programs. Appropriate syntactic means will be presented in Chapter 4. Here we simply follow this approach in an informal manner.

The issue raised here is not a limitation of formal methods but a subtle point related to the semantics of programs: in spite of appearances we have to be careful not to confuse Boolean expressions occurring in tests with logical expressions occurring in assertions. Therefore in the following, we distinguish the logical constants **f** and **t**, used in formulas, and the Booleans `false` and `true`, used in programming.

2.4.4.4 Another Program. Since the occurrence of $P(x)$ in the exit test of the loop is harmful, let us remove it. What is left is while $x \neq q$ do *body to be determined*. But, when exiting the loop we would necessarily have $x = q$, which is not what we expect.

A basic technique which turns out to be useful in this kind of situation is to replace a constant (the only one we have in the test is q) by a variable, say y. When exiting a loop while $x \neq y$ do etc. we have $x = y$, and we want one of the following assertions to be true:

---

[7]It is at least the most meaningful behavior (except when we only want to read a value for which we proved, as here, that its value is irrelevant). In most cases, allowing the execution to continue leads to unpredictable results often difficult to analyze. Languages such as C make this unfortunate choice.

- either $x = y < q$, if $P(x)$ (line 5 of the specification),
- or $x = y = q$, if there is no satisfactory element in the table (line 6 of the specification).

We will naturally test $P(x)$ in the body of the loop, with the intention of exiting the loop in the case of success; then we have to equate x and y, without modifying x since x contains the value we are looking for: hence we consider y:=x. If the test fails, x is incremented as in the previous program. If the successive tests always fail, y must behave like q in the previous program, q is then a good candidate for the initial value of y. Hence we have an elegant program which may escape even our experienced programmer:

```
1    x:=p ; y:=q ;
2    while x≠y  do
3        if P(x) then y:=x else x:=x+1 done ;
```

The correctness proof is performed as above. We just have to add in the invariant, that y is between x and q (see I_1) and that x satisfies P when y is strictly smaller then q (I_3):

$$I \stackrel{\text{def}}{=} I_1 \wedge I_2 \wedge I_3 \; ,$$

$$I_1 \stackrel{\text{def}}{=} \underbrace{x \in \mathbb{N} \wedge y \in \mathbb{N} \wedge p \leq x \wedge x \leq y \wedge y \leq q}_{\text{domain of } x \text{ and of } y} \; ,$$

$$I_2 \stackrel{\text{def}}{=} \underbrace{\forall i \in \mathbb{N} \, (p \leq i \wedge i < x) \Rightarrow \neg P(i)}_{\text{unsuccessful exploration}} \; ,$$

$$I_3 \stackrel{\text{def}}{=} \underbrace{y < q \Rightarrow P(x)}_{\text{success}} \; .$$

Now we can check that at the entry point of the loop body, we have $x \neq y$, hence $x < q$ because of I_1. Then $P(x)$ can be easily computed.

For loop termination concerns, we can take $y - x$ as the variant; details are left as an exercise.

2.4.5 Discussion

This little example illustrates a tricky point that occurs in programming, in formal specification, and in logic as well: handling *partial functions*.

A **partial function** is a function which is not defined everywhere. For example, if we consider functions over real numbers, $1/x$ is not defined for $x = 0$, and $\sqrt{x}$ is not defined for $x < 0$. For an example over $\mathbb{N}$, the square root function is only defined over $\{0, 1, 4, 9 \ldots\}$. Basic notions of functions are recalled in § 3.4.3 on page 48.

An array can be regarded as a partial function which is defined over an interval of integers, i.e. a (special) subset of $\mathbb{N}$. It may happen that computing a function which is described in a programming language either loops or aborts for particular values of its arguments; then we still have a partial function.

We already have a problem at the notation level: what is denoted by $f(x)$ when f is partial and is not defined for x? The matter would be simpler if we could tell in advance whether or not f is defined for x. But in the general case such knowledge cannot be provided by *mechanical* methods, if x is the result of a computation. We will consider three main approaches to this issue, one based on classical two-valued logic, one using a third truth value and one based on types.

It is important to keep in mind that notations coming from mathematics often take a slightly different meaning in programming. This was illustrated in § 2.4.4.3 on P ∧ Q and on P(x).[8] There is another pitfall with the concept of a variable. The concept we use in a programming language like C is quite different from the concept we use in mathematics: it is essentially a memory address, and generally corresponds to values that are difficult to predict because of aliasing phenomena, that is, when two names refer to the same piece of memory.

2.5 Summary

Considering the right abstraction level is essential for writing precise specifications without getting lost in the details. Logic turns out to be an excellent tool in this area. This chapter also introduced, in a semi-formal way, a specification technique based on logical assertions as well as simple reasoning about them. Reasoning obviously lies within the realm of logic.

We also observed the ambiguity of informal text, and that such ambiguities can be overlooked at first sight: recall the two interpretations proposed on page 17 for a specification based on a precondition and a postcondition. Moreover, similarities between mathematical notations and programming languages may cause a number of confusions: program variables are not exactly mathematical variables; Boolean expressions cannot always be considered to be predicates; partial functions have a somewhat different status.

Our example for illustration purposes was very simple. What happens when we consider real large-scale software? The risk of lapses, ambiguities and inconsistencies increases dramatically. Formalizing (parts of) the specification becomes more difficult. However, it should be noted that, during the lifecycle of a software, we always have at least one formalization step: encoding in a programming language. Moreover it is better to formalize our knowledge as early as possible, so we can then derive information about the behavior of the system under consideration, compare the latter with desired properties, and make more accurate design decisions. At the same time, it is important not to freeze implementation choices too early. In this respect, good abstraction mechanisms are essential.

Hence, powerful and expressive languages endowed with a precise semantics turn out to be very useful. Again, logic provides essential tools. However,

---

[8] Recall how P(x) was introduced in our toy programming language on page 23.

they have to be chosen very carefully. The difficulty of this task should not be underestimated.

2.6 Semantics

Real software is written in programming languages, then compiled and executed on real (or virtual) machines. A complete guarantee of their behavior would require exhaustive verification right down to the hardware level. This is of course a gigantic task, but one against which we are not, however, entirely powerless. We will not consider here the application of formal methods to hardware specification and verification, although they are used in that arena at least as much as for software.

In contrast let us say a few words on programming languages. Reasoning about concrete programs is legitimate provided that the language used is endowed with a well-understood formal semantics. There are several kinds of semantics. Among them, the most important are denotational semantics, axiomatic semantics and operational semantics.

Denotational semantics aims at giving programs a mathematical meaning which is independent from computations on particular machines, including abstract machines. In most cases this mathematical meaning takes the form of a *function* covering an appropriate domain. In contrast, **operational semantics** defines the behavior of a program by its effect on an abstract machine. Finally, **axiomatic semantics** tells us the effect of each program statement on assertions over the state of an abstract machine.

Each semantics has its uses. Denotational semantics provides a better representation of the very nature of a program. Operational semantics may form the basis of the design of a compiler. Reasoning rules to be applied to concrete programs are based on axiomatic semantics. The preferred situation is when all three type of semantics are available and when each one is consistent with the others.

2.7 Notes and Suggestions for Further Reading

Many textbooks on formal specification techniques (e.g. [PST91, Jon90, Mor90, Wor92, WL88]) provide an easy-to-read introduction to logic and set-theoretic concepts used in techniques such as Z, B or VDM.

The idea of reasoning about programs seems to be as old as programming itself. It was mentioned in the 1940s by the logician Alan Turing, who invented the concept of a universal machine (a machine where the program is registered in memory). Logical assertions were introduced in flow charts in 1967 by Floyd [Flo67], then in structured programming languages following the seminal work of Hoare [Hoa69] and Dijkstra, we return to these in Chapter 4.

There are several introductory textbooks on programming language seman- tics. Hanne and Flemming Nielson's book [NN92] present the main approaches clearly. The short book by Gordon [Gor79] and the reference book by Stoy [Sto77] are more specifically devoted to denotational semantics. One may con- sult [Sch88] for a further study.

The bounded linear search algorithm comes basically from textbooks by [Coh90] and [Kal90] on Dijkstra-style approaches. The starting point of the authors is a specification similar to the last one in § 2.4.4.1, where $x = q$ encodes the failure of the search. This specification is simple to understand and perfectly relevant if the problem to be solved is the search for an element in an *array*. Why did we dismiss $T \cup \{failure\}$ in a first stage and finally come back to it? Precisely because our initial problem was to search for an element in a *table*, whatever the actual detailed structure of the table. The concept of sum introduced in § 2.3.4 perfectly fits our requirements for a high-level specification. Actually, most implementations considered in programming turn out to use data structures with two distinct variants.

3. A Presentation of Logical Tools

Mathematical logic has spread out in a variety of ways — model theory, proof theory, set theory, computability — according to Barwise's classification [Bar77]. To this taxonomy we can add type theory, which has become more important since the time of Barwise's overview. From our point of view, the importance of logic can be summarized as follows:

- it provides a natural framework for precisely constructing and expressing various concepts in computing;
- it lends itself well to formalization.

The first of these points has been described in Chapter 2. The properties of a program are quite naturally expressed in logic. The language of sets also finds many applications in this domain. Variables manipulated by programs range over a state space that is nothing more than a set defined by composing particular basic sets (specifically, integers, characters, etc.) by means of set operations (for example, the `record` construct of the Pascal language or the `struct` construct of C are both a form of Cartesian product). In other respects, computability theory makes us aware of the existence of unrealizable specifications.[1] Finally, type correctness makes programming more accurate and more secure.

Returning to the second point, above, our interest in formalization is twofold. On the one hand, the rigor of our specification texts and our reasoning about them is increased, since this is based on the manipulation of symbols that may be easily verified; on the other hand, the effort may be automated, or at least aided, by computer. It must be noted that the complete formalization of proofs, whether in software development or in a mathematical context, has a tendency to submerge the principal ideas under a plethora of more or less trivial lemmas. For such an approach to be viable, at least a partial automation proves to be indispensable in practice.[2] Proof theory provides essential tools in this respect.

On a practical level, mathematical logic aids in developing specification languages. An intuitive understanding of concepts, such as we acquire in school and in college, is often sufficient. Certain specification languages such as Z or B transform the language of sets and logic to accommodate the organizational needs of computing by means of adequate structuring mechanisms.

---

[1] Not because they are contradictory, but more subtly because no program can be derived to compute the desired function.

[2] An alternative point of view is presented in § 9.6.

Knowledge of certain more advanced aspects of logic is often very useful. This will be illustrated in § 3.1. Section 3.2 will give an overview of the historical context of mathematical logical. We will describe the different branches in § 3.3. Basic mathematical terms will be recalled in § 3.4. We will end with more technical discussions on well-founded relations and ordinals from § 3.5 — these concepts play a key role in issues of termination and computability in § 3.7. The last two sections may be omitted on a first reading.

3.1 Some Applications of Logic

3.1.1 Programming

Let's take a piece of paper on which are drawn some ordinary figures, and try to determine if a given point is inside this figure, or if a given line cuts that figure. In three dimensions, this presents a very concrete problem of aerial control. The reader is invited to spend a few minutes considering a solution in the programming language of his or her choice.

Do we, for example, construct some form of structured variables for each basic form? Do we try to combine everything into a tree structure? We must consider every possible interaction.

It's much more simple: we use the *characteristic function* of the figure under consideration, that is a function that for every point returns the value *true* if the point belongs to the figure, and the value *false* otherwise. The reader should be able to easily express the characteristic function of basic figures (discs, rectangles, etc.) in the programming language of choice. But this representation doesn't really catch our interest unless we can construct new figures from known figures. For example, the intersection of two figures represented by f and g is a function which, when applied to the point p, returns *true* if and only if $f(p) = true$ and $g(p) = true$. The function that computes the intersection is very general, and makes a total abstraction from the particulars of the figures themselves. Other forms of composition (complement, union) are also easy to obtain, as are transformations such as translations, symmetries or rotations.

Everything rests on one essential ingredient: the ability to pass functions as parameters and return functions as a result. What programming language should we choose? At first sight we find the concept of a pointer to a function, widely used in the C programming language, to be convenient. In reality, this is only sufficient to cover the case when the functions used are finite in number and are known in advance. The problem with not perceiving these limitations is that we may hope to be able to resolve the problem by taking a sufficiently shrewd approach. In reality, only the functional languages, based on the λ-calculus (see later) such as Scheme, ML or Haskell, provide a sufficiently general mechanism.

The underlying problem is to know if functions are considered as objects that can be manipulated in the same way as data structures. This is not a trivial question. We will see that in set-based specification techniques, we regularly manipulate binary relations, functions being a particular case of relations. These relations are intended to be implemented with data structures (tables, pointers, etc.) or algorithms (procedures, functions). Choosing the right solution is delicate. If the development is undertaken unadvisedly, or rashly, it may well end up with an inefficient or overly complex implementation — or just fail.

3.1.2 Sums and Unions

Let us examine some other constructs used in formal languages. The reader probably knows already how to use symbols such as $\cup$ and can associate it with a simple intuitive interpretation — combining the elements of two sets. This notation is generally used to combine sets of the same "kind". For example we can state:

$$\{x \in \mathbb{R} \mid 1 \le x \le \pi\} \cup \{x \in \mathbb{R} \mid 2 \le x \le 2\pi\}$$
$$= \{x \in \mathbb{R} \mid 1 \le x \le 2\pi\} \ .$$

We don't feel the need to combine dissimilar sets, for example a set of integers, a set of couples and a set of sets:

$$\{1,2,3\} \ , \quad \{\langle 1,2 \rangle, \langle 3,4 \rangle\} \quad \text{and} \quad \{\{1,3,4\},\{1,5\}\}$$

which would yield:

$$\{1, \ 2, \ 3, \ \langle 1,2 \rangle, \ \langle 3,4 \rangle, \ \{1,3,4\}, \ \{1,5\}\} \ ;$$

but after all, nothing is impossible. We actually often need to mix heterogeneous data in computing. For example, in protocols, when we want to manipulate messages having different formats in a uniform way. Or in parsers, when we construct a syntax tree: a node corresponding to a statement can have two children if it represents the sequential composition of two statements, three children if it represents an if-then-else statement, etc.; moreover we see that nodes can represent statements or expressions. A data structure representing elementary geometric figures, say circles or triangles, would have, respectively, two fields (the center, which is a point, and the radius, which is a distance) or three fields (the vertices, which are points). A more elaborate example is the set of finite integer sequences, which can be seen as an infinite union:[3]

$$\{\varnothing\} \cup \mathbb{N} \cup (\mathbb{N} \times \mathbb{N}) \cup (\mathbb{N} \times \mathbb{N} \times \mathbb{N}) \cdots$$

However, mixing heterogeneous objects is not harmless. It is plainly meaningful to reject, at compile time, a test like $a = b$ if a and b have different types. The usual interpretation is that a and b take their values from two different

---

[3]We need a singleton for representing the empty sequence. The usual set-theoretical trick is to take $\{\varnothing\}$.

sets A and B, say floats and strings. But we could just as easily agree that a and b take their values from the same set: $A \cup B$! And let us stress that we cannot just disallow $A \cup B$, as this notion is needed in the previous examples.

How can we get the flexibility that we need while simultaneously controlling the coherency of data and operations? The concept of sum introduced in 2.3.4 is just the ticket. In a good type system, A, B and $A + B$ can be distinguished.

A sum is dealt with using an operator able to check whether a given element s comes from an element a of A or from an element b of B, and then to direct the computation appropriately; the computation depends on a in the first case and on b in the second case. Such constructs are available in modern languages like ML. In Pascal (or C) it is possible to emulate a sum using a record construct with variants and a switch field, but it is the responsibility of the programmer to ensure that a variant is always used in a way consistent with the switch field. Note that during the initial design of ASN1, a standardized language for describing the format of data exchanged in protocols, sums were not recognized as a primitive concept, leading to many complications.

In ASN1, the expression CHOICE { a A, b B } yields a value whose type is either A or B. Switch fields (like a or b) are mandatory only since 1994. Before this date, they were confused with labels, which are integers encoding the type of the fields of a compound value. They are clumsy and cannot solve the ambiguity which appears if A and B happen to represent the same type.

3.1.3 Chasing Paradoxes Away

Let us again consider the example of sequences. They can be characterized by the following property: "to be empty or an integer or a pair of integers or etc.".

We often need to form sets from elements satisfying a given property — such a set is defined by *comprehension*. In this way we enter into the realm of the first version of set theory, where every collection made of objects characterized by a given property is a set. This so-called "naïve set theory" turned out to be inconsistent! Technically, an inconsistent system is a system where one thing and its contrary can be proved (formally: $P \wedge \neg P$) or, equivalently, everything can be proved.

Let us consider one of the simplest paradoxes, called Russell's paradox. In general, a set is not a member of itself. For instance, we have $\neg(\mathbb{B} \in \mathbb{B})$ because $\mathbb{B}$ is *not* a Boolean. Could we imagine a set which is a member of itself? Yes, though we have to think a bit.[4] Anyway, what matters is not whether such sets exist or not, but that we consider the property $x \in x$ and its negation.

Let us define by comprehension $R \overset{\text{def}}{=} \{x \mid \neg(x \in x)\}$. If $R \in R$, R must satisfy the characteristic property of members of R, that is, $\neg(R \in R)$. If $\neg(R \in R)$, R possesses the characteristic property, hence $R \in R$. If we define $P \overset{\text{def}}{=} R \in R$, we have P and $\neg P$ at the same time, which is inconsistent.

---

[4]Consider, for instance, the set of sets which can be defined with less than a hundred English words.

◯   Formally we have just shown that $P \Rightarrow \neg P$ and $\neg P \Rightarrow P$. By the equivalence (3.6) on page 47, the first implication yields $P \Rightarrow (P \Rightarrow \mathbf{f})$, which by (3.11) boils down to $(P \wedge P) \Rightarrow \mathbf{f}$, then to $P \Rightarrow \mathbf{f}$ which we use twice. First, it can be written $\neg P$, and we deduce P from the second implication. Second, combined with P we get $\mathbf{f}$.

The same paradox arises if one accepts too broad a concept of "property" (instead of set), more specifically if one accepts that the scope of a property may extend to all objects, including properties. Just replace every set by its characteristic property in the above reasoning. We then consider properties A which are false when applied to themselves and we define: $R(A) \stackrel{\text{def}}{=} \neg A(A)$, which has $\forall A\, R(A) \Leftrightarrow \neg A(A)$ as a consequence. Taking $A = R$ we deduce the absurd $R(R) \Leftrightarrow \neg R(R)$.

We will see in the following that several solutions have been proposed in order to avoid paradoxes. For the moment, let us just mention that the most celebrated in mathematics is the axiomatic set theory of Zermelo–Fraenkel. However, as it is an untyped theory, it is not well suited to computer science. This explains why specification languages based on set theory, such as Z and B, introduce an additional typing mechanism.

In summary, logic provides concepts and tools that allow us to understand the benefits, limitations and design issues of specification and programming languages. One has to pay attention to two pitfalls:

- a lack of expressiveness may lead to complications in using a language; for instance, it is sometimes just impossible to state the properties we wish to verify;
- conversely, some powerful constructs which seem correct at first sight may turn out to be much too powerful; that is, in the case of a property language, the underlying logic may become inconsistent; or, in the case of a programming language, they may lead to run-time errors which are difficult to analyze.

3.2 Antecedents

From an historical perspective, mathematical logic emerged a century ago for the purposes of precisely and rigorously constructing the foundations of mathematics. It was known, since the times of Dedekind and Cantor, that all mathematical objects (numbers, functions, vectors and so on) could be constructed from natural integers using only set-theoretic operations. However, those operations, when defined in an intuitive way, allowed one to derive paradoxes such as Russell's paradox. The whole mathematical edifice was threatened, leading to the "foundation crisis", and then to an intensive activity aiming at establishing common reasoning principles, such as deductive or inductive reasoning, on firm ground. This was one of the main motivations for David Hilbert to put forward his well-known programme, that would (in principle) reduce mathematics to finite manipulation of symbols.

A number of techniques invented in this framework happen to fit well with the needs of computer science, because, on the one hand, symbol manipulation plays a central role and, on the other hand, manipulated objects (both programs and data) are of finite or countable size (see § 3.4.6). Among theories born at that time, and which are of interest to us, we can cite predicate logic, type theory, axiomatic set theory, the λ-calculus, and intuitionistic logic. If we add the works of the 1930s on proof theory (Gentzen and Herbrand) we can see that the foundations of modern programming were largely available before the birth of the first computer!

On the mathematical side, things took an unexpected path in 1931, when Kurt Gödel proved his famous incompleteness theorem for arithmetic, sounding the death-knell of Hilbert's programme. To put it in a concrete way, it means that the most secure and restrictive reasoning forms are not strong enough to justify the principle of induction, not even to mention the stronger axioms contained in Zermelo–Fraenkel set theory. However, the latter turned out to be sufficiently powerful to serve as a basis for all known mathematics, and it is unlikely that an inconsistency will be discovered in it. The Zermelo–Fraenkel system remains the most commonly used nowadays.

3.3 The Different Branches of Logic

3.3.1 Model Theory

There are basically two complementary ways of writing a specification:

– describing the properties of a system;
– providing a model of the system by means of built-in constructs.

One sometimes uses the terminology **property oriented** and **model oriented** formal specification. Properties are expressed by logical axioms whereas models are derived with the help of set-theoretic operations. This duality is already present in mathematical logic, where we have a syntax for expressing logical properties and a semantics describing what we are talking about. This aspect of logic is called **model theory**. One distinguishes, on the one hand, the concept of a logical statement built upon a formal language, for example:

$$\forall x \exists y (y > x) \ , \tag{3.1}$$

and on the other hand the concept of a model satisfying this statement; for instance, (3.1) admits, among other models, $\mathbb{N}$ endowed with the relation "greater-than", $\mathbb{R}$ endowed with the relation "less-than" and $\mathbb{N}$ endowed with the relation "is-a-multiple-of".

A fundamental concept of model theory is the relation called **logical consequence** or **semantic consequence**. A sentence E is a semantic consequence of the sentences $A, B, C...$ if *every* model having the properties $A, B, C...$ has also the property E. This is a very concrete relation. Let us consider, for instance, the three properties "every terminal is a piece of equipment", "every piece of equipment possesses a registration number" and "this phone is a terminal". A practical consequence, of interest to the department in charge of inventories, is that in any situation where the above three properties hold true, we have, systematically, "this phone possesses a registration number". The concept of model is represented here by what we just called a situation.

3.3.2 Proof Theory

However, the concept of semantic consequence suffers from a big handicap: it is very difficult or even impossible to check it directly, because we must consider every possible model and there is, in general, an infinite number of them. This is why one may prefer to use another relation called **provability**. We say that a sentence E is provable from the sentences A, B, C... if we can construct a formal proof of E using only hypotheses A, B, C... in combination with axioms and the rules of logic. E is **refutable** if its negation is provable.

Of course, the logician must ensure that those formal manipulations respect the semantics, hence the concept of **soundness**. The converse property (every semantic consequence is provable) is a form of **completeness**. Another kind of **completeness** relates a collection of formulas Γ with *one* intended model $\mathcal{M}$, stating that the latter is completely characterized by Γ, i.e. every true (respectively false) formula in $\mathcal{M}$ is provable (respectively refutable) from Γ.

If we consider the formal specification of a piece of software, we can easily admit a specification to be incomplete at a high level stage. We only expect that the operations of our software respect a number of constraints, expressed by the means of logical formulas, but we may want to leave several options open. For instance, if we specify the calculation of $\sqrt{2}$ with a tolerance of 10^{-3}, the programmer is free to provide an implementation computing any result between $\sqrt{2} - 10^{-3}$ and $\sqrt{2} + 10^{-3}$. In many protocol specifications, some messages have to be answered in a very precise manner while others are considered less important. Sometimes we cannot afford incompleteness: in security software, all possible cases must be handled.

Apart from the links between semantic consequence and provability, there are interesting issues concerning provability alone. For example: if we know that E is provable, can we find a proof of E using only sub-formulas of E? If the answer is yes, the proof search space can be significantly restricted. This is especially important for automated proof tools. The study of axiom sets and logical rules, seen as formal calculations (by this we mean purely syntactic manipulations where we forget how formulas are interpreted) and their relationship with the concept of semantic consequence are the realm of **proof theory**.

In model theory, the semantics of logical sentences is provided by truth values. This is sometimes called the **Tarskian** tradition, in honor of the logician Tarski who deeply clarified its basis. Proof theory provides a different semantic perspective, which is in some sense more accurate, where logical sentences are associated with a set of proofs that conclude to these sentences, instead of to a simple value (true or false). This set of proofs can also be seen as a set of algorithms. This tradition is sometimes called **Heytingian** [GLT89].

The aim of Heyting was to interpret **intuitionistic logic** invented by Brouwer in a formal manner, during the foundation crisis. (At the same time normal logic was termed **classical logic**.) Intuitionistic logic contests the validity of a number of laws. The most well-known of these is the law of the excluded middle, which is formally stated as $p \lor \neg p$. Let us first point out that some consequences of this law are somewhat unexpected, for instance: "when you cast a dice, if you get an even result then it is smaller than three, or conversely". Formally, $p \Rightarrow t \lor t \Rightarrow p$ is accepted by classical logic but rejected by intuitionistic logic. We will see in § 3.7.3 another surprising example which is related to recursive functions. More deeply, the excluded middle is rejected because of a new interpretation of disjunction: in order to accept $p \lor q$, intuitionists want to know which proposition is provable amongst p and q. More precisely, it is enough for them to have the capability to compute the answer to that question. Then they can accept some instances of $p \lor \neg p$, but not any one.

In order to illustrate the difference in points of view, let us take a situation x in a game of chess and let $r(x)$ denote the fact that the black king is in check and in the situation x. An intuitionist can accept the sentence $r(x) \lor \neg r(x)$ because, by a mechanical application of the rules of the game (the explicit definition of $r(x)$) we can know whether the black king is in check in the situation x. Such reasoning remains valid in classical logic, of course. But in this framework we can also conclude this immediately using the law of the excluded middle. We can see that the explanation required by the intuitionist provides much more information.

The existential quantifier is interesting as well. In order to prove $\exists x \, P(x)$, the intuitionist wants to know, or to be able to compute a witness, x satisfying $P(x)$. A proof that the hypothesis $\neg \exists x \, P(x)$ leads to a contradiction, for instance, is not sufficient.

Simple common situations, where the law of the excluded middle is rejected by intuitionists, can be expressed in the form $(\exists n \, p(n)) \lor \neg (\exists n \, p(n))$, or, equivalently, $(\exists n \, p(n)) \lor (\forall n \, \neg p(n))$, where p is a property of natural numbers for which it is unknown whether, or not, there exists an n such that $p(n)$. Even if we have a mechanical procedure for deciding, for any given n, whether $p(n)$ holds or not — formally: even if we know $\forall n \, p(n) \lor \neg p(n)$ — the obvious algorithm for testing $(\exists n \, p(n)) \lor (\forall n \, \neg p(n))$, which successively checks whether, or not, $p(0)$, $p(1)$, ..., would involve an infinite number of tests if p happens to be false everywhere. As this algorithm may not terminate, it cannot be considered as reliable for providing an answer to our question. Suitable properties p can be constructed from unsolved mathematical conjectures. So-called Brouwerian arguments use, typically, the existence of 100 consecutive '9's arbitrarily far into the decimal expansion of the number π.

Intuitionistic logic has important uses in computer science because of its constructive features. In particular, there is a close relationship with type systems which we consider in Chapter 11.

3.3.3 Axiomatic Set Theory and Type Theory

A real model, like the one considered above in the example of telephone equipment, is not quite conventional in model theory. One merely considers mathematical structures, that is, sets endowed with particular operations. The same is true in computer science: in a model-oriented technique, models are written using set-theoretical constructs, though they are much less sophisticated than in model theory. For instance, one would consider an abstract set of equipment, having the same relation with reality as data structures of the corresponding software.

In order to be able to reason in a safe manner, building blocks for such sets need to be well defined. However, we know that the problems raised are not trivial. Several solutions have been proposed for eliminating the paradoxes of "naïve" set theory.

3.3.3.1 Typing Formulas. The most ambitious solution was proposed by Bertrand Russell [vH67, p. 199]. His idea was to introduce *types* in order to prohibit expressions like $x \in x$, or any expression which would yield the latter after a calculation. Actually **type theory** was not an attempt to save set theory or to reconstruct it on safe ground, but rather a new approach to establishing the foundations of mathematics. The first versions of type theory turned out to be unsatisfactory because they imposed inconvenient restrictions and some axioms were ad hoc. The idea has been significantly reshaped since then, expecially following the work of Martin-Löf [ML84], and a fair amount of mathematics can now be developed in a typed framework.

Ideas progressed in a similar way in computer science, and even more successfully: the first typing systems, for languages such as Pascal, proved to be too restrictive. But, subsequent progress led to programming languages that are both convenient in practice and strongly typed (notably, languages of the ML family).

A number of important ideas came to light with typing, such as the idea of **stratification**. Typing, at least in its most elementary form, stratifies sets (and properties) in distinct layers: at layer 0, individuals; at layer 1, sets of individuals (and properties about individuals); at layer 2, sets of sets of individuals (and properties of sets of individuals); and so on. Distinguishing first-order logic, second-order logic, etc. (see below) comes directly from this idea. This kind of typing is called **predicative**, which means that in order to define a concept, only concepts defined in lower layers can be used.

We find something analogous in computer science, when a software system is structured into layers. A function or a procedure which is defined using only previously defined functions and procedures can also be qualified as predicative. Note that in computer science we generally use the terminology **recursive** instead of impredicative: saying that a recursive function is defined as "a function of itself", a paradoxical way of stating things, is precisely recognizing that this function has an impredicative definition. There is a clear motivation to use only predicative definitions in logic: paradoxes like Russell's are then avoided. Note that in our presentation in § 3.1.3, the set R is impredicatively defined.

3.3.3.2 Axiomatizing Set Theory. The other attempt to suppress paradoxes consisted of defining **set theory** using an axiomatic form, in the framework of predicate logic. The main inventors were Zermelo,[5] Fraenkel and Skolem. In Shoenfield's presentation in [Bar77, ch. B.1], the idea of stratification appears quite clearly. This can explain why the well-known paradoxes could not be reproduced. One of the most important points concerns the definition of a set by comprehension, that is, by the means of a characteristic property of its elements. An axiom, called the separation axiom, states that we can form a set by comprehension *only* if we first have a sufficiently large set where we take elements having the desired property. As a consequence, we cannot directly define $A \cup B$ as the set of elements x such that $(x \in A) \lor (x \in B)$.

Thanks to this axiomatization, it proved possible to retrieve the ingredients provided by the "naïve" theory of Cantor, that were needed for developing the desired mathematical concepts, and hence its quick operational success.

3.3.4 Computability Theory

A last part of logic is the study of **computable functions**, that is, functions which can actually be defined by computations. This is an intuitive concept which must be formalized in order to become workable. Several proposals were made in the 1940s, among others: Turing machines, λ-calculus (Church) and recursive functions[6] (Gödel, Herbrand). Each of these approaches is a way of formalizing the concept of an algorithm, and in essence, defines a primitive programming language.

A simple reasoning on set sizes shows that many functions are not computable.[7] Moreover, it turned out that all aforementioned formalisms represent exactly the same class of functions: for instance we can encode any partial recursive function with a Turing machine and vice versa. The concept of a computational process seems then to be faithfully represented by any of these formalisms. This postulation is known as the **Church thesis**. To date, it has never been shown to be wrong.

---

[5]Zermelo's first paper on this topic was published the same year as the one by Russell on type theory, cf. [vH67].

[6]Note that the meaning of "recursive" in logic is precise but, unfortunately, different from its meaning in computer science. We saw that the latter corresponds rather to "impredicative". The definition of "recursive" is given at the end of this chapter.

[7]If we restrict ourselves to functions over natural integers (without loss of generality, because all useful data structures can be encoded by integers), the set of functions from N to $\{0, 1\}$ — and *a fortiori* to N — is not countable, whereas the set of functions defined by the means of a language having a finite or countable vocabulary is countable.

As a consequence, a programming language is said "to have the power of Turing machines" if it has the maximal expressive power we can expect — but it does not tell us whether this language is easy or difficult to use. Informally, a **Turing machine** is composed of an internal state, a tape with an infinite number of squares, a read-write head and instructions used to move the head and/or to write on the current square according to the current state and the symbol present on the current square. All common programming languages, including assembly languages, have the power of Turing machines. Among formalisms which do *not* have the power of Turing machines, we can cite finite state automata (which can parse or generate regular languages) and push-down automata (which can parse or generate context-free languages). Roughly, in order to get the power of Turing machines, the key ingredients are:

- basic arithmetic operations (addition, compare to zero);
- a notion of loop where the exit condition is computed at each iteration (e.g. the `while` of Pascal in comparison with the `for`);
- unbounded memory space; note that only a finite amount of memory is available on real computers, but the difference is hardly perceptible in practice.

Once this class of functions came to light, a number of fundamental questions could be asked and sometimes solved. The most well known of them is **the halting problem of Turing machines**: can we *mechanically* and in a *finite* number of steps, decide whether an arbitrary Turing machine running on arbitrary input data will eventually reach the state "computation end"? To put it in other words, can we know in advance — say, at compile time — whether or not the execution of a program will end, or whether a partial recursive function is defined on a given input data? It can be shown that this problem is actually **undecidable**, which means that no Turing machine can compute the answer to this question. As a practical consequence, a computer that could tell in advance whether an arbitrary program "loops" or not is definitely magical.

Notes. When we try to prove the correctness of algorithms, proving their termination is a crucial issue. The aforementioned result does not prevent us from doing it, it just states limitations on the extent of the help that we can expect from automation.

Note that this undecidability result came after the incompleteness theorem of Gödel; it is, moreover, proven along similar lines. Decidability and completeness are actually strongly related questions.

Computability (or recursion) theory comprises many other technical results that are not covered in this book. Their impact on formal methods is, in any case, quite weak nowadays.

3.4 Mathematical Reminders

We recall here useful basic concepts of set theory, logic and algebra.

3.4.1 Set Notations

In the following, A, B, C... denote "sets" whereas a, b, c... denote "elements"; we use quotes because the concepts of "elements" and sets are in fact relative, the members of a set can quite acceptably be sets themselves.

A **singleton** is a set having exactly one element, such as $\{a\}$. An **unordered pair** is a set having exactly two elements, such as $\{a, b\}$. The set with no elements is denoted by $\emptyset$. Two sets are **disjoint** if their intersection is empty.

We say that A is **included** in B, and written $A \subset B$, if every element of A is also an element of B. In particular, we have $A \subset A$ and $\emptyset \subset A$ for any set A. If A is included in B, we also say that A is a **subset** of B and that B is a **superset** of A. Two sets A and B are **equal** if they contain exactly the same elements. Hence $A = B$ if and only if $A \subset B$ and $B \subset A$. A is **strictly included** in B if $A \subset B$ and $A \neq B$. Then A is also called a **proper subset** of B. The set of subsets of A is called the **powerset** of A, it is denoted by $\mathcal{P}(A)$ or 2^A.

The **union**, the **intersection** and the **Cartesian product** of two sets were previously introduced on page 20. The **Cartesian square** of A is $A \times A$. The **difference** $A - B$ is the set of elements which are members of A but not of B. The **symmetric difference** $A \setminus B$ is the set of elements which are members of either A or B (but not A and B). Thus $A \setminus B = (A \cup B) - (A \cap B)$.

The set A^n denotes the Cartesian product $A \times A \ldots \times A$ (with n occurrences of A), i.e. the set of n-tuples $\langle a_1, \ldots, a_n \rangle$ such that $a_i \in A$. A^1 is identified with A. We agree that A^0 is the singleton $\{\emptyset\}$ — another singleton would do the job just as well, this one is the most simple we can construct in a universe where no element is known *a priori*.

Besides definitions by extension introduced on page 20, it is possible to define a set by **comprehension**, i.e. by providing a characteristic property of its elements. We use $\{x \mid P(x)\}$ to denote the set of elements x such that $P(x)$, and $\{x \in E \mid P(x)\}$ to denote the set of elements x which are members of E and such that $P(x)$. The second form is better because the first can lead to paradoxes.

3.4.2 Logical Operators

Tables 3.1 and 3.2 summarize the intuitive meaning of logical operators as well as their relation to set-theoretic operations. These intuitions will be developed and explained in subsequent chapters.

The meaning of conjunction $\wedge$ and of negation $\neg$ is just the one you would expect. The same is true of disjunction $\vee$ as well, but be aware that we have a *non exclusive or*. Interpreting implication $P \Rightarrow Q$ must be done with greater caution: nothing tells us that there is an actual causality relation between P and Q. We can only say

Table 3.1

t	true
f	false
¬	not
∧	and
∨	or
⇒	implies
⇔	is equivalent to
∀	for all
∃	exists

Table 3.2

E		$x \in E$
A, B		P, Q
$A \cap B$	inter	$P \wedge Q$
$A \cup B$	union	$P \vee Q$
$A - B$	minus	$P \wedge \neg Q$
$A \setminus B$	symmetric difference	$\neg(P \Leftrightarrow Q)$
$\varnothing$	empty set	f

that Q happens to be true when P is true. Thus $\forall x\, R(x) \Rightarrow S(x)$ means that all x verifying R verify S as well. If no x verifies R, we agree that the formula $\forall x\, R(x) \Rightarrow S(x)$ is true. We have, therefore, in this case $\forall x\, \mathbf{f} \Rightarrow S(x)$ and, as $S(x)$ may be true or false, we see that both $\mathbf{f} \Rightarrow \mathbf{t}$ and $\mathbf{f} \Rightarrow \mathbf{f}$ are true.

The logical equivalence $P \Leftrightarrow Q$ is an abbreviation for the conjunction $(P \Rightarrow Q) \wedge (Q \Rightarrow P)$. It behaves like an equality; hence we can replace P with Q when $P \Leftrightarrow Q$. Table 2 above can read: $x \in A \cap B \Leftrightarrow x \in A \wedge x \in B$, etc., $x \in \varnothing \Leftrightarrow \mathbf{f}$.

Numerous logical laws can be stated using equivalences. For instance, consecutive conjunctions can be reordered with $P \wedge Q \Leftrightarrow Q \wedge P$ and $(P \wedge Q) \wedge R \Leftrightarrow P \wedge (Q \wedge R)$. The same is true for disjunction. We have also $P \wedge \neg P \Leftrightarrow \mathbf{f}$ and $P \vee \neg P \Leftrightarrow \mathbf{t}$.

The constants $\mathbf{t}$ and $\mathbf{f}$ can be eliminated using $P \wedge \mathbf{t} \Leftrightarrow P$, $P \vee \mathbf{f} \Leftrightarrow P$, $P \wedge \mathbf{f} \Leftrightarrow \mathbf{f}$ and $P \vee \mathbf{t} \Leftrightarrow \mathbf{t}$. Hence we see that $x \in \varnothing \Leftrightarrow \mathbf{f}$ boils down to $x \in \varnothing \Rightarrow \mathbf{f}$.

Here are other very useful identities:

$$P \vee (Q \wedge R) \Leftrightarrow (P \vee Q) \wedge (P \vee R) \tag{3.2}$$
$$P \wedge (Q \vee R) \Leftrightarrow (P \wedge Q) \vee (P \wedge R) \tag{3.3}$$
$$\neg\neg P \Leftrightarrow P \tag{3.4}$$
$$P \Rightarrow Q \Leftrightarrow \neg P \vee Q \tag{3.5}$$
$$P \Rightarrow \mathbf{f} \Leftrightarrow \neg P \tag{3.6}$$
$$\neg(P \wedge Q) \Leftrightarrow \neg Q \vee \neg P \tag{3.7}$$
$$\neg(P \vee Q) \Leftrightarrow \neg Q \wedge \neg P \tag{3.8}$$
$$\neg\forall x\, P(x) \Leftrightarrow \exists x\, \neg P(x) \tag{3.9}$$
$$\neg\exists x\, P(x) \Leftrightarrow \forall x\, \neg P(x) \tag{3.10}$$
$$(P \wedge Q) \Rightarrow R \Leftrightarrow P \Rightarrow (Q \Rightarrow R) \tag{3.11}$$
$$(P \wedge U) \Rightarrow Q \Leftrightarrow P \Rightarrow (\neg U \vee Q) \tag{3.12}$$

For example, using (3.6), the last line of Table 2 is equivalent to $\neg x \in \varnothing$: as expected, no element can be a member of $\varnothing$. The laws (3.7) to (3.10), called De Morgan's laws, allow us to distribute negation across other connectives. The equivalence (3.11) provides two ways for expressing "if I have P, if I have Q then I have R". This can also be written $P \Rightarrow Q \Rightarrow R$. Using (3.5) we get the equivalence (3.12) that allows us to move a formula U to the opposite side of an implication at the price of a negation.

> EXERCISE. Show the equivalence $(P \lor Q) \land \neg(P \land Q) \iff \neg(P \iff Q)$. Justify $A \setminus B \iff \neg(P \iff Q)$, where $P \overset{\text{def}}{=} x \in A$ and $Q \overset{\text{def}}{=} x \in B$, from $A \setminus B = (A \cup B) - (A \cap B)$.

> EXERCISE. Find the logical laws used in the reasoning on page 29 for proving partial correctness of the first bounded linear search program.

3.4.3 Relations and Functions

A (binary) **relation** R from A to B is a subset of $A \times B$. Its elements, which are ordered pairs $\langle a, b \rangle$ with $a \in A$ and $b \in B$, are also denoted by $a \mapsto b$. Then we say that a **is related to** b or that a **maps to** b by R. We often use the infix notation $a R b$ instead of $\langle a, b \rangle \in R$.

A simple example of a relation is the **identity relation** on a set A, which is the set of all ordered pairs $a \mapsto a$ such that $a \in A$.

A relation R on A is **reflexive** if for all x in A, $x R x$.

It is **symmetric** if $\forall x, y \in A$, $x R y \Rightarrow y R x$.

It is **anti-symmetric** if $\forall x, y \in A$, $(x R y \land y R x) \Rightarrow x = y$.

It is **transitive** if $\forall x, y, z \in A$, $(x R y \land y R z) \Rightarrow x R z$.

An **equivalence relation** is a reflexive, symmetric and transitive relation. An **order** is a reflexive, anti-symmetric and transitive relation. An order is **total** when two elements can always be compared: $\forall x, y \in A$, $x R y \lor y R x$. In the opposite case (or if we don't know) we have a **partial** order.

If R is an order on A and if B is a subset of A, an element m of A is a **lower bound** (respectively an **upper bound**) of B if $\forall b \in B \; m R b$ (respectively $\forall b \in B \; b R m$).

A relation R from A to B is **defined at** a with $a \in A$, if there exists an ordered pair $a \mapsto b$ in R, i.e. if a is mapped to an element of B by R. The **domain** of R is the set of elements a such that R is defined at a. R is a **total** relation if its domain is A. In the opposite case (or if we don't know) we say that R is **partial**. The set of total functions from A to B is denoted by $A \to B$.

A **function** f from A to B is a relation such that if $x \mapsto y_1$ and $x \mapsto y_2$ are members of f, then $y_1 = y_2$ (intuitively, applying a function to a given element always yields the same result). If $x \in A$ and if x is in the domain of f, we denote fx or $f(x)$ the unique element y of B such that $x \mapsto y$ is a member of f.

The **composition** of two functions f and g from B to C and from A to B, respectively is the function from A to C denoted by $g \circ f$ such that $(g \circ f) x = g(fx)$. This definition generalizes if f and g are relations. In that case $x \mapsto z$ is a member of $g \circ f$ if and only if there exists a y in B such that $x \mapsto y$ is a member of f and $y \mapsto z$ is a member of g.

The first **projection** p_1 is the function from $A \times B$ to A defined by $p_1\langle a, b\rangle = a$. Similarly the second projection p_2 is defined by $p_1\langle a, b\rangle = b$. More generally, the ith projection is the function from $A_1 \times \ldots A_i \times \ldots$ to A_i defined by $p_i\langle a_1, \ldots a_i, \ldots\rangle = a_i$.

A function is **injective** if distinct elements are mapped to distinct elements. A function f from A to B is **surjective** if all elements of B are mapped by f. A **bijection** is a total, injective and surjective function.

3.4.4 Operations

An **operation** $\star$ on the set A is a total function from $A \times A$ to A. It is **commutative** if for all x, y of A we have $x \star y = x \star y$. It is **associative** if for all x, y, z of A we have $(x \star y) \star z = x \star (y \star z)$.

The element e is called an **left identity** element of $\star$ (respectively a **right identity** element) if for all x in A we have $e \star x = x$ (respectively $x \star e = x$). The element a is called a **left absorbing** element of $\star$ (respectively a **right absorbing** element) if for all x in A we have $a \star x = a$ (respectively $x \star a = a$). The element x' is called a **left inverse** (respectively **right inverse**) of x if $x' \star x = e$ (respectively $x \star x' = e$). An **identity element** (respectively an **inverse**, an **absorbing element**) is a left and right identity (respectively inverse, absorbing) element.

Notation: when the underlying operation $\star$ is clear from the context, it is often omitted: one writes xy instead of $x \star y$. If $\star$ is associative, one also writes x^n for $x \star \ldots \star x$ (with n occurrences of x). The inverse of an element x (when it exists) is denoted by x^{-1}.

Example: given a set A, let $\mathcal{R}_A$ denote the set of relations on A. Then $\circ$ is an operation on $\mathcal{R}_A$, with the identity relation as an identity element. The **inverse** of a relation R (written R^{-1}) is then the set of ordered pairs $y \mapsto x$. such that $x \mapsto y$ is in R. A function is injective if and only if the inverse relation is a function. A function is surjective if and only if the inverse relation is total. A function is bijective if and only if the inverse relation is a total function.

An element x is said to be **idempotent** if $x \star x = x$. The operation $\star$ is **idempotent** if all elements of A are idempotent.

EXERCISE. The connectives $\wedge$, $\vee$, $\Rightarrow$ and $\Leftrightarrow$ can be seen as operations on $\mathbb{B}$ (see § 5.1.3). Which of them are commutative? Associative? Idempotent? Which ones possess an identity element? An absorbing element? Invertible elements? Do not neglect $\Leftrightarrow$.

3.4.5 Morphisms

Let us consider the set of natural integers endowed with addition and the identity element 0 on the one hand, the set of natural integers endowed with multiplication and the identity element 1 on the other hand. The function φ which maps n in $\mathbb{N}$ to 3^n preserves the identity element and the operation in the following sense: $\varphi(0) = 1$ and $\varphi(m + n) = \varphi(m) \times \varphi(n)$. We say that φ is a morphism from $\langle \mathbb{N}, +, 0 \rangle$ to $\langle \mathbb{N}, \times, 1 \rangle$.

Let us consider a more general case. We take a set E endowed with a function f, an operation $\star$, and a relation R. This structure is denoted by a 4-tuple: $\langle E, f, \star, R \rangle$. Let us take a similar structure $\langle E', f', \star', R' \rangle$. A morphism of $\langle E, f, \star, R \rangle$ to $\langle E', f', \star', R' \rangle$ is a function φ from E to E' which preserves the structure in the following sense. Let x, y, z be arbitrary elements of E and let x', y', z' their respective targets by φ: thus we have $x' = \varphi(x)$, $y' = \varphi(y)$ and $z' = \varphi(z)$. The function φ is a morphism if:

- φ preserves the function: if $y = f(x)$, then $y' = f'(x')$;
- φ preserves the operation: if $z = x \star y$ then $z' = x' \star' y'$;
- φ preserves the relation: if $x R y$ then $x' R' y'$.

An **isomorphism** is a bijective morphism. Two structures are **isomorphic** if they are related by an isomorphism. Intuitively, we can to a fair extent agree that they are identical because they have exactly the same properties.

3.4.6 Numbers

Common number sets ($\mathbb{N}$, $\mathbb{Z}$, $\mathbb{Q}$ and $\mathbb{R}$) are recalled on page 22. Natural numbers can be generated from the empty set using the following encoding: 0 is encoded by $\{\} = \varnothing$, 1 is encoded by $\{0\} = \{\varnothing\}$, 2 is encoded by $\{0, 1\} = \{\varnothing, \{\varnothing\}\}$, $\ldots$ n is encoded by $\{0, \ldots n - 1\}$.

It is not as obvious as it may seem to define what is a finite or an infinite set. A first idea could be to count its elements and to say that the set E is infinite if there is an injection (an injective function) from $\mathbb{N}$ to E. In fact, the "axiom of infinity" stated in Chapter 7 says that there is a set containing $\mathbb{N}$. We can avoid the reference to $\mathbb{N}$ in the following way: a set E is said to be **infinite** if and only if there is a bijection from E to a proper subset of E.

A set E is **countable** if there exists a sequence (u_n) of elements of E covering E, or, equivalently, if there exists a surjective function from $\mathbb{N}$ to E (intuitively: we can count the elements of E). For example, finite sets, $\mathbb{N}$ itself, $\mathbb{Z}, \mathbb{Q}$ $\mathcal{P}_F(\mathbb{N})$ (the set of **finite** subsets of $\mathbb{N}$) are countable. Among sets that are not countable we have $\mathbb{R}$ and $\mathcal{P}(\mathbb{N})$ (the set of all subsets of $\mathbb{N}$). Here is an important example for computer science: a set whose elements can always be denoted by a finite sequence of characters taken in a finite alphabet is countable. In particular, the set of programs defined in all programming languages is countable, whereas the set of functions on natural numbers is not countable.

A collection where the element can be repeated is called a **family** or a **multiset**. Formally, if E is a set, a family of elements of E is a total function from E to $\mathbb{N}$.

3.4.7 Sequences

A sequence $u_0, u_1, \ldots u_n, \ldots$ of elements u_i of E is a total function from $\mathbb{N}$ to E: u_n is just another notation for $u(n)$. A sequence can be defined directly (for example $v_n = n^2$) or by **induction**, by providing the value of u_0 and a function yielding the value of u_{n+1} from u_n (for example $u_0 = 0$ and $u_{n+1} = u_n + 2n + 1$). The typical way of proving properties of such sequences is through proof by induction. On the last example it is easy to prove: $\forall n \, u_n = v_n$.

We sometimes need to talk about sequences that are finite or infinite. We mean, total functions from A to E, where A is either a subset of $\mathbb{N}$ of the form $\{n \in \mathbb{N} \mid n < a\}$ for a given natural number a, or $\mathbb{N}$ itself. We will then use the explicit terminology "finite or infinite sequence", "finite sequence" when A has the first form and "infinite sequence" when A has the second form. In other contexts "sequence" will always denote an infinite sequence.

3.5 Well-founded Relations and Ordinals

3.5.1 Loop Variant and Well-founded Relation

We have seen in § 2.4.1.2 that the termination of a program can be studied by considering a quantity v that decreases at each step while staying in $\mathbb{N}$. Let us emphasize the last point. It is not enough to ensure that the variant v is a decreasing number:

- an integer can decrease *ad vitam æternam* by taking arbitrarily large negative values;
- a positive rational or real number can decrease while approaching a lower limit without reaching it.

The point is that v must take a *finite* number of values. Reasoning with a "decreasing number" is of course an incorrect wording, which has to be formalized with a finite or infinite sequence $v_0, v_1, \ldots v_n, \ldots$ as we will see below.

In order to model the problem of termination, let us first consider the set S of the values that can be taken by the state of a program.[8] The change in this state is observed at certain points between which we admit that nothing important can happen.[9] Each execution step corresponds to a state transition which is modeled as an ordered pair $\langle s_i, s_f \rangle$ where s_i and s_f, the value of the state respectively at the beginning and at the end of the transition, are

---

[8]For the sake of completeness we should include in the state a component for the program counter and another for the execution stack. We proceed in this manner in order to define an operational semantics.

[9]We can choose fine grain observation, corresponding to elementary instructions or large grain observation, corresponding to blocks of such instructions: the point is that executing those "grains" always terminates.

members of S. We then introduce the set of transitions $\mathcal{T}$, which is a relation on S.

When we reason with a variant v, the latter is a function of the state s. Each transition $\langle s_i, s_f \rangle$ at the level of states corresponds to a transition $\langle v(s_i), v(s_f) \rangle$ at the level of the variant. The general situation is then captured by a set S endowed with a relation $\mathcal{T}$.

The changes of the state during an execution beginning at initial state s_0 are then modeled by a finite or infinite sequence $s_0, s_1, ...s_n, ...$ such that two consecutive elements s_k and s_{k+1} are always related by $\mathcal{T}$. Ensuring the termination of the program boils down to prohibiting the sequence from being infinite. For example, in the case of natural integers, there is no infinite sequence $v_0, v_1, ...v_n, ...$ such that $v_0 > v_1 > ... > v_n > ...$, which allowed us to justify the technique of the variant on page 24. When no such sequence exists the relation is said to be **Noetherian**. We can similarly consider the inverse relation (recall that, for instance, $<$ and $>$ are inverse relations). We then have a well-founded relation. Let us develop this concept.

Let E be a set and R a relation on E. Let x and y be two elements of E, we say that x is a **predecessor of y for R** if $x R y$. When there is no ambiguity we simply say that x is a predecessor of y. A **chain** is a finite or infinite sequence $e_0, e_1, ...e_n, ...$ of elements of E such that e_{n+1} is always a predecessor of e_n: $\forall n \in \mathbb{N} \; e_{n+1} R e_n$. R is a **well-founded** relation if R contains no infinite chains.[10]

The concept of predecessor that we use here generalizes from the usual one on integers: just take for R the relation noted R_1 below. For an arbitrary relation R, the predecessor of an element, when it exists, need not be unique.

In summary, expressing that a program terminates boils down to saying that the underlying transition relation $\mathcal{T}$ is Noetherian, or that the inverse relation $\mathcal{T}^{-1}$ is well founded. In practice, instead of reasoning directly on the set of states S endowed with $\mathcal{T}^{-1}$, it is worth considering a simplified view E of S endowed with a corresponding relation R, which must be well founded as well. The loop variant presented in the above example amounts to taking $\mathbb{N}$ for E and $<$ for R.

3.5.2 Examples

The relation $<$ is well founded on $\mathbb{N}$, but is not well founded on $\mathbb{Z}$, nor on any interval of $\mathbb{R}$ or of $\mathbb{Q}$. Any relation included in a well-founded relation is also well founded. Hence all sets of ordered pairs of natural integers $\langle m, n \rangle$ verifying $m < n$ are well founded. Here are three examples:

$$R_1 \stackrel{\text{def}}{=} \{\langle n, n+1 \rangle \mid n \in \mathbb{N}\}$$

---

[10]Nothing prevents the repetition of an element in a sequence. If x is such that $x R x$, the sequence $x, x, ... x, ...$ is then an infinite chain. If x and y satisfy $x R y$ and $y R x$, the sequence $x, y, ... x, y, ...$ is an infinite sequence as well.

$R_2 \overset{\text{def}}{=} \{\langle n, 2n+\varepsilon\rangle \mid n \in \mathbb{N} \wedge n > 0 \wedge (\varepsilon = 0 \vee \varepsilon = 1)\}$

$R_3 \overset{\text{def}}{=} \{\langle n, n+2\rangle \mid n \in \mathbb{N}\}$

The relation $>$ is not well founded on $\mathbb{N}$, but it becomes so on a finite subset of $\mathbb{N}$. As a consequence, relations having the form $R_4(q)$ are well-founded:

$R_4(q) \overset{\text{def}}{=} \{\langle n+1, n\rangle \mid n \in \mathbb{N} \wedge n < q\}$

Here is a very important example. Let R be a well-founded relation on E and let S be a well-founded relation on F, the relation defined over $E \times F$ by

$$R_5 \overset{\text{def}}{=} \{\langle\langle x, u\rangle, \langle x', v\rangle\rangle \mid x, x' \in E \wedge u, v \in F \wedge x R x'\} \ \cup$$
$$\{\langle\langle x, u\rangle, \langle x, u'\rangle\rangle \mid x \in E \wedge u, u' \in F \wedge u S u'\}$$

is well founded. This construction corresponds to the **lexicographic ordering** used by all of us when consulting a dictionary.[11] This example is more subtle than the previous ones. If we consider the relation $<$ on $\mathbb{N}$ (or its subsets R_1, R_2 and R_3), we have already observed that all decreasing sequences are finite. But additionally, we know an upper bound on the length of such sequences as soon as we know the first element (the latter is such an upper bound). In contrast, if we take the structure $\langle \mathbb{N}, < \rangle$ or even $\langle \mathbb{N}, R_1 \rangle$ for $\langle F, S \rangle$ in R_5, it is no longer possible to give an upper bound for decreasing sequences starting from $\langle x_0, n_0 \rangle$ if there is no x_1 in E such that $x_1 R x_0$. In that case there exist an infinite number of finite decreasing sequences starting from $\langle x_0, n_0 \rangle$, and their length is arbitrarily large.

The lexicographic ordering on the Cartesian product of two or of any finite number of well-founded sets is well founded. Note however, that the lexicographic order on words, that is, arbitrarily large finite sequences of elements of a well-founded set E, is *not* well founded. For instance, with $E = \{0, 1\}$ and $0 < 1$, we have the infinite decreasing chain 1, 01, 001, 0001, etc.

Generalizing the technique of loop variants with well-founded relations can be useful in two ways:

1. We can acquire a knowledge of the number of iterations performed when executing a loop.
2. We can cope with more complex situations involving several loops, whether embedded or not.

3.5.2.1 Counting Iterations in a Loop. First recall that the number of iterations n_i depends on the initial value v_0 of the variant. In general, the latter depends in turn on a preliminary computation or on an external event — reading a number for example — and is then essentially unpredictable. In contrast we can ask how n_i depends on v_0.

---

[11] One should pay attention to the following technical point: a well-founded relation like R or S is not an order because it cannot be reflexive. We come back to the links between these concepts in § 3.5.4.1.

Let us take N as the domain of the variant. If the well-founded relation at hand is $<$, we only know that $n_i \leq v_0$. If the relation is R_1, we have $n_i = v_0$. If the relation is R_2, we know that n_i is close to the base 2 logarithm of v_0.

3.5.2.2 Using more Complex Well-Founded Relations.
In order to study the termination of programs composed of several loops using only one well-founded relation, the domain E we have to consider for the latter has to be larger[12] than N. Here we content ourselves with the simple case of a program made of a first loop, followed by the computation or the reading of an arbitrarily large positive integer L and finally a second loop.

Let us first consider each loop separately. Assume that the variant of the first is v in N endowed with R_1 whereas the variant of the second is w also in N endowed with R_1. For all initial values v_0 and w_0, it is intuitively clear that the program terminates since each loop terminates. If we knew in advance the value of w_0, we could take $u = v + w$ as the global variant, in the same domain N endowed with R_1. To be more precise, u would be defined as $v + w_0$ in the first loop, as w_0 between the two loops and as w in the second loop. But we cannot proceed in this way if the value L taken by w_0 is unknown in advance and arbitrarily large.

A satisfactory solution is to take for E the sum of two copies of N or, equivalently, the Cartesian product[13] $\{0,1\} \times$ N. The variant u is $\langle 1, v \rangle$ in the first loop, $\langle 1, 0 \rangle$ between the two loops (let us call this element ω) and $\langle 0, w \rangle$ in the second loop. Our well-founded relation $R_{1,1}$ is defined by $\langle i, n \rangle R_{1,1} \langle i, n+1 \rangle$ (intuitively it behaves like R_1 on each copy of N) and $\langle 0, n \rangle R_{1,1} \omega$. $R_{1,1}$ is contained in the relation R_5 above, where we take $E = \{0,1\}$, $R = \{\langle 0,1 \rangle\}$, $F = $ N and $S = R_1$.

Let us point out that, in contrast with most relations presented so far, the element $\langle \omega \rangle$ admits an infinite number of predecessors in $R_{1,1}$. However, a decreasing sequence starting from any element of $\{0,1\} \times$ N is necessarily finite.

A relation like $R_4(q)$ can be convenient in practice. For instance, $R_4(N)$ may be used for a direct termination proof of the bounded linear search program instead of reasoning on the difference $N - x$, as we did on page 25.

We also remark on R_2, R_3 and R_4 that it is not required that only one value (0) has no predecessor, even if we consider only natural (i.e. non-negative) numbers:[14] in R_2, we have 0 and 1; in R_3 we have 0 and all odd natural numbers; in $R_4(q)$ we have all numbers greater or equal to q. This is reflected in the loop invariant and in the exit test. For example, with R_4, we have to ensure that, at the beginning of the loop, the variant v is strictly less than q (condition $(V_<)$ on page 24, reshaped with R_4, tells us that during an iteration v is necessarily incremented by 1); in this situation, we are led to put $v \leq q$ in the invariant,

---

[12]In a sense coming from the theory of ordinal numbers, see below.

[13]Technically we can also represent N + N by N (consider even and odd numbers). But it would only make the definition of the well-founded relation more complicated with no compensation in the reasoning. The concept of ordinal presented below clarifies the situation.

[14]We choose to keep $0 \leq v$ in the invariant.

and then to take $v = q$ as the exit condition. With R_3 the exit test would correspond to $v = 0$ and the invariant would entail that v is even.

3.5.3 Well-founded Induction

Given a well-founded relation R on a set E, we can prove that a property P is true on all elements of E by showing the following proposition (H) which tells us, in familiar terminology, that P propagates:

> given any element x of E,
> if P is true on all predecessors of x,   (H)
> then P is true on x.

In particular, we have to show that P is true on all x without a predecessor, which corresponds to the base cases.

This kind of reasoning is called **well-founded induction**. Usual induction on $\mathbb{N}$ is a (simple) special case of well-founded induction, where the relation considered is R_1. Assume that, despite the fact that (H) has been shown, we have an element e_0 where P is not true; e_0 has at least one predecessor, since P is true for all elements without a predecessor; by (H) we also know that P is false on at least one of the predecessors of e_0; let e_1 be one of them. Repeating the process would then yield an infinite decreasing chain $e_0, e_1, \dots e_n, \dots$, which is impossible because R is well-founded.

The previous reasoning implicitly uses a principle called the *axiom of choice*, which will be introduced in Chapter 7. Indeed, in order to construct the chain $e_0, e_1, \dots e_n, \dots$ we simultaneously construct the infinite family $P_0, P_1, \dots P_n, \dots$ where P_i is the non-empty set of predecessors of e_i. At each step, we have to choose e_{i+1} in P_i.

The rule of the loop is an application of well-founded induction. Let us illustrate what happens with the relation R_3. This corresponds to a loop B where the initial value of v is even:

while $v \neq 0$ **do** ...$v := v - 2$... **done**

We then have to show the property $P(n)$ defined by $\{v = n \wedge I\} \, B \, \{I\}$, where I is the loop invariant. We distinguish the "true" base case $n = 0$ (corresponding to a successful exit test) from the "false" ones (odd values of v). In the latter cases $P(n)$ is trivially true by reduction to the absurd, provided we put "v is even" in the invariant.

3.5.4 Well Orders and Ordinals

We can present well-founded induction from special order relations. Here are some preliminary definitions. The main point to remember is that two isomorphic ordered sets are essentially the same up to the name of their elements. A set E endowed with an order R will be denoted by a 2-uple $\langle E, R \rangle$.

Let $\langle E, R \rangle$ and $\langle F, S \rangle$ be two ordered sets. A function f from E to F is **monotone** if the order is preserved by f:

$$\forall x, y \in E \quad xRy \Rightarrow f(x) S f(y) \ .$$

An **isomorphism** is a monotonic bijection. Two ordered sets $\langle E, R \rangle$ and $\langle F, S \rangle$ are **isomorphic** if there is an isomorphism from $\langle E, R \rangle$ to $\langle F, S \rangle$.

3.5.4.1 Well Orders. Let E be a set endowed with an order R. Given a subset A of E, a **minimum** of A, if it exists, is an element a of A such that there is no predecessor different from a in A: if $xRa \wedge x \in A$ then $x = a$ (R is reflexive!). If R is total, a minimum of A must be unique.

We say that R is a **well order** if R is total and if every subset of E possesses a minimum. Note that E possesses a unique minimum m in that case.

Let us note $R_{\neq}$, the relation defined by $x R_{\neq} y$ if and only if xRy and $x \neq y$. If R is a well order, $R_{\neq}$ is a well-founded relation. Conversely it is possible to construct a well order from a well-founded relation. But beware: a given well order can come from several well-founded relations.

The concept of a well-founded induction is defined as in § 3.5.3 if we replace R with $R_{\neq}$. The base case concerns only m. This principle can be justified as follows. Suppose that the set A of elements e which do not verify P is not empty, A possesses a minimum a which must be different from m; the predecessors of a are not members of A, hence they verify P, but with (H) we then have that P is also true of a, so a cannot be a member of A, a contradiction.

Some well orders are especially important: ordinals.

3.5.4.2 Ordinals. Let E be a set endowed with the well order R. The **section** X_a determined by an element a of E is defined as the set of elements x which are smaller than a:

$$X_a \stackrel{\mathrm{def}}{=} \{x \in E \mid xR_{\neq}a\} \ .$$

E endowed with the well order R is an **ordinal** if for all a of E we have $X_a = a$. Thus, to verify that 3 is an ordinal, we just have to remember that in set theory $3 \stackrel{\mathrm{def}}{=} \{0, 1, 2\}$, which actually yields $2 \stackrel{\mathrm{def}}{=} \{0, 1\} = X_2$. The first ordinals are exactly $\varnothing$, $\{\varnothing\}$, $\{\varnothing, \{\varnothing\}\}$, etc., where the order is inclusion or, equivalently, membership (the two relations happen to coincide on ordinals).

Given an arbitrary ordinal x we can construct its successor $x \cup \{x\}$. We then start from $\varnothing$ and we construct all natural numbers step by step. The next step consists of taking $\mathbb{N}$ itself (it can be shown that $\mathbb{N}$ satisfies the required properties). $\mathbb{N}$ is traditionally noted ω in this context.

The process carries on in the same way: ω, $\omega \cup \{\omega\}$ (noted $\omega+1$), etc. Apart from 0 only two cases can occur for an ordinal: either it contains a greatest element, it has then the shape $x \cup \{x\}$ and it is called a **successor ordinal**; or, it does not contain a greatest element and it is called a **limit ordinal**.

The first limit ordinal is ω. The next one, noted 2ω, is the limit of $\{0, 1, \dots \omega, \omega+1, \dots\}$. Carrying on this process we define $3\omega, \dots n\omega, \dots \omega^2, \dots \omega^\omega, \dots \omega^{\omega^\omega}$,

... until a new limit ordinal ϵ_0 which verifies $\omega^{\epsilon_0} = \epsilon_0$. There are still many other ordinals. Ordinals up to ϵ_0 are used in the automated proof assistant of Boyer–Moore and in PVS in order to formalize termination arguments [Rus93].

An important theorem about ordinals states that a well order is always isomorphic to an ordinal. Ordinals can then be used for measuring the complexity of termination proofs of algorithms. Let us also remember that the most general form of induction is well-founded induction, because the concept of a well-founded relation is finer than the concept of well order.

3.5.4.3 Ordinals and Cardinals. Cardinals are another concept of set theory that can be used for measuring the size of a set. We will not go into detail here. We say that two sets have the same cardinality if there exists a bijection between them. Finite sets have a cardinal 0, 1, 2, ... n with $n \in \mathbb{N}$.

Next we have $\mathbb{N}$ itself, whose cardinality is denoted $\aleph_0$ (pronounced *aleph zero*). We already know from § 3.4.6 that many infinite sets are countable: in other words, their cardinality is $\aleph_0$.

Another important point is the following. If the cardinal of a set E is α, then the cardinal of $\mathcal{P}(E)$ is strictly greater than α.

All ordinals presented so far are countable. A better wording is: the underlying sets of those ordinals are countable. We must remember that what matters in an ordinal is the corresponding order. Indeed, there are many (non-isomorphic) ways to order the elements of $\mathbb{N}$, and each of them corresponds to a different ordinal.[15] However, the order is completely irrelevant for cardinals.

In contrast to ordinals, cardinals don't seem to have applications in formal methods. Note, however, that they play an important role in set theory.

One of the first questions raised at the very beginning of development in set theory was the following: let c be the cardinal of $\mathbb{R}$; c is also the cardinal of $\mathcal{P}(\mathbb{N})$, thus we have $c > \aleph_0$; but is there an intermediate cardinal? Cantor thought that the answer should be no — this is called the *continuum hypothesis* — but the question turned out to be arduous. Gödel showed in the 1930s that this hypothesis is consistent with (i.e. cannot be disproved from) the axioms of set theory, while conversely Cohen showed in 1963 that it cannot be proven in set theory. This reveals the somewhat arbitrary character of set theory. We come back to this point at the end of Chapter 7.

3.6 Fixed Points

Let $\mathcal{E}$ be a set and f be a function from $\mathcal{E}$ to $\mathcal{E}$. A fixed point of f is an element x of $\mathcal{E}$ such that $x = f(x)$. For example 1 and 5 are fixed points of the function

---

[15]For example, if the order we consider is $<$, the corresponding ordinal is ω. However, let us consider the order R, defined by $x R y$ if $x < y$ and $x \neq 0$, and by $x < 0$ for all x: the corresponding ordinal is $\omega + 1$. Intuitively, in the latter case, natural integers are put in the following order: 1, 2, ... 0. The two relations $<$ and R are not isomorphic since only the second one possesses a greatest element.

on $\mathbb{R}$ that maps x to $(x^2 + 5)/6$. The theorem of Knaster–Tarski states that under quite general conditions, f is guaranteed to have a least or a greatest fixed point. This allows us to *define* x by a fixed-point equation.

We suppose that (1) $\mathcal{E}$ is ordered by a relation $\leq$; (2) f is monotone, that is, $x \leq y \Rightarrow f(x) \leq f(y)$; (3) all non-empty subsets A of $\mathcal{E}$ have a greatest lower bound $glb(A)$ (it is not necessary that $glb(A)$ is a member of A); and (4) $post_f = \{x \in \mathcal{E} \mid f(x) \leq x\}$ is non-empty (elements of $post_f$ are called post-fixed points of f). In our example we have $4 \in post_f$. Then f possesses a least fixed point which is $\mu_f = glb(post_f)$.

Indeed — let us remove the index f — as μ is a lower bound of post-fixed points, we have $\mu \leq x$ for all x such that $f(x) \leq x$, then, as f is monotone: $f(\mu) \leq f(x) \leq x$; then $f(\mu)$ is also a lower bound of $post$. As μ is greater than all lower bounds, we get $f(\mu) \leq \mu$. By monotony $f(f(\mu)) \leq f(\mu)$, hence $f(\mu) \in post$, then $\mu \leq f(\mu)$ since μ is a lower bound of $post$. By anti-symmetry of $\leq$ we have that $\mu = f(\mu)$.

Symmetrically, if all non-empty subsets A of $\mathcal{E}$ have a least upper bound $lub(A)$ and if the set $pre_f = \{x \in \mathcal{E} \mid x \leq f(x)\}$ of pre-fixed points of f is non-empty, then f possesses a greatest fixed point $\nu_f = lub(pre_f)$.

The least fixed point can also be reached from below when $\mathcal{E}$ possesses a least element $\bot$ (take $\mathcal{E} = [0, +\infty[$ in the previous example): we construct the monotonic sequence $(u)_\alpha$ with $u_0 = \bot$, $u_{\alpha+1} = f(u_\alpha)$ and $u_{\lim(\alpha_n)} = lub\{u(\alpha_n)\}$. The process ends at the first limit ordinal ω if f is continuous, i.e. $f(lub\{x_i\}) = lub\{f(x_i)\}$ for all monotonic sequences $(x_i)_{i \in \mathbb{N}}$. For the greatest fixed point, one would proceed symmetrically from a greatest element $\top$ in E.

The relation $\leq$ is not required to be total here. We can then apply the previous results with the inclusion relation on a set of sets, for example $\mathcal{E} = \mathcal{P}(E)$: $\varnothing$ plays the role of $\bot$, $glb(A)$ is the intersection of elements of A, $lub(A)$ is the union of elements of A and E plays the role of $\top$.

3.7 More About Computability

Here we give more precise definitions for the concepts of computability mentioned above [Gir87b, Bar90]. Here, unless we explicitly write *partial recursive function*, a *recursive function* will mean a *total recursive function*, according to the original definition of Gödel and Herbrand. Note that, following the work of Kleene, many textbooks use the opposite convention.

Let us consider a problem P. If we have a search process for solutions of P at our disposal which (i) succeeds if a solution exists, and (ii) answers "no" in the converse case, this process is called a **decision procedure**. If condition (i) only is satisfied, i.e. if the process may go on looking indefinitely for a solution where no solution exists, it is called a **semi-decision procedure**. To summarize what follows, a decision algorithm is a recursive function, while a semi-decision procedure is a partial recursive function.

For the remainder of this chapter, the functions considered are arithmetic functions. By that we mean functions that take natural integers as input and that return a natural integer. For the sake of uniformity, constants are considered to be functions of arity 0. In order to lighten the notation, applying a function f to n arguments $x_1 \dots x_n$ is denoted by $f(\vec{x})$, where $\vec{x}$ is seen as the n-tuple $\langle x_1, \dots, x_n \rangle$ — the value of n is the arity of f.

In order to formalize the concept of an algorithm, we need a formal language capable of expressing algorithms, and we have to stipulate the computations associated with legal expressions. This can be done with very low level constructs, but it is more convenient to use functions directly. It is easy to understand, for example, how to compute the composition of two functions provided one knows how to compute each of them separately. We proceed by introducing primitive recursive functions, then recursive functions and finally partial recursive functions, which correspond to progressively larger classes of algorithms.

It is important to keep in mind the distinction between the function which is computed, that is, a set of ordered pairs (the **extension** of the function), and the algorithm which performs the computation: two different algorithms may independently and correctly compute the same function f; for example one of them could be primitive recursive while the other is not. According to the following definition, f is then considered as primitive recursive. Indeed, the word *function* below takes its extensional meaning — though the underlying computation remains crucial in the rules (R_i) given below.

It may transpire that the most efficient algorithm that computes a given primitive recursive function is not primitive recursive. For instance, the obvious primitive recursive way for computing the minimum of two integers m and n is not symmetrical: it takes e.g. m steps, while a better algorithm would take $\min(m, n)$ steps. Indeed, a result due to Loïc Colson shows that the latter algorithm cannot be encoded using primitive recursion. Recursion theory is then an important theoretical tool, but the light shed on the concept of expressivity is limited.

3.7.1 Primitive Recursion

The **initial functions** are:
- the constant 0;
- the successor function $S(n) = n + 1$;
- the projections $\mathrm{pr}_i^n(x_1, \dots, x_n) = x_i$, $\quad 1 \le i \le n$.

We then consider the formation rules:

(R_1) **composition** rule: take $k + 1$ functions $h_1, \dots h_k$ and g already constructed and construct the function f defined by $f(\vec{x}) = g(h_1(\vec{x}), \dots, h_k(\vec{x}))$;

(R_2) **primitive recursion** rule: take two functions g and h already constructed and construct the function f defined by

$$\begin{cases} f(\vec{x}, 0) = g(\vec{x}) \\ f(\vec{x}, n+1) = h(\vec{x}, n, f(\vec{x}, n)) \end{cases}.$$

A **primitive recursive presentation** or, a **primitive recursive algorithm**, is an expression constructed only from initial functions and by application of rules (R_1) and (R_2). A **function** f is **primitive recursive** if there exists a primitive recursive presentation which computes f.

The occurrence of f on the right of "=" in (R_2) is not that problematic. Indeed it is clear that $f(\vec{x}, 0)$ is defined for all $\vec{x}$, then $f(\vec{x}, 1)$, and so on. The function f can be regarded as a sequence defined by induction but parameterized by $\vec{x}$: $f(\vec{x})_0, \ldots f(\vec{x})_n, f(\vec{x})_{n+1}, \ldots$

Examples. Addition is primitive recursive, as it can be defined by $\mathrm{add}(m, 0) = m$ and $\mathrm{add}(m, n+1) = S(\mathrm{add}(m, n))$. Multiplication is defined in a similar way. We can then define the factorial function ($\mathrm{fact}(0) = 1$ and $\mathrm{fact}(n+1) = \mathrm{mult}(n+1, \mathrm{fact}(n))$) subtraction (see below), the exponential function, and many other functions over integers. The linear search of an integer n such that $P(n) = 0$, *is not* primitive recursive even if P is:

$$R = R'(0)$$
$$R'(n) = \text{if } P(n) = 0 \text{ then } n \text{ else } R'(n+1) .$$

There is no way to define this function using only the previous rules. By contrast, there is a primitive recursive presentation of bounded linear search between p and q similar to the program given in § 2.4.4.

$$R = R'(q - p)$$
$$\begin{cases} R'(0) = q \\ R'(n+1) = h(n, R'(n)) \end{cases}$$
$$h(n, r_2) = \mathrm{tzer}(q - (n+1), r_2, P(n+1))$$
$$\begin{cases} \mathrm{tzer}(r_1, r_2, 0) = r_1 \\ \mathrm{tzer}(r_1, r_2, n+1) = r_2 \end{cases}.$$

Note that testing the equality to zero, realized by tzer, makes use of rule (R_2), with $g = \mathrm{pr}_1^2$ and $h = \mathrm{pr}_2^4$.

In a programming language like Pascal, we get primitive recursive functions if we restrict iterative control structures to **for** loops (general **while** loops have to be prohibited[16]): in **for** loops, the number of iterations is computed (at run-time, however) *before* the loop. One of the main properties of primitive recursive functions is that they are total, in other words the corresponding programs terminate in all cases.

---

[16] **goto** statements and "recursive" (!) procedures must also be prohibited, as it is clear that such mechanisms are at least as powerful as the **while** loop.

A number of functions over natural integers, like subtraction, are usually not defined everywhere. As a consequence of the last remark, their primitive recursive presentation extends them over the whole set $\mathbb{N}$. The default value is often 0. Thus the usual primitive recursive definition of the predecessor function is $P(0) = 0$ and $P(n + 1) = n$ — using (R_2) with $g = 0$ and $h(n, a) = n$, that is, $h = \mathrm{pr}_1^2$. We get subtraction by iteration of P.

There are total functions that cannot be defined by a primitive recursive presentation, but they are not that easy to find. One of the simplest is the Ackermann function:

$$\left\{ \begin{array}{l} A(0, n) = n + 1 \\ A(m + 1, 0) = A(m, 1) \\ A(m + 1, n + 1) = A(m, A(m + 1, n)) \ . \end{array} \right.$$

It can be shown that this function grows faster than all primitive recursive functions. Its termination can be proven by well-founded induction using a lexicographic ordering based on relation R_5 of § 3.5.2, with $E = F = \mathbb{N}$ and $R = S = R_1$.

3.7.2 Recursion, Decidability

The previous examples clearly show that primitive recursive functions do not exhaust intuitively computable functions. In order to enrich our set of functions, let us introduce the following rule:

(R_3) **minimalization** rule: take a function g already constructed such that

$$\forall \vec{x} \, \exists m \, g(\vec{x}, m) = 0 \tag{3.13}$$

and construct the function f that maps $\vec{x}$ to the smaller m such that $g(\vec{x}, m) = 0$, denoted by $f(\vec{x}) = \mu m[g(\vec{x}, m) = 0]$.

Intuitively, a way to compute this function is by a linear search program: successively try $m = 0$, $m = 1$, etc. until an m satisfying $g(\vec{x}, m) = 0$ is found.

A **recursive presentation**, also called an **algorithm**, is an expression constructed only from initial functions and by application of rules (R_1), (R_2) and (R_3). A **function f is recursive** if there exists a recursive presentation which computes f.

For example, the linear search program R given on page 60 is encoded by a trivial application of (R_3): by hypothesis there exists an n such that $P(n) = 0$, where P is primitive recursive; then we take simply $R = \mu m[P(m) = 0]$.

Again, recursive functions are total functions: requiring condition (3.13) amounts to ensuring a priori that the previous linear search program terminates. In other words, intuitively, an algorithm is a program which provides an answer for all input data. We then get a precise formal definition for the intuitive concept of an algorithm. This formal definition may be considered as arbitrary. However, as in physics, experience decides the matter.

Here we encode a predicate P by a function f_P from tuples of integers to $\{0, 1\}$. A **predicate** is **recursive** if the corresponding function f_P is recursive. We can also define a **recursive set** E as a set (of integers, or of tuples of integers) having a recursive characteristic function. It means that we have at our disposal an algorithm for deciding, given any tuple $\vec{x}$, whether or not $P(\vec{x})$, or equivalently, whether or not $\vec{x}$ is a member of E. We say that a **problem** is **decidable** if the corresponding predicate is recursive. In the opposite case we say that the problem is **undecidable**.

3.7.3 Partial Recursion, Semi-Decidability

In practice and in logic as well, we need to consider programs which do not always terminate. Thus we are led to weaken the rule (R_3) by relaxing condition (3.13).

(R_3') **partial minimalization** rule: take a function g already constructed and construct the function f such that, if there exists an m such that $g(\vec{x}, m) = 0$, returns $f(\vec{x}) = \mu m[g(\vec{x}, m) = 0]$, or else is not defined.

The new rule (R_3') allows one to construct partial functions. Therefore, we now agree that our rules construct partial functions from partial functions.

A **partial recursive presentation** is an expression constructed only from initial functions and by application of rules (R_1), (R_2) and (R_3'). A function f is **partial recursive** if there exists a partial recursive presentation which computes f. Here is another definition: a partial recursive function is a function which can be encoded using a Turing machine.[17] The *Church thesis for partially computable functions* states that the class of partial recursive functions formalizes the intuitive concept of program.

Roughly, we can say that a recursive function is a partial recursive function whose termination is proven in all cases. Let us consider the linear search program given on page 60, where we add an integer parameter x in the search criterion P:

$$R(x) = R'(x, 0)$$
$$R'(x, n) = \text{if } P(x, n) = 0 \text{ then } n \text{ else } R'(x, n + 1) \ .$$

In general, the search succeeds only for special values of x. For example, if we want to search the smaller n such that $2n = x$, we can choose for $P(x, n)$ the expression $(x - 2n) + (2n - x)$ (pay attention to the definition of subtraction!); then it is clear that the computation terminates for even values of x and for no others.

Here is another example of a partial recursive function, sometimes called the Syracuse function. It can only return 1, and in all known experiments

---

[17]We don't present a formal definition of Turing machines here: it is a bit long but raises no difficulty.

it does return. But termination for all inputs remains an open problem so far, thus we don't know if this function is recursive.

$$\begin{cases} U(0) = U(1) = 1 \\ U(n) = U(\frac{n}{2}) & \text{if } n \text{ is even and } n > 1 \\ U(n) = U(3n+1) & \text{if } n \text{ is odd and } n > 1 \ . \end{cases}$$

What is the status of the function t_U that returns 1 if U is total and otherwise returns 0? This presentation of t_U is not recursive. However, the function k_i, which returns a fixed i, is recursive; then t_U is recursive as well, since t_U is either k_0, or k_1, though we don't know which one. We conclude that a computable function, as formally defined in recursion theory — a classical theory admitting the excluded middle principle, is not quite the same as a function we know how to compute.

The last important basic concept we present here is the concept of a recursively enumerable predicate or set. As suggested by the name, it is a set which can be completely covered by application of a calculable function on 0, 1, 2, etc. Equivalently, we can say that membership of this set is a **semi-decidable** problem.

We say that a **set** is **recursively enumerable** if it is the domain of a partial recursive function. We say that a **predicate** P is **recursively enumerable**:

- if it is the characteristic predicate of a recursively enumerable set;
- or, equivalently, if the function g defined by $g(x) = 0$ for all x such that $P(x)$, and undefined elsewhere, is partial recursive;
- or, equivalently, if there exists a recursive function f such that, for all y verifying $P(y)$, there exists x such that $y = f(x)$.

Let x be an integer and let P and Q be partial recursive predicates. The functions computing $P(x) \wedge Q(x)$, $P(x) \vee Q(x)$ and $\neg P(x)$ are partial recursive. If P and Q are recursive, these functions are recursive as well, which allows us to determine if $P(x)$, $Q(x)$, $P(x) \wedge Q(x)$, $P(x) \vee Q(x)$ and $\neg P(x)$ are true. If P and Q are only recursively enumerable, we are only able to determine if $P(x)$, $Q(x)$, $P(x) \wedge Q(x)$ and $P(x) \vee Q(x)$ are true. We have also the following theorem:

Theorem 3.1
A predicate (respectively, a set) is recursive if and only if itself and its negation (respectively, its complement) are recursively enumerable.

3.7.4 A Few Words on Logical Complexity

If the predicate P is recursively enumerable, then so are the predicates $P(x) \wedge Q(x)$, $P(x) \vee Q(x)$ $\forall x < n\ P(x)$, $\exists x < n\ P(x)$ and $\exists x\ P(x)$. However, $\neg P(x)$ and $\forall x\ P(x)$ are not always recursively enumerable.

The intuitive idea is that it is possible to encode the search for an x satisfying $P(x)$ by pseudo-simultaneously checking $P(0)$, $P(1)$, etc., but checking $\forall x\, P(x)$ would in general require an infinite number of verifications. As a consequence, a formula including unbounded quantifiers and (partial) recursive predicates specifies a relation between its free variables, but we may not have any algorithm for computing it. A relation thus specified is called **arithmetical**.

Kleene established that arithmetical relations can be classified according to the **arithmetical hierarchy**, which measures their logical complexity. Formulas are put under the form $\forall x_n \exists x_{n-1}...\Delta$ or $\exists x_n \forall x_{n-1}...\Delta$, where the predicate Δ is primitive recursive. Each class is characterized by the first quantifier and the number of quantifier *alternations*. Formulas of the first kind are designated by Π_n^0, formulas of the second kind by Σ_n^0. For example $\exists n\ n^2 = 25$ is Σ_1^0, while $\forall c \exists r\ r^2 \le c \wedge c < (r+1)^2$ is Π_2^0. The reader can consult [vL90a, Cou91, Sho93] for a rigorous definition.

There is a tight link between complexity of program termination proofs, ordinals and logical complexity [Gir87b, CW97, Wai91, Wai93].

3.8 Notes and Suggestions for Further Reading

The reader interested in the sources of mathematical logic can find the texts of founding fathers edited and commented on by J. van Heijenoort in [vH67]. The *Handbook of Mathematical Logic* [Bar77] is a reference book for specialists. However, a number of chapters are very accessible, notably: the first, which is a good introduction to model theory; the chapter written by Shoenfield is a good introduction to set theory; and the chapter written by Rabin includes many decidability results.

The example of geometric figures comes from a contest organized by the US Air Force. Two teams, championing a functional language (Haskell, in fact), submitted similar solutions based on the principles[18] indicated in § 3.1.1. There are many introductory books on functional programming, for instance [Pau91], [BW88], [CMP02] and [CM98].

---

[18]They beat all other approaches hands down, which came as a surprise because traditionally favorite domains for functional languages were compilation or theorem provers.

4. Hoare Logic

The techniques to be discussed in this chapter are aimed at reasoning about algorithms. We first introduce the traditional notation for annotating a program with assertions. This yields a special kind of proposition and we give the logical rules which govern them — specifically, Hoare logic. Finally, we show another interpretation of these rules, due to Dijkstra, which leads to a technique allowing one to *calculate* a program that establishes a given assertion.

4.1 Introducing Assertions in Programs

Chapter 2 showed how to specify what we expect from a program or from a piece of code, using assertions which are logical formulas over the input and output data of this program. It turned out to be useful to put assertions inside a program, because (among other reasons) instructions sometimes make sense only if they are executed from a suitable state. This state is itself defined by the value of all program variables at a given time.

For instance, let us suppose that the state is defined by three numerical variables x, y and z, and that the program consists of a sequence of instructions:

S_1 ; S_2 ; S_3 ; z:=2/(y-x) ; S_5

Just before the fourth instruction, x has to be different from y. Such conditions are traditionally inserted at the relevant point in the code between curly (or, set) brackets:

S_1 ; S_2 ; S_3 ; { ¬(x=y) } z:=2/(y-x) ; S_5 .

We could then complete the table search program of page 31 as follows:

```
1    x:=p ; y:=q ;
2    while x≠y  do { p≤x<q }
3       if P(x) then y:=x else x:=x+1 done ;
```

Here is the program together with its complete specification, derived from the last specification (page 28). Recall that line 8, which is in the form $A \Rightarrow B \Rightarrow C$, reads $A \Rightarrow (B \Rightarrow C)$, that is, "$A$ implies that B implies C", or in other words, "if I have A and I have B then I have C".

```
1   (p ∈ N) ∧ (q ∈ N) ∧ p≤q ,
2     P: predicate defined for all elements of [p..q[ }
3   x:=p ; y:=q ;
4   while x≠y  do { p≤x ∧ x<q }
5     if P(x) then y:=x else x:=x+1 done ;
6   { x ∈ N ∧ p≤x ∧ x≤q
7       ∧ x<q ⇒ P(x)
8       ∧ x=q ⇒ (∀ i ∈ N) (p≤i ∧ i<q) ⇒ ¬P(i) }
```

We have to deal with *two* concepts of a **variable**. The concept we use in programming is a name that concretely denotes a piece of memory, or, more abstractly, a portion of the state whose contents varies in the course of execution. This is, for instance, the case with x and y in the above program. In addition, we have logical variables which are used to construct logical formulas. Such variables were used informally throughout Chapter 3, for example x on pages 42 and 44. They will be formally introduced in Chapter 5. They represent a value that does not depend on execution but rather on external considerations. However, we need to mention program variables in logical formulas — assertions — and consequently mix these two kinds of variables! We already did that with x and N on page 25. Any effective use of rule (4.4) below mentions program variables within I and the logical variable V.

Fortunately, the confusion can be tolerated to an extent. The key point is not to fall into the pitfall of aliasing, as mentioned on page 32. In brief, we can agree that the state assigns a value to logical as well as program variables, but that logical variables can be considered as constants during execution. Note that, in our table search program, p and q are also arbitrary constants. We consider this point again in connection with the semantics of logical formulas (see § 5.2.3).

In Chapter 8 we will take an additional view point, where the semantics of program variables itself is manipulated: they are regarded as fields (or more mathematically: projections) of the state.

4.2 Verification Using Hoare Logic

The correctness proof of a program will be structured according to the structure of the program itself. Let us first analyze the latter. A program is composed of program elements (sequence, alternative constructs, loops) which are themselves composed of smaller and smaller elements, until we have the simplest ones, that is, assignment. Each program element (including the whole program itself) can be considered separately: it performs its own action on the state, which is also formalized by a relation between a precondition and a postcondition. For example, the above program is the sequential composition of:

- line 3, which is itself the sequential composition of:
  - the assignment x:=p,
  - the assignment y:=q,

– lines 4 and 5: a loop whose body is:
  – line 5: an alternative between
    – the assignment y:=x,
    – the assignment x:=x+1.

4.2.1 Rules of Hoare Logic

The relation between the precondition and the postcondition of a compound element depends only on the components and on the kind of composition. Hence we can construct the proof incrementally. The simplest examples are the empty statement **skip**, which establishes the postcondition P from the precondition P, and the sequential composition S; S'.

On page 27 we introduced the notation $\{P\}\, S\, \{Q\}$ for "S establishes the postcondition Q from the precondition P". The effect of **skip** is then axiomatized as

$$\{P\} \, \mathbf{skip} \, \{P\} \; . \tag{4.1}$$

On the other hand, it is clear that, if $\{P_1\}\, S\, \{P_2\}$ and if $\{P_2\}\, S'\, \{P_3\}$ then the sequence S; S' establishes the postcondition P_3 from the precondition P_1. This deduction rule is given as:

$$\frac{\{P_1\}\, S\, \{P_2\} \qquad \{P_2\}\, S'\, \{P_3\}}{\{P_1\}\, S\,;\, S'\, \{P_3\}} \; . \tag{4.2}$$

The rule for alternation, the "if-then-else" statement, is not very difficult either. Premises read: "S_1 (respectively S_2) establishes Q from the precondition P in the case when C is true (respectively false)":

$$\frac{\{P \wedge C\}\, S_1\, \{Q\} \qquad \{P \wedge \neg C\}\, S_2\, \{Q\}}{\{P\} \, \mathbf{if}\ C\ \mathbf{then}\ S_1\ \mathbf{else}\ S_2\, \{Q\}} \; . \tag{4.3}$$

The rule for the loop involves an invariant denoted by I, which occurs in the precondition and which must be preserved by the body of the loop when the input condition C is true, and a natural integer v — the variant — which decreases at each execution of the body of the loop ($<$ can be replaced with another well-founded relation). Then termination is guaranteed and both I and $\neg C$ are true at the exit of the loop:

$$\frac{(I \wedge C) \Rightarrow v \in \mathbb{N} \qquad \{I \wedge C \wedge v = V\}\, S\, \{I \wedge v < V\}}{\{I\} \, \mathbf{while}\ C\ \mathbf{do}\ S\, \{I \wedge \neg C\}} \; . \tag{4.4}$$

We have another useful rule, which tells us that we can strengthen the precondition and weaken the postcondition:

$$\frac{P' \Rightarrow P \qquad \{P\}\, S\, \{Q\} \qquad Q \Rightarrow Q'}{\{P'\}\, S\, \{Q'\}} \; . \tag{4.5}$$

We are left with the rule for assignment, which may seem surprising at first sight because it works backwards. It is actually an axiom, that is, a rule without a premise, since an assignment is not composed of simpler program elements:[1]

$$\{[x := E]P\} \; \texttt{x:=E} \; \{P\} \; . \tag{4.6}$$

The formula $[x := E]P$ represents P where E is substituted for x. This axiom goes from the postcondition to the precondition: it states that every property which is true for x after the assignment must be true for E before the assignment. For example, if $x > 5$ is the postcondition of $\texttt{x:=x+1}$, Intuitively, we had $x > 4$ before this assignment; we get an equivalent precondition if we replace x with $x + 1$ in the postcondition: $x + 1 > 5$.

An axiom such as $\{P\} \; \texttt{x:=E} \; \{P \wedge x = E\}$ would be unsatisfactory for at least two reasons:

1. x may occur in P, then we cannot keep the same P in both the precondition and the postcondition;
2. x may also occur in E; for example $\texttt{x:=x+1}$ certainly does not establish the postcondition $x = x + 1$.

4.2.2 Correctness of the Bounded Linear Search Program

Now we have all the ingredients we need for concocting a formal correctness proof of the program on page 66. Five formulas (not counting the specification itself) to prove the correctness of a three-line program may seem like rather a lot. However, our example happens to concentrate all fundamental constructs into a small space.[2] Rules (4.1) to (4.6) are sufficient for proving the correctness of arbitrarily complex algorithms. Of course, we also need normal laws of logic, for example the laws recalled in § 3.4.2 or others which are explained in forthcoming chapters.

Let us now show that our program is correct. We analyze its structure. First we have a sequential composition of two consecutive assignments followed by a loop. Then we apply rule (4.2), where P_1 and P_3 are, respectively, the precondition and the postcondition of the specification. We have to find P_2, which is also the loop invariant I according to (4.4). Following the idea explained in our informal reasoning on pages 29 and 31 we consider:

$$I \;\overset{\text{def}}{=}\; I_1 \wedge I_2 \wedge I_3 \; ,$$

$$I_1 \;\overset{\text{def}}{=}\; \underbrace{x \in \mathbb{N} \wedge y \in \mathbb{N} \wedge p \le x \wedge x \le y \wedge y \le q}_{\text{domain of } x \text{ and of } y} \; ,$$

$$I_2 \;\overset{\text{def}}{=}\; \underbrace{\forall i \in \mathbb{N} \, (p \le i \wedge i < x) \Rightarrow \neg P(i)}_{\text{unsuccessful exploration}} \; ,$$

---

[1] For the sake of simplicity, we agree that expressions on the right-hand side of an assignment don't have side effects.

[2] The programming language we consider here has the power of Turing machines.

$$I_3 \stackrel{\text{def}}{=} \underbrace{y < q \Rightarrow P(x)}_{\text{success}} \ .$$

Rule (4.5) is often used in the following way: in order to prove $\{P'\}\, S\, \{Q\}$, take a precondition P such that $\{P\}\, S\, \{Q\}$ is easy to show, and verify that P is a consequence of P'. This strategy works when S is an assignment and Q is known: a simple reading of (4.6) provides a good candidate for P.

We verify easily that I is true after the two assignments of line 3, by an application of (4.6), and taking the precondition into account. Indeed, the latter entails $[x := p]\,[y := q]I$.

The loop variant is $v = y - x$. We still have to verify

$$\{I \wedge x \neq y \wedge y - x = V\}\ S\ \{I \wedge y - x < V\}$$

where S is an alternative. Note that the assertion we introduced in line 4 of the program is a consequence of $I \wedge x \neq y$. We apply rule (4.3). It is easy to verify that the variant decreases in the two branches; we now consider invariant preservation.

In the first branch, after simplification, $[y := x]I_1$ yields $x \in \mathbb{N} \wedge p \leq x \wedge x \leq q$ which is a consequence of I_1; $[y := x]I_2$ yields exactly I_2; $[y := x]I_3$ is in the form $A \Rightarrow P(x)$ which is satisfied since $P(x)$ plays the role of C in (4.3).

The precondition of the second branch contains I_1 and $x \neq y$, thus it implies $x < y$; that is (in $\mathbb{N}$) $x + 1 \leq y$, hence $[x := x + 1]I_1$ is satisfied. On the other hand, as $\neg P(x)$ is initially true, we also have $\neg y < q$ taking I_3 into account; then $[x := x+1]I_3$ is in the form $A \Rightarrow P(x+1)$ where A is false. We are left with $[x := x + 1]I_2$ which can be decomposed in $I_2 \wedge \neg P(x)$ and is clearly satisfied.

Finally, the postcondition we look for is a consequence of $I \wedge \neg C$, that is, $I \wedge x = y$ here: we just use ordinary logical manipulations.

4.3 Program Calculus

The use of Hoare logic we just considered requires that we look *a posteriori* for intermediate assertions, such as loop invariants. This may turn out to be crippling. Other researchers, notably Dijkstra, advocate a different, *constructive*, approach whereby a program is designed together with its correctness proof. In short, one has to start from a given postcondition Q and then look for a program that establishes Q from the precondition. Often, analyzing Q provides interesting hints to finding the program.

4.3.1 Calculation of a Loop

Let us again consider bounded linear search. The postcondition is:

$$x \in \mathbb{N} \wedge p \leq x \wedge x \leq q \tag{4.7}$$

$$\wedge \quad x < q \Rightarrow P(x) \tag{4.8}$$

$$\wedge \quad x = q \Rightarrow \forall i \in \mathbb{N}\, (p \leq i \wedge i \leq q) \Rightarrow \neg P(i) \ . \tag{4.9}$$

Our idea is, of course, to use a loop. Its postcondition, given (4.4), is a conjunction $I \wedge \neg C$. The first strategy we can try is to share out the conjuncts (4.7) to (4.9) among I and $\neg C$. Assertions about the domain of x in (4.7) fall clearly within the invariant. The assertion (4.9), which involves a quantifier, is too complicated for a test. Let us then envisage $\neg C = x < q \Rightarrow P(x)$, that is $C = x < q \wedge \neg P(x)$. This leads us to a program having the following shape:

```
1    x:=p ;
2    while x<q ∧ ¬P(x) do ... done ;
```

The body of the loop — x:=x+1 — can be guessed at without calculation. We then get a variant of the first algorithm for linear bounded search given on page 28, as well as a good approximation to the invariant to be used in its correctness proof. This is not so bad, although this program requires P to be defined over q. A derivation of the second program is explained in [Coh90].

4.3.2 Calculation of an Assignment Statement

A striking example for the synthesis of an assignment statement, inspired by [Coh90], is the computation of the cube of a natural integer N where the only allowed arithmetical operation is addition. The first postcondition we consider is $c = N^3$.

Aiming at a loop, a technique already mentioned (page 30) consists of replacing a constant with a variable. The effect of this transform is to put the postcondition in the form $I \wedge \neg C$. Here the only available constant is N, hence we put the postcondition in the form $c = x^3 \wedge x = N$. Then we look for a program having the following shape:

```
1    establish I ;
2    while x≠N do
3        preserve I while making x closer to N done ;
```

where the loop invariant is $I \overset{\text{def}}{=} c = x^3$.

An obvious way to establish I at the beginning of the loop is to take $x = c = 0$. We can partially guess the body of the loop: increment x, with the aim of successively computing 1^3, 2^3, 3^3, etc. The loop variant is $N - x$, and we will leave this unchanged.

The loop body contains x:=x+1 and an assignment to c such that the invariant is preserved. Here reasoning is made easier if we consider a *simultaneous assignment*: the sequential composition of x:=x+1 and c:=... would introduce a cumbersome intermediate state. The shape we envisage for line 3 is then:

```
3        x,c := x+1,E done ;
```

where E is an expression that is yet to be found, and we want (invariant preservation and assignment rule):

$$I \wedge x \neq N \Rightarrow [x, c := x + 1, E] I . \tag{4.10}$$

We get an *equation where the unknown is the program*, or at least a part of the latter: the expression E. In order to solve (4.10) we calculate:[3]

$[x, c := x + 1, E] I$

$=$ {definition of I}

$[x, c := x + 1, E](c = x^3)$

$=$ {simultaneous substitution}

$E = (x + 1)^3$

$=$ {arithmetic}

$E = x^3 + 3x^2 + 3x + 1$

$=$ {use of the hypothesis I, that is $c = x^3$}

$E = c + 3x^2 + 3x + 1$.

The expression $3x^2 + 3x + 1$ raises a problem: it is not a sum of known quantities. Let us introduce d and assume, at the same time, that $d = 3x^2 + 3x + 1$. We can complete the previous calculation:

$E = c + 3x^2 + 3x + 1$

$=$ {use of the hypothesis $d = 3x^2 + 3x + 1$}

$E = c + d$.

To summarize, we have:

$$(c = x^3 \wedge d = 3x^2 + 3x + 1) \Rightarrow$$
$$[x, c := x + 1, c + d](c = x^3) , \qquad (4.11)$$

to be compared with (4.10). Then we actually consider $I \stackrel{\text{def}}{=} I_1 \wedge I_2$ with $I_1 \stackrel{\text{def}}{=} c = x^3$ and $I_2 \stackrel{\text{def}}{=} d = 3x^2 + 3x + 1$. The implication (4.11) can then be written

$$I \Rightarrow [x, c := x + 1, c + d]I_1 .$$

Note that, if $[S]$ is a substitution, $[S](I_1 \wedge I_2) = [S]I_1 \wedge [S]I_2$, we still have to establish that I_2 is invariant; that is, to find an appropriate assignment for d. Then we calculate (E' is an expression to be found):

$[x, c, d := x + 1, c + d, E'] I_2$

$=$ {definition of I_2, substitution, arithmetic}

$E' = 3(x^2 + 2x + 1) + 3(x + 1) + 1$

$=$ {use of the hypothesis I_2}

$E' = d + 6x + 6$

$=$ {invention of e satisfying $I_3 \stackrel{\text{def}}{=} e = 6x + 6$}

---

[3]The format we use is explained in § 9.6.2.

$$E' = d + e \ .$$

We repeat the process in order to make I_3 invariant:

$$[x, c, d, e := x + 1, c + d, d + e, E''] I_3$$

$=$ {definition of I_3, substitution, arithmetic}

$$E'' = 6(x + 1) + 6$$

$=$ {use of the hypothesis I_3}

$$E'' = e + 6 \ .$$

We just have to initialize the loop by means of a simultaneous assignment, that is, to (easily) find C, D and E such that:

$$[x, c, d, e := 0, C, D, E] I \ .$$

This leads us to the following nice program:

```
1    x,c,d,e := 0,0,1,6 ;
2    while x≠N do
3        x,c,d,e := x+1,c+d,d+e,e+6 done ;
```

4.3.3 Weakest Precondition

Given two assertions A and B, we say that A is **stronger** than B, and that B is **weaker** than A, if $A \Rightarrow B$.

The process illustrated in § 4.3.2 rests on a calculation of expressions having the shape $[S]P$ where S is a substitution and P is a predicate — an assertion which depends on a number of variables. This process can be generalized if, for each program element S, we have at our disposal a simple means to calculate the **weakest precondition** P such that $\{P\} S \{Q\}$. The latter is denoted[4] by $[S]Q$.

$[S]$ is called a **predicate transformer**: when applied to Q, it returns the weakest P such that $\{P\} S \{Q\}$:

$$\{P\} S \{Q\} \iff P \Rightarrow [S]Q \ . \tag{4.12}$$

We have for example :

$$[\mathbf{skip}]Q \overset{\text{def}}{=} Q \ , \tag{4.13}$$

$$[\mathbf{x:=E}]Q \overset{\text{def}}{=} [x := E]Q \ , \tag{4.14}$$

$$[S\,;\,S'] \overset{\text{def}}{=} [S] \circ [S'] \ . \tag{4.15}$$

It turns out to be convenient to generalize the classical construct **if B_1 then** S_1 **else** S_2 to a non-deterministic choice:

---

[4]This notation, used in the B language (see Chapter 6), is inspired by the notation of substitutions. Dijkstra's original notation is $wp.S.Q$.

if $B_1 \rightarrow S_1$ □ $B_2 \rightarrow S_2$ **fi**

where B_2 does not need to be the negation of B_1. The corresponding weakest precondition is:

$$[\textbf{if } B_1 \rightarrow S_1 \ \square \ B_2 \rightarrow S_2 \ \textbf{fi}] Q \ \stackrel{\text{def}}{=} \ \begin{cases} B_1 \vee B_2 & \wedge \\ B_1 \Rightarrow [S_1]Q & \wedge \\ B_2 \Rightarrow [S_2]Q & . \end{cases} \qquad (4.16)$$

This construct fits better with program calculation, as well as multiple assignment with relation to sequential composition of assignments. Note that it is easy to translate an algorithm written with non-deterministic choices and multiple assignments into a programming language with usual alternative constructs and sequential composition of single assignments. We don't give further details here; the ideas are explained and illustrated with many examples in [Dij76, Coh90, Kal90]. The above constructs (sequential composition, multiple assignment, **skip**, loop, choice expressed with □) make up the language of **guarded commands** devised by Dijkstra.

4.4 Scope of These Techniques

Hoare logic has been used in a number of industrial projects, to provide guarantees on critical programs following their realization. A notable example is the railway signaling software for line A of RER in Paris. However, it turned out to be difficult to transfer the results to versions of the software implemented for other towns.[5]

The techniques *à la* Dijkstra allow skilled people to design algorithms which can be surprisingly subtle and elegant. Large-scale programming, however, is not within the scope of these techniques. Structuring mechanisms, such as subroutines, modules and so on are needed for more realistically sized systems. Normal programming languages include somewhat complex features, such as recursive procedures with side effects, pointers, dynamic data structures, etc. But it is not that simple to define and to use an axiomatic semantics for them. Apart from algorithm design, the techniques considered in this chapter apply mainly to small subsets of common programming languages. It is interesting to remark that such subsets fit well with the programming standards used for critical software. Moreover, recall that "complete" C and languages derived from it are seriously disadvantaged compared to languages provided with a clear formal semantics, such as ML.

In any case, methods and techniques introduced in this chapter are useful for everyday programming. Even an informal use of invariants and variants makes the design of a loop significantly easier. For example, who never hesitated when considering initial or terminal values of a loop index?

---

[5]Development teams decided then to switch to B, which is quite similar in some respects, but offers techniques and tools that are useful for maintenance.

Let us also remark that in a programming language, such as Eiffel [Mey88], using assertions is explicitly and strongly encouraged. They can be checked at run-time and are linked to the exception mechanisms, providing a valuable aid to debugging. Similar features are also available in Objective Caml (a version of ML) and even, to some extent, in C. Note that only computable assertions (in particular, without unbounded universal quantifiers) make sense in this context.

4.5 Notes and Suggestions for Further Reading

Introducing assertions in programs is an idea dating back at least to Floyd [Flo67]. It has been structured under what is now called Hoare Logic in [Hoa69], and applied to Pascal in [HW73]. The language of guarded commands and Dijkstra's approach to the design of correct-by-construction sequential programs are both presented by their author in [Dij76], and in various textbooks, e.g. [Kal90] and [Coh90].

Among recent innovations, a number of researchers have provided automated support for Hoare-style proof of imperative programs in a general framework. For example, such ideas are developed and implemented by J.-C. Filliâtre [Fil99] for Coq — the version of type theory that we consider in Chapter 12 — and PVS — which is also discussed in this chapter.

As mentioned earlier, the techniques considered in this chapter are essentially relevant when one considers programming-in-the-small. An important technique for dealing with larger-scale software development is refinement. The basic idea consists of relating concrete specifications to abstract specifications, so that we can reason about high-level properties of a system without being hampered by unnecessary low-level details. We will say more about this in Chapter 6. The interested reader may also consult the article by Gardiner [GM91] and the book by de Roever [dRE98].

5. Classical Logic

Logic provides a syntax for expressing properties. A "meaning" of these expressions and their compositions is defined by the concepts of an interpretation and of a model. We begin by introducing the most simple of these expressions, called propositions. We then present the general case of formulas, which are expressions that depend on the value of parameters called variables, or which can themselves be variables. These formulas may be quantified using ∀ (for all) and ∃ (there exists).

In this chapter we examine different logics: the logic of propositions (§ 5.1), first-order logic (§ 5.2), and higher-order logic (§ 5.5), along with a variant of first-order logic, which we will examine as part of a discussion of partial functions (§ 5.4). Equality and arithmetic are tackled in § 5.3. We conclude with basic concepts of model theory (§ 5.6).

5.1 Propositional Logic

5.1.1 Atomic Propositions

We assume a collection of elementary expressions called **atomic propositions**, which are application dependent. These atomic propositions may then be combined by means of logical connectors (and, or, not, etc.). There are two possibilities:

1. We do not need to break down these expressions. In this case we represent them by a letter identifier (for example, P, Q, etc.); if we need to better express the ideas we are trying to represent, we may use a longer identifier, for example it_is_sunny; these symbols are called **proposition symbols**;
2. The atomic expression is structured. In this case the interpretation depends on the subject and there are as many possible interpretations as there are subjects. For example, we can consider the individuals denoted by Claudio, Elliot, John, and construct three expressions stating the fact that Claudio, Elliot and John are telephone subscribers in the same way as follows: is_a_subscriber(Claudio). We employ a functional notation that is justified by the fact that is_a_subscriber will be interpreted by a

function from the set of people to {*true*, *false*}. We call is_a_subscriber a **one-place predicate symbol**, or simply **a one-place predicate**.[1]

Similarly, we can introduce predicates with any number of places. For example, to express that Claudio rents a given telephone we introduce the constants tel1, tel2, etc., as well as a two-place predicate rents; now we can write:

$$\text{rents(Claudio, tel27)} . \tag{5.1}$$

Then, an individual can be expressed as a function of one, or several, other individuals. For example, we can introduce the functions denoted by father_of, which allows us to express the fact that Elliot's father (the father of Elliot) rents telephone number 5:

$$\text{rents(father\_of(Elliot), tel5)} . \tag{5.2}$$

Note. The first situation is a particular case of the second: the proposition symbols (P, Q, it_is_sunny) can be considered to represent zero-place predicates. On the other hand, the difference between the first type of expression and the second is superficial for now. For example, the collection of expressions above can be replaced by Claudio_is_subscribed, ..., Claudio_rents_tel27, father_of_Elliot_rents_tel5, etc. The advantage of a structured representation of atomic propositions is that it allows for the synthesis of a great number of them in a systematic way.

⬭   This comment suggests that the predicate calculus can be reduced to the propositional calculus (see the definitions below) provided that quantifiers can be eliminated. In fact, Herbrand showed that every first-order logic proof, within a sufficiently general class of formulas, may be transformed to a proof in the logic of propositions; Herbrand even provided an algorithm to perform this transformation. This has had important consequences in automatic programming and the development of Prolog. We will return to this in Chapter 9. Henkin also used processes aimed at reducing propositions to first-order, thereby establishing results of completeness.[2] See [Bar77, ch. 1] and [Gal86].

5.1.2 Syntax of Propositions

Atomic propositions are the building blocks of propositions. It is convenient to have two predefined atomic propositions, **t** and **f**, representing the proposition

---

[1]It should be noted that this is an abuse of terminology. This will be made more clear when we address semantics.

[2]It is a little unusual to present propositional logic by introducing function and predicate symbols straightaway. These symbols are essential only in first-order logic. They are already useful, however, and we can see a continuity between propositional logic and first-order logic. Really what distinguishes between them is the use of variables and quantifiers.

that is always true and the proposition that is always false, respectively. To illustrate the terminology, consider expression (5.3):

$$\mathtt{rents(father\_of(Elliot), tel5)} \wedge \mathtt{rents(Claudio, tel27)} \ . \qquad (5.3)$$

In this example:

- (5.3) is a proposition ;
- `rents(father_of(Elliot),tel5)` and `rents(Claudio,tel27)` are also propositions, more precisely atomic propositions;
- `father_of(Elliot)`, `Elliot`, `tel5`, `Claudio` and `tel27` are terms, the last four being simply constant symbols; as no variable is used so far they are, in fact, constant terms;
- `rents` is a predicate symbol;
- `father_of` is a function symbol.

Propositions are defined as follows:

1. Every atomic proposition is a proposition;
2. If A is a proposition, its negation, written $\neg A$ (pronounced "not A") is a proposition;
3. If A and B are propositions, $A \vee B$, $A \wedge B$, $A \Rightarrow B$ and $A \Leftrightarrow B$ (pronounced " A and B", "A or B", "A implies B" and "A equivalent to B", respectively) are propositions;
4. There are no other propositions other than those constructed via the preceding three rules.

Notes.

(1) This definition gives only the essentials of propositions, the *abstract syntax* in computer science terminology. To reduce ambiguities in a concrete expression such as $P \wedge Q \vee R$, it is convenient to introduce priority levels for the operations $\wedge$, $\vee$, etc., as well as parentheses when necessary. In this book, we use conventional parentheses "(" and ")" for this, as well as square brackets "[" and "]"
(2) Often the symbols "$\rightarrow$" and "$\supset$" are used in place of "$\Rightarrow$", and "$\equiv$" in place of "$\Leftrightarrow$".

Atomic propositions are formally constructed by combining two ingredients — predicate symbols and **constant terms**, the latter being themselves constructed by means of the constant symbols (such as `tel2`) and the function symbols (such as `father_of`) that we assumed initially:

1. A proposition symbol is an atomic proposition;
2. If P is an n–place predicate symbol, and if $t_1, \dots t_n$ are constant terms, then $P(t_1, \dots t_n)$ is an atomic proposition;
3. Every constant symbol is a constant term;
4. If f is an n–place function symbol, and if $t_1, \dots t_n$ are constant terms, then $f(t_1, \dots t_n)$ is a constant term;

5. There are no other atomic propositions or constant terms other than those constructed via the preceding four rules.

⊜   A zero-place predicate can be viewed as an atomic proposition, in which case the first rule is a special case of the second. Similarly, a given symbol can be considered to be a zero-place function symbol, in which case the third rule is a special case of the fourth.

5.1.3 Interpretation

The approach to interpreting the preceding notions is as follows. First we assume the set $\mathbb{B} = \{true, false\}$; *true* and *false* are called **truth values**. We then consider a universe of discourse $\mathcal{D}$ (more formally, a *non-empty* set of constants called a **domain**), satisfying certain properties. We then establish a correspondence between the symbols, individual people, and these properties.

In our example we create a correspondence between given names and real people, let's say Claudio with Abbado, John and Elliot with Gardiner,[3] the symbols tel1, tel2, etc. with actual telephones and the symbol father_of with the function that associates an individual with his/her father. To every atomic proposition we attach a truth value; for example *true* for P, it_is_sunny and Claudio_is_subscribed, *false* for Q, John_is_subscribed and Elliot_is_subscribed; if we prefer the structured representation, this amounts to associating the function {Abbado ↦ *true*, Gardiner ↦ *false*} with the predicate symbol is_a_subscriber.

The general case offers no surprises: constant symbols represent constants, function symbols represent functions, and so on. The only point that warrants particular attention is that all represented functions are *total* (they are defined for all values of the domain). We will return to this later.

An **interpretation** $\mathcal{I}$, therefore, is a correspondence that assigns:

- an element $c_{\mathcal{I}}$ from the domain $\mathcal{D}$ to every constant symbol c;
- a total function $f_{\mathcal{I}}$ from $\mathcal{D}^n$ to $\mathcal{D}$ to every n–place function symbol f;
- an element $P_{\mathcal{I}}$ of $\mathbb{B}$, **t** and **f** being necessarily interpreted by *true* and *false* respectively, to each proposition symbol P;
- a total function $P_{\mathcal{I}}$ from $\mathcal{D}^n$ to $\mathbb{B}$ to each n–place predicate symbol.

⊜   If E is a set, by convention E^0 denotes a singleton, let's say $\{1\}$; that allows us to identify E^{0+n} to E^n by means of a natural bijection $(1, x) \mapsto x$. Then every total function from E^0 to F may be identified as an element of F (the image of 1). Taking $E = \mathcal{D}$ and $F = \mathbb{B}$ (respectively, $F = \mathcal{D}$) the assimilation of propositions as zero-place predicates (respectively, constants as zero-place functions) is justified.

---

[3]Every constant symbol must correspond to an individual, but there is nothing to prevent two different symbols from relating to the same individual.

We see that an interpretation allows the assignment of a value in $\mathcal{D}$ to each constant term t; it is sufficient on each occurrence of a function symbol f in t, to apply the corresponding function $f_\mathcal{I}$ to the value of its arguments. Similarly, every atomic proposition $P(...)$ has a truth value obtained by applying $P_\mathcal{I}$ to the value of its possible arguments.

The same approach allows for the assignment of a truth value to all propositions. The connectives $\neg$, $\vee$, $\wedge$, $\Rightarrow$ and $\Leftrightarrow$ are associated with $\mathbb{B}$ to $\mathbb{B}$ and $\mathbb{B} \times \mathbb{B}$ to $\mathbb{B}$ functions defined via well-known truth tables (Figure 5.1).

P	Q	$P \vee Q$	$P \wedge Q$	$P \Rightarrow Q$	$P \Leftrightarrow Q$	$\neg Q$
false	false	false	false	true	true	true
false	true	true	false	true	false	false
true	false	true	false	false	false	
true	true	true	true	true	true	

Figure 5.1: Truth tables.

Observe that $P \wedge Q$ is true if and only if P and Q are both true. Nevertheless, it is unsatisfactory to present the semantics of $\wedge$ based solely on the usual meaning of the word *and*, because there are many such meanings! We can see three here:

- I took my hat and my coat
  (concept of a collection or grouping)

- I took my hat and I left
  (close to logical conjunction but with a concept of a temporal ordering) and

- See Naples and die
  (concept of a permission and of a succession).

The other connectives present similar ambiguities. The use of truth tables avoids this pitfall by invoking a clear mathematical concept, the application of a function to arguments.

5.2 First-order Predicate Logic

The language we've considered thus far, the propositional logic, doesn't allow us to express relatively simple facts, for example:

- if Claudio rents telephone 2, then Claudio is a subscriber.

It is clearly desirable to be able to capture more general properties, such as:

- every individual who rents a telephone is a subscriber.

To this effect, we first need parameterized propositions, for example:

– if x rents telephone y, then x is a subscriber.

A parameterized proposition is called a **formula**. The next step consists of quantifying formulas. Universal quantification over x in

– if x rents a telephone, then x is a subscriber

expresses:

– for all x, if x rents a telephone, then x is a subscriber.

In plain English:

– every individual who rents a telephone is a subscriber.

This expression can be viewed as a potentially infinite conjunction:

– if Claudio rents a telephone, then Claudio is a subscriber *and* if John rents a telephone, then John is a subscriber *and* etc.

Finally, existential quantification of y in

– x rents telephone y

is:

– there exists y such that x rents telephone y.

In plain English:

– x rents a telephone.

In the same way, this expression can be viewed as a disjunction:

– x rents telephone 1 *or* rents telephone 2 *or* ... etc.

The logic employed here is **predicate logic**, or more precisely **first-order** predicate logic because the variables considered throughout are drawn from a domain of constants, $\mathcal{D}$, and cannot represent functions over $\mathcal{D}$ nor propositions.

5.2.1 Syntax

We need to complete the notions of a term and of a proposition that we introduced earlier. We introduce into the language a set of **variables** $\mathcal{V} = \{x, y, ...\}$ and two symbols $\forall$ (for all) and $\exists$ (there exists), also called **quantifiers** (respectively, the **universal** and **existential** quantifiers). The constants, functions, variables and predicates that we assumed form what is termed a **first-order language**.[4]

Terms, **atomic formulas** and **formulas** are then defined by replacing *constant term* with *term*, *atomic proposition* with *atomic formula* and *proposition* with *formula* in the previous definitions. We add the following rules:

---

[4]What we often call a language is really a set of terms and formulas. That amounts to the same thing since the language here is completely determined by variables and the constant, function and predicate symbols.

- every variable is a term;
- if P is a formula and if x is a variable then $\forall x P$ and $\exists x P$ are formulas. x need not occur in P, although in practice this is often the case.

By convention, a quantifier extends as far as possible, taking any parentheses into account. For example, $\forall x\ P \Rightarrow Q$ does not represent $(\forall x P) \Rightarrow Q$, but rather $\forall x\ (P \Rightarrow Q)$.

Example. Every x that rents some thing (y) is a subscriber:

$$\forall x [\exists y\ \mathtt{rents}(x, y)] \Rightarrow \mathtt{is\_a\_subscriber}(x)\ . \tag{5.4}$$

Comment. The expected interpretation here is that every person who rents a telephone is a subscriber, but in the given formula there is nothing that requires that x must denote a human being and that y must denote telephone equipment. In contrast to college mathematics, quantifiers are not constrained to a domain of definition (i.e., the set of human beings or the set of telephones, in the previous example):

$$\forall x \in \mathtt{humans},\ (\exists y \in \mathtt{tels},\ \mathtt{rents}(x, y)) \Rightarrow \mathtt{is\_a\_subscriber}(x)\ .$$

Writing the formula in such a way uses the concept of sets *within* the language, something that we have carefully avoided in this section. That doesn't constitute a reduction in the expressive power of the logic, as the same effect is obtained by representing not sets (such as humans) but characteristic predicates:

$$\begin{aligned}\forall x\ \mathtt{is\_a\_human}(x) \Rightarrow\\ [(\exists y\ (\mathtt{is\_a\_tel}(y) \wedge \mathtt{rents}(x, y))) \Rightarrow \mathtt{is\_a\_subscriber}(x)]\ .\end{aligned} \tag{5.5}$$

The concepts of a set, a function, etc., have only been used in an informal manner and in the metalanguage, so that the syntax and the necessary material for interpretation could be described. The syntax of logic itself does not include the symbol $\in$. There is, however, an important first order language that uses $\in$ — axiomatic set theory. We note that the use of a symbol denoting a set is subject to certain restrictions.

In set-based specification languages, quantifiers are necessarily constrained: quantified formulas are of the form $(\forall x \in E)P$ or $(\exists x \in E)P$, (also written $\forall x \in E \bullet P$ or $\exists x \in E \bullet P$), and the rules employed guarantee that E exists. But to justify the correctness of mechanisms employed, a well-developed theory of sets must be available beforehand.

5.2.2 Example of the Table

In the example of searching for an integer between two bounds, terms represent natural numbers. These are constructed from a constant symbol 0 and a one-place function symbol S. The latter represents the successor function; for example, the integer 2 is represented as S(S(0)). Other symbols representing addition, multiplication, and other operations on integers are useful but not necessary.

An almost omnipresent predicate is that of equality. We introduce the two-place predicate (symbol) equal, but we will use the usual infix notation $x = y$ instead of equal(x, y). Similarly, for comparisons, we will write the predicates $<$ and $\leq$ in an infix manner. Moreover, we will consider the three-place predicate between, the intended meaning of between(a, b, c) being: b is contained between a (inclusive) and c.

Let us suppose that we wish to find an element divisible by 37 in the interval $[p..q[$, where p and q are variables.[5] We introduce the predicate symbol div37; the integer x to be found must satisfy the formula:

$$(\text{between}(p, x, q) \,\wedge\, \text{div37}(x))$$
$$\vee \quad (x = q \,\wedge\, \forall i \ \text{between}(p, i, q) \Rightarrow \neg\text{div37}(i)) \ . \tag{5.6}$$

5.2.3 Interpretation

How do we interpret a formula depending on x? Consider, for example, the formula is_a_subscriber(x). It is clear that its value, *true* or *false*, depends *a priori* on the value of x. We had a similar situation for is_a_subscriber (without "(x)"), which was interpreted by a function from $\mathcal{D}$ to $\mathbb{B}$. Here, we introduce the concept of an assignment, which is a function from a set of variables $\mathcal{V}$ to $\mathcal{D}$. Let us fix an assignment Γ, the value given to is_a_subscriber(x) is then is_a_subscriber$_\mathcal{I}(\Gamma(x))$. More generally, the value of a term and the truth value of a formula over $\mathcal{D}$ depends on the interpretation $\mathcal{I}$ *and* on the assignment Γ.

To interpret a quantified formula such as $\exists y \ \text{rents}(x, y)$, it should be noted that its truth value depends only on x and not on the quantified variable y: suppose that $\mathcal{D}$ contains only two constants c_1 and c_2, this formula has the same value as $\text{rents}(x, c_1) \vee \text{rents}(x, c_2)$. Note that we could just as easily have written $\exists z \ \text{rents}(x, z)$. We have uncovered the phenomenon of dummy variables, well known in mathematics, in expressions such as $\sum_{y=1}^{n} f(y)$ or $\int f(y) \, dy$.

In logic, we use the term **free** or **bound** variable. For example, in the formula $\exists y \ \text{rents}(x, y)$, x is free while y is bound. Only free variables can be viewed as parameters of a formula.

One must be conscious of the fact that in the same formula a variable x can have both free and bound occurrences; for example x in $P(x, y) \wedge \forall x Q(x, y)$. The free occurrences of x are defined by: (1) every occurrence of x in a term or an atomic formula is free; (2) every free occurrence of x in P is also free in $\neg P$; (3) every free occurrence of x in P is also free in $P \vee Q, P \wedge Q, P \Rightarrow Q, P \Leftrightarrow Q$; *idem* for every free occurrence of x in Q; (4) no occurrence of x in $\forall x P$ or in $\exists x P$ is free.

---

[5] Following the convention of Chapter 2, the value returned is q if no value divisible by 37 is contained in $[p..q[$.

The **substitution** of c for x in R, where R is a term or a formula, is defined by replacing all *free* occurrences of x in R with c. We will write this $[x := c]R$. In the following definition, c will represent a constant and we will assume without loss of generality a constant symbol c_v for every value v of the domain $\mathcal{D}$. When c is not a constant but rather a term possessing free variables, we must first rename all quantified variables of R — $[x := y](\exists y(y > x))$ is not $\exists y(y > y)$ but $\exists y_1(y_1 > y)$.

To be completely rigorous it is necessary to mathematically define the concepts of a term, of an occurrence and of a substitution. That is done by defining a concept of a **tree domain** — intuitively, an address space structured in the form of a tree; a term is defined as an application of such a space to the set of constant and function symbols used. That is purely technique, and gives the results one expects for justifying practical manipulations. The reader seeking a more rigorous exposition is directed to [Gal86].

We can now give the definition of the **interpretation** $\mathcal{I}$ of a formula in the assignment Γ:

- the interpretation of constant, function and predicate symbols is the same as in the propositional case (assignment makes no change);
- if x is a variable, its interpretation $x_{\mathcal{I}}$ is $\Gamma(x)$;
- the connectors $\neg, \wedge$, etc. are interpreted as before;
- $\forall x P$ is interpreted by *true* if for every value v of $\mathcal{D}$, $[x := c_v]P$ has the value *true*, and by *false* otherwise;
- $\exists x P$ is interpreted by *true* if there exists a value v of $\mathcal{D}$ for which the formula $[x := c_v]P$ has the value *true*, and by *false* otherwise.

Overall, the truth value of a formula containing n free variables $x_1, \ldots x_n$ depends on $\Gamma(x_1), \ldots \Gamma(x_n)$. It may be useful to consider that this formula is interpreted by a function from $\mathcal{D}$ to $\mathbb{B}$.

We already pointed out in § 4.1 that the variables used in programs represent "state portions" whose value varies during the course of an execution. Let us fix a program with its variables y_i; we can formalize it by the means of a set of states S and of an appropriate projection p_{y_i} for each variable y_i of the program, *provided there is no aliasing*. The value represented by the variable y_i in the state s is then $p_{y_i}(s)$. We will proceed in this way in Chapter 8.

Symmetrically, we can consider that in each state s, we have a function Γ_s such that $\Gamma_s(y_i)$ provides the value of the variable y_i in the given state. Indeed, Γ_s is an assignment in the sense given above. Then we can reason in a formal way about a program by representing its variables by logical variables and each state by an assignment defined over these variables and over other regular logical variables as well.

Let us for instance interpret the formula $x \leq N$ of page 25. We represent an execution by a sequence of assignments $\Gamma_0, \Gamma_1, \ldots$ where $\Gamma_i(x)$ varies according to the evolution of x allowed by the program, whereas $\Gamma_i(N)$ remains fixed: N is not part of the program.

The expressive power of first-order logic is considerably greater than that of the propositional logic, because one can potentially achieve infinity using a finite number of formulas. For example:

$$\text{int}(0) \wedge (\forall x \ \text{int}(x) \Rightarrow \text{int}(S(x)))$$

has as a consequence

$$\text{int}(\underbrace{S(...S(\,0)...))}_{n}$$

where n is arbitrarily large. To obtain the same result in the propositional logic, we would straightaway need to express an infinite number of propositions such as

$$\text{int}(S(S(0)))\ .$$

Note: as soon as we have at least one constant symbol and one function symbol, the possible combinations enable us to conceive of an infinite number of propositions, even if we cannot express them explicitly.

5.3 Significant Examples

Most applications require the use of at least integers and equality. For this reason, we introduce the necessary symbols and what we refer to as their theory, made up of logical formulas called axioms. The interested reader may wish to refer to more precise definitions of these concepts in § 5.6.1.

5.3.1 Equational Languages

A language $\mathcal{L}$ is said to be **equational** if it contains the binary predicate $=$. This predicate, if it is to behave as equality, must always implicitly satisfy the following three axioms:

- the fact that $=$ is an equivalence relation (3 axioms);[6]
- the principle of substitution of equals for equals, that is, the Principle of Leibniz. For every n-ary function symbol f, n axioms are required:

$$\forall x_1 ... \forall x_n \forall y_i \ x_i = y_i \Rightarrow$$
$$f(... x_{i-1}, x_i, x_{i+1} ...) = f(... x_{i-1}, y_i, x_{i+1} ...)\ ,$$

likewise for every n-ary predicate symbol P:

$$\forall x_1 ... \forall x_n \forall y_i \ x_i = y_i \Rightarrow$$
$$P(... x_{i-1}, x_i, x_{i+1} ...) \Leftrightarrow P(... x_{i-1}, y_i, x_{i+1} ...)\ .$$

This symbol is always interpreted by the equality over the domain of interpretation $\mathcal{D}$.

---

[6]In fact, reflexivity is sufficient; it, combined with the Principle of Leibniz, allows us to use symmetry and transitivity also.

⭕   In fact, the axioms allow for the interpretation of $=$ by any equivalence relation compatible with the operations of the language $\mathcal{L}$ (that is, a *congruence*). But it is also possible to consider the quotient of $\mathcal{D}$ by the relation $\mathcal{D}'$, which provides an interpretation under which "$=$" is indeed equality.

Algebraic specification languages are equational languages. Most theories of mathematics are equational and, generally, model theory considers equational languages. On the other hand, basic proof theory generally does not address equality, which poses specific problems. While axioms are just equations, we must resort to the theory of rewriting systems.

⭕   For more general axiomatizations, combining logical connectors and equality, an important technique employed in automatic proof is paramodulation [RW69]. We will not address that here, but the interested reader will find a good description in [CL73].

Comment. If we consider second-order logic, equality can be *defined* as the second-order predicate that expresses the fact that x and y are equal if they have exactly the same properties:

$$x = y \stackrel{\text{def}}{=\!=} \forall P\ P(x) \Leftrightarrow P(y) \ .$$

5.3.2 Peano Arithmetic

A particularly important theory, due to Peano, is one which formalizes **arithmetic**. This is a first-order equational theory over the language composed of the constant 0, the unary function symbol S (representing the *successor* function), the binary function symbols "+" and ".", and the relation $<$. These operations are written here in the infix form, following common usage. The integer n is represented by $\underbrace{S(...S(0)...)}_{n}$.

5.3.2.1 Axioms of Peano Arithmetic. The axioms are as follows.

No two integers are the same:

$$\forall x \quad \neg(0 = S(x)) \ ,$$
$$\forall x \forall y \quad S(x) = S(y) \Rightarrow x = y \ .$$

Axioms of addition:

$$\forall x \quad x + 0 = x \ ,$$
$$\forall x \forall y \quad x + S(y) = S(x + y) \ .$$

Axioms of multiplication:

$$\forall x \quad x.0 = 0 \ ,$$
$$\forall x \forall y \quad x.S(y) = x + (x.y) \ .$$

Axioms of comparison:

$$\forall x \quad \neg(x<0) \ ,$$
$$\forall x \forall y \quad x<S(y) \iff x<y \ \lor \ x=y \ .$$

Note that the axioms of addition, multiplication and comparison are constructed by systematically considering the possible patterns of the second argument, which is either 0 or $S(y)$.

Our last axiom is actually a collection of axioms, because ϕ represents an arbitrary first-order formula having x as a free variable. A collection of axioms defined in this way is called a **schema**. We then have an infinite number of possible instances for a schema. The key point is that they may be recognized by an algorithm: we say that Peano arithmetic is **recursively axiomatizable**.

Induction schema:

$$\phi(0) \ \land \ [\forall x \ \phi(x) \Rightarrow \phi(S(x))] \ \Rightarrow \ \forall x \ \phi(x) \ .$$

We can, for example, take the formula $x < S(x+x)$ for $\phi(x)$, signifying that x is less than or equal to $2x$. The principle of induction in this case is:

$$0 < S(0+0) \ \land \ [\forall x \ x < S(x+x) \Rightarrow S(x) < S(S(x)+S(x))]$$
$$\Rightarrow \forall x \ x < S(x+x) \ .$$

There is nothing to stop us from taking a generally false formula such as $x=0$, for $\phi(x)$:

$$0=0 \land [\forall x \ x=0 \Rightarrow S(x)=0] \ \Rightarrow \ \forall x \ x=0 \ ;$$

but of course there is no hope of proving the second premise!

The formula ϕ can be more complex, for instance it can depend on other free variables and use logical connectors. Moreover, it is acceptable to choose variables other than x for the induction. An interesting example is the following:

$$\phi(x,y) \ = \ x<y \Rightarrow S(x)<S(y) \ .$$

Taking y as the inductive variable, we obtain the axiom:

$$(x<0 \Rightarrow S(x)<S(0)) \ \land$$
$$(\forall y \ (x<y \Rightarrow S(x)<S(y)) \ \Rightarrow \ [x<S(y) \Rightarrow S(x)<S(S(y))]) \qquad (5.7)$$
$$\Rightarrow \forall y \ x<y \Rightarrow S(x)<S(y) \ .$$

5.3.2.2 Application to the Table Example. For the table example that we described in § 5.2.2, we use the language of arithmetic augmented with two predicate symbols, between and div37. These symbols do not represent arbitrary predicates, but are linked to $<$ and $=$. We wish to define between(x,y,z) by $x\leq y \land y<z$, but $\leq$ does not exist in our language. We can introduce it and state the following axiom:

$$\forall x \forall y \quad x \leq y \iff x < y \lor x = y \ .$$

Another possibility is to note that, thanks to the second axiom of comparison, one can always replace $x \leq y$ by $x < S(y)$. We therefore can avoid "$\leq$" and state the following axiom about between:

$$\forall x \forall y \forall z \ \text{between}(x, y, z) \iff x < S(y) \land y < z \ . \tag{5.8}$$

We don't really have a need to axiomatize div37, since it has no effect on the criteria for searching in a table. If it were necessary we could introduce a constant thirty_seven, with the axiom:

$$\text{thirty\_seven} = \underbrace{S(...(0)...)}_{37} \ .$$

The axiom of div37 would then be:

$$\forall x \ \text{div37}(x) \iff \exists y \ x = y.\text{thirty\_seven} \ .$$

5.3.2.3 Models of Arithmetic. This section refers to the concepts coming from model theory as described in § 5.6.

It is intuitively clear that the set of natural numbers $\mathbb{N}$ together with obvious functions is a model of Peano arithmetic, which we call the **standard model**. But is it the only one? We can find others, such as the set of even integers where $+$, 0 and $<$ are interpreted without change, while S and "." are interpreted, respectively, by $n \mapsto n + 2$ and $m, n \mapsto mn/2$. In fact, these two models are identified by the isomorphism $n \mapsto 2n$.

We obtain a much more unexpected result by applying the theorem of compacity and the theorem of Löwenheim (cf. § 5.6.2) [Gal86]: Peano arithmetic admits a countable model non-isomorphic to $\mathbb{N}$. The existence of such models, which we call non-standard models, shows that $\mathbb{N}$ *is not* entirely characterized by the axioms of Peano. We will see in § 9.8.2 that this fact may be established by other means, and in a stronger manner through Gödel's theorem of incompleteness. On the one hand, Gödel's proof, contrary to that of the theorem of Löwenheim, uses only the finite processes recommended by Hilbert and accepted by the intuitionists; on the other hand, it shows that the introduction of supplementary axioms to fill the gap serves no purpose.

5.4 On Total Functions, Many-sorted Logics

Function symbols are interpreted by total functions, whereas one might want to model partial functions. Let's take the function father as an example; if we interpret it over the concrete set $\mathcal{A}$ of inhabitants of London, it is clear that this function is far from being total. We are then driven to taking for the interpretation of father a function from $\mathcal{A}$ to $\mathcal{A}$ associating with every person a his/her legal father if the latter is in the set $\mathcal{A}$, otherwise any value (for example, a itself). It would be more judicious to name this predicate father_if_he_exists. This modeling must be completed by introducing a predicate has_a_father, which characterizes those persons whose father is also in the domain.

In general, a partial function can be modeled by a total function and the characteristic predicate of its domain of definition. We use formulas that simultaneously combine both of these aspects of the function, for example:

$\forall x$ has_a_father$(x)\Rightarrow$
    [is_subscribed(father$(x)) \Rightarrow$ is_subscribed$(x)]$.

The need for characteristic predicates is far more obvious when the domain $\mathcal{D}$ mixes elements of different types, for example people and telephones (cf. the comment on page 81).

The interpretation by total functions can be attacked as being artificial and redundant. In our example, it assigns a value *a priori* to father(tel3) or to rents(tel1,tel2), even though this value has no influence. But, blindly replacing total functions by partial functions brings its own complications. In particular, this can lead to the introduction of a third truth value $\perp$, pronounced **undefined**. In fact, there are *many* three-valued logics, which have different properties and are less straightforward than ordinary logic. The specification languages VDM, Raise-SL and Abel use different three-valued logics. Typically it is less easy to reason with them; for example, in Abel, implication is not reflexive; in VDM, the deduction theorem (see § 9.1) does not hold; in Raise, conjunction and disjunction are not commutative.

There are, nonetheless, some interesting compromises, consisting of fixing *a priori* the domains of definition of functions used. The most simple (multisorted logic) consists of decomposing the domain of interpretation $\mathcal{D}$ into several disjoint domains $\mathcal{D}_1, \dots \mathcal{D}_i, \dots$ Every n-ary function symbol is interpreted by a total function $\mathcal{D}_{i_1} \times \dots \times \mathcal{D}_{i_n} \to \mathcal{D}_{i_0}$. The key is that this partitioning of $\mathcal{D}$ can be expressed in the syntax and then checked statically: for each symbol used, we declare a signature using sorts (i.e., domain symbols), father : person $\to$ person, for example.

The interpretation naturally assigns one $\mathcal{D}_i$ to each sort. Interpretations obtained in this way are **heterogeneous algebras** or **Σ-algebras**, and they play a fundamental role in algebraic specification languages.

In passing, we describe a concept used in Chapter 11: given a vocabulary Σ of function symbols $f_1, f_2 \dots f_n$, the **initial algebra** over Σ is the set of closed terms formed with $f_1, f_2 \dots f_n$. To be more precise, the concept of a morphism introduced in § 3.4.5 must be used: an algebra is initial as long as there exists a unique morphism between it and every other algebra over Σ. Every algebra isomorphic to the algebra of closed terms is initial. Let us take, for example, Peano arithmetic omitting addition, multiplication, comparison and induction: all that remains is the set of terms generated by 0 and S, which signifies on the one hand that every natural number is represented by a term of the form S(...(0)...), and on the other hand that two terms having a different number of applications of S represent different integers. But if we add an axiom such as $S(S(S(0))) = 0$, with the intention of defining modulo 3 arithmetic, the algebra we get is no longer initial in the class of algebras over Σ.

Frequently, the domains of definition of certain operations are distinct, but not disjoint. For example, addition is defined over $\mathbb{N} \times \mathbb{N}$, while division is only defined over $(\mathbb{N} - \{0\}) \times \mathbb{N}$; the push operation is defined for all stacks, while the pop operation is only defined for non-empty stacks. In these two examples, we would like to express that for two domains $\mathcal{D}_i$ and $\mathcal{D}_j$, we have $\mathcal{D}_i \subset \mathcal{D}_j$. For that, certain specification languages such as OBJ [GM00, JKKM92] permit the declaration of an order between sorts. The underlying theory becomes more complex, and static verification may become impossible: determining that an expression has a non-null value is an undecidable problem in the general case. This leads to a restriction in the use of logical connectors.

On the other hand, these extensions do not increase the expressive power of the ordinary (mono-sorted) first-order logic, in which all the restrictions mentioned are expressible by well-chosen characteristic predicates. In fact, different logical connectors offer a great richness of expression which can be used profitably in defining a varied range of characteristic predicates.

In summary, amongst the formalisms mentioned here, first-order logic offers the greatest expressive power, while multi- or order-sorted languages permit more static checking and ease of formulation. In order to achieve more expressive power, we must go beyond the first-order.

5.5 Second-order and Higher-order Logics

While expressing specifications and reasoning about their properties, we may end up introducing mathematical functions whose logical complexity is arbitrarily great. This is particularly the case if we wish to express general principles in a uniform manner.

First-order quantification holds only over variables from the domain of constants $\mathcal{D}$. This does not allow for the expression of properties or of functions ranging over other functions or properties. Let us consider, for example, the composition of two functions. As we all know, this is defined as $(g \circ f)(x) = g(f(x))$. This seems simple, yet the following assertions are not expressible in first-order logic.

$$\forall f \, \forall g \quad \forall x \ (g \circ f)(x) = g(f(x)) \ ,$$
$$\forall f \quad f \circ \mathrm{Id} = f \ ,$$
$$\forall f \, \forall g \, \forall h \quad h \circ (g \circ f) = (h \circ g) \circ f \ ,$$

Here are other examples:

− if a property holds for 0 and if it is true for an integer then it is true for its successor, then it is true for all integers:

$$\forall P \ [P(0) \wedge \forall n P(n) \Rightarrow P(n+1)] \Rightarrow \forall x P(x) \ .$$

Similar inductive principles can be written for a large range of data types in computer science.

- An injective function has a left inverse:

$$\forall f \ (\forall x \forall y \ x \neq y \Rightarrow f(x) \neq f(y)) \Rightarrow \exists g \forall x \ g(f(x)) = x \ .$$

This property can be useful in refining to doubly-linked data structures.[7]

- Given P, Q, ..., properties about individuals x, y, ..., we define $P\&Q$ as the property of x which is true if and only if x satisfies properties P and Q:

$$(P\&Q)(x) \ \stackrel{\text{def}}{=} \ P(x) \wedge Q(x) \ .$$

This concept of a conjunction is used in temporal logic, see § 8.5.1.

- If P is a hereditary property of x and if y is a descendant of x, then y also has property P:

$$\forall P \ \mathtt{hereditary}(P) \Rightarrow [\forall x \forall y P(x) \wedge \mathtt{descendant}(x,y) \Rightarrow P(y)] \ .$$

We note that ∘ takes two functions as its arguments and returns a function, that & takes two predicates as its arguments and returns a predicate, and that **hereditary** is a predicate over predicates. This feature is extremely interesting as it permits the expression of general reusable principles within very varied contexts. But here we must consider Russell's paradox (see page 39, for a discussion of its second version) if we begin to write formulas such as **hereditary(hereditary)**. To avoid this, Russell proposed a distinction between two kinds of predicates: first-order predicates over first-order terms, and second-order predicates over first-order predicates (such as **hereditary**). Similarly, ∘ is a second-order function.

Second-order logic introduces, in addition to first-order predicates, functions and variables, second-order predicates, functions and variables which may be universally or existentially quantified. These quantifiers are sometimes written $\forall^2$ and $\exists^2$ to distinguish them from first-order quantifiers. Second-order variables are interpreted by functions from $\mathcal{D}^n$ to $\mathcal{D}$ or from $\mathcal{D}^n$ to $\mathbb{B}$. Second-order predicates and functions may take first-order predicates and functions as arguments.

Repeating this process, we derive third, fourth and higher-order logics. In **higher-order logic**, we have variables, functions, predicates, and quantifiers of order n, for every integer n. We can refine this concept of an "order" and get type systems, as shown in Chapter 11.

---

[7]For example, in a hotel reservation system where every reservation r is for a room $f(r)$, one could specify that two different reservations are for two different rooms. This fact can be used at the specification level to talk about *the* reservation for a given room p, knowing that there is at most one. During a refinement, this reservation might be named $g(p)$, representing f by a pointer to a room and g by an inverse pointer.

◯   If Prop denotes the type of propositions, the type of predicates over, say, the natural numbers is nat → Prop, while the type of predicates over such predicates is (nat → Prop) → Prop. Therefore, it is no longer possible to express Russell's paradox within a typed environment. Ensuring the total absence of paradoxes in a practical type system is not trivial, but has been done for the most common ones.

In the semantics of programming languages, we often use higher-order functions or properties. This is typically the case in denotational semantics where the meaning S_P of a program, or a program element, P, is a function from the initial state to the final state. To give the semantics of language constructors which form complex elements E, starting with simple elements E_1, E_2, ..., we are naturally inclined to consider functions giving S_E from S_{E_1}, S_{E_2} ... One can also give the semantics of a program not as a transformation of states, but as a transformation of predicates expressed over the state. This approach, advocated by Dijkstra, for the specification and construction of correct programs, is also the basis of the B method.

These logics are considerably more expressive than first-order logic, but certain properties of decidability, which are useful in automatic proof, are lost. Interactive proof-assistant software has been developed using these logics, see Chapter 12.

◯   We mention here that second-order monadic logic (in which it is possible to quantify over unary predicates) possesses interesting properties of decidability relevant to computing science, especially automata theory. In this logic, we distinguish individual variables x, y, ... and unary predicate variables X, Y, ... which allows us to write formulas such as $X(x)$.

Equivalently, we can consider that second-order monadic logic is first-order logic augmented with set variables X, Y, ...; instead of $X(x)$ we then write $x \in X$. These variables are interpreted by parts of $\mathcal{D}$.

Weak second-order monadic logic is defined with the same language, but the variables X, Y, ... are interpreted by *finite* parts of $\mathcal{D}$. As a practical application, let us mention MONA [KM01], an environment using weak second-order monadic logic as its specification language.

5.6 Model Theory

Model theory [CK90, Bar77] has seen substantial mathematical developments, but seems to have little utility in the area of formal specification. On the other hand, the underlying ideas are often used, and are recalled here. We are concerned with completing the vocabulary introduced above with the idea of interpretation. We conclude with an illustration of two theorems of model theory.

5.6.1 Definitions

We are given a first-order language $\mathcal{L}$ (most of the following definitions apply to languages of any order).

A given interpretation $\mathcal{M}$ determines if an expression *without free variables* P of $\mathcal{L}$ is true or false. We say that $\mathcal{M}$ is a **model** of P, or that $\mathcal{M}$ **satisfies** P if P has the value *true* in $\mathcal{M}$. We write this $\models_{\mathcal{M}} P$.

In the following, we use the expression **closed formula** to refer to a formula without free variables. We note that a proposition is a closed formula without quantifiers. A **theory** is a collection of closed formulas.

Let T be a theory over $\mathcal{L}$. An interpretation $\mathcal{M}$ is a **model** of T, written $\models_{\mathcal{M}} T$, if $\mathcal{M}$ is a model of every formula of T. A theory T is said to be **satisfiable** if it possesses a model, and **unsatisfiable** otherwise.

A key idea in logic is the relation of consequence. The fact that a closed formula is a consequence of other closed formulas does not depend on the interpretation.

Given a closed formula P, and a collection of closed formulas Γ, we say that P is a **logical consequence** or a **semantic consequence** of Γ if every model of Γ is also a model of P. We write this $\Gamma \models P$.

The relations $\models$ and $\models_{\mathcal{M}}$ are easily distinguished: $\models$ expresses a relationship between formulas, while $\models_{\mathcal{M}}$ expresses a relationship between a (mathematical) model and a formula.

Here are several properties of $\models$:

- if $\Gamma \models P$, *a fortiori* $\Gamma, Q \models P$;
- if $\Gamma \models P$, and if $P \models Q$, then $\Gamma \models Q$;
- $\Gamma \models$ expresses that Γ is unsatisfiable; if $\Gamma \models P$, then $\Gamma, \neg P \models$.

The consequences of Γ form a set of formulas called the **theory generated** by Γ. The elements of Γ are called **axioms** of this theory. For example, the theory generated by the axioms (5.1), (5.2) and (5.4) comprises the formula `is_subscribed(Claudio,tel27)`.

A statement such as (5.4) is not true in all interpretations; however, it is the case of statements such as:

$$(P \wedge Q) \Rightarrow P \ ,$$
$$(\exists x \forall y P(x,y)) \Rightarrow (\forall y \exists x P(x,y)) \ .$$

A closed formula T which is true in every interpretation is said to be **valid**, written $\models T$; the intuitive meaning is that T is a semantic consequence without assumption. A valid proposition is called a **tautology**. We note that a valid formula is a semantic consequence of any theory; it is therefore not useful to introduce valid formulas amongst axioms of a theory.

5.6.2 Some Results of Model Theory; Limitations of First-Order Logic

The activity of modeling, whether in mathematics or computer science, often necessitates the search for a system of axioms characterizing the model under consideration. Occasionally, such a system does not necessarily exist within the given logic, typically first-order logic. Model theory provides tools which enable the detection of this sort of situation.

To illustrate this proposition, here is a simple example drawn from commutative group theory. We consider first the axioms, over the equational first-order language formed from the constant 0_g and the binary function $+$, written in infix form:

$$\forall x \forall y \quad x + y = y + x \ ,$$
$$\forall x \forall y \forall z \quad (x + y) + z = x + (y + z) \ ,$$
$$\forall x \quad x + 0_g = x \ ,$$
$$\forall x \exists y \quad x + y = 0_g \ .$$

The following are two properties of commutative groups, based on the concept of a divisor, that we would like to characterize axiomatically: we say that x is a **divisor** of y of order n if $\underbrace{x + ... + x}_{n \ times} = y$.

A commutative group is of **finite order** if every element is a divisor of 0. A commutative group is **divisible** if every element possesses a divisor of order n, for all n. These concepts can be axiomatized in second-order logic, quantifying over the integer n. We can take

$$\forall x \ \exists n \ \texttt{nat}(n) \wedge (\texttt{times}(n, x) = 0_g \ ,$$
$$\forall n \ \texttt{nat}(n) \Rightarrow \forall x \exists y \ \texttt{times}(n, x) = y \ ,$$

respectively for the axioms. The function $\texttt{times}$ can be axiomatized by:

$$\forall x \ \texttt{times}(0, x) = 0_g \ ,$$
$$\forall n \ \texttt{nat}(n) \Rightarrow \forall x \ \texttt{times}(\mathsf{S}(n), x) = x + \texttt{times}(n, x) \ .$$

The problem is that the predicate $\texttt{nat}$ is not first-order: the first-order axioms $\texttt{nat}(0)$ and $\forall n \ \texttt{nat}(n) \Rightarrow \texttt{nat}(\mathsf{S}(n))$, express that $0, \mathsf{S}(0), \mathsf{S}(\mathsf{S}(0))$, etc. are natural numbers, but it must be added that these are the only ones. We have the following negative results:

– it is impossible to characterize the class of divisible groups by means of a finite number of first-order axioms;
– it is impossible to characterize the class of finite order groups by means of a set (even an infinite set) of first-order axioms.

Moreover, we cannot axiomatize the real numbers in first-order logic. The proof of these results (see [Bar77, Ch. 1]) involves the following two theorems, which no longer hold true at second or higher orders.

Theorem 5.1 (Löwenheim)
Let T be a countable set of axioms; if there is a model of T, then there is a model of T whose set of elements is countable.

Theorem 5.2 (compacity)
A first-order theory T admits a model if and only if every finite part of T admits a model.

○   We can adapt this reasoning for various data structures of computer science and obtain similar results of impossibility, expressing that these structures cannot be characterized by a finite number of first-order axioms. A simple example mentioned in [Jon90] is Veloso's stack. It has been known for a long time that normal first-order logic is not suitable for systems having only finite models [AU79]. Logics with the concept of a *fixed point* were conceived to remedy this.

5.7 Notes and Suggestions for Further Reading

Propositional logic, first-order logic, and other issues discussed in this chapter are introduced in a number of texts. Particularly useful are the two volumes by Cori and Lascar [CL00, CL01], which are centered around the concept of a model. For a more detailed presentation of multi-sorted logic, see [Lal93] and [Gal86]. [GG90] address the issue from a philosophical point of view.

Reference works on model theory include [CK90]. A good introduction to this topic can also be found in the first two chapters of [Bar77].

6. Set-theoretic Specifications

This chapter is devoted to formal methods based on set theory. In set theory, a system is modeled using sets which are either considered to be primitive sets (for instance, sets of individuals, of books, of keyboards, etc.) or constructed by means of combinations of primitive subsets using set-theoretic operations. Specific languages can be distinguished from each other according to the way set-theoretic concepts are used, their underlying logic or how they assist in the production of programs from specifications. In this chapter we will introduce some well-known formal notations representative of the approach: Z, which appeared in the 1970s, VDM, which was born in the 1960s, and B, which was developed in the 1990s.

6.1 The Z Notation

Z can be roughly described as a syntactic envelope built on top of usual classical set-theoretic notations. The concept of a set is used as a universal means of expression. A first, and distinct, advantage of this approach is uniformity: the state space of a system is modeled as a set, types are sets, even operations are sets. Indeed, the latter are modeled as relations, that is, subsets of the Cartesian product of the set of states. Z provides symbols for various kinds of relations (functions, injections, partial injections, etc.) and a number of operators allowing one to construct relations from previously known relations.

6.1.1 Schemas

In Z, the state space and the operations of a system are declared by means of tables called **schema**. A schema is made of two parts. In the first part, we declare fields much as we would declare variables in a language like Pascal. Each field has a type which is constructed from built-in sets (e.g. the set of integers) and the usual set-theoretic operators (union, Cartesian product, etc.). The second part of the schema states constraints on the possible values of the fields by means of logical assertions.

A schema is surrounded with a frame. Its name is written in the first line of that frame. A horizontal line separates the declaration part and the predicate part. When several predicates are present, they are implicitly connected by a conjunction.

```
_____Example_ schema_____
x, y : Z
_____
x ≥ 0
y ≥ 5
x + y = 10
```

This simply denotes the definition of a set by comprehension, the usual mathematical notation is:

$$\{\langle x, y \rangle \in \mathbb{Z} \times \mathbb{Z} \mid x \geq 0 \wedge y \geq 5 \wedge x + y = 10\} \; ;$$

However, the schema notation becomes more interesting when the number of fields and the volume of assertions increase.

Z provides mechanisms for schema composition that allow one to structure a specification. For instance, the previous schema can be obtained through the composition of the two next schema.

```
_____first_ piece_____        _____second_ piece_____
x : Z                              y : Z
_____                      _____
x ≥ 0                              y ≥ 5
```

More precisely, we get the first schema by adding a constraint on both x and y.

```
_____Schema_ example_____
first_ piece
second_ piece
_____
x + y = 10
```

This schema can also be regarded as a subset of:

$$\{x \in \mathbb{Z} \mid x \geq 0\} \times \{y \in \mathbb{Z} \mid y \geq 5\} \; ,$$

that is, a relation between *first_ piece* and *second_ piece*. We can of course introduce a schema that expresses the last constraint separately:

```
_____constraint_____
x, y : Z
_____
x + y = 10
```

The conjunction of our three schema can simply be written:

Schema_ example $\stackrel{\text{def}}{=}$
    first_ piece $\wedge$ *second_ piece* $\wedge$ *constraint* .

Other logical operators are allowed as well. Thus

$other\_schema \stackrel{\mathrm{def}}{=}$
  $(first\_piece \lor second\_piece) \land constraint$

represents:

```
┌─────────other_schema──────────────
│  x, y  :  Z
│ ─────────────────────────────
│  x ≥ 0 ∨ y ≥ 5
│  x + y = 10
└───────────────────────────────────
```

Those combinations constitute the **schema calculus**. If S_1 and S_2 are two schema and $\star$ is a logical operator ($\lor$, $\land$, etc.) the expression $S_1 \star S_2$ represents the schema whose first part is the juxtaposition of declarations of S_1 and S_2, and whose second part is $P_1 \star P_2$ where P_1 (respectively P_2) is the predicate present in the second part of S_1 (respectively S_2). For the first clause to make sense, we must have no clash between the two declarations: common identifiers must have the same type.

6.1.2 Operations

The schema introduced up to now allow one to specify the state of a system. In order to describe an operation, two versions of the state are related: the state just before the operation and the state just after the operation. Z uses the following convention: if the first state is defined by variables x, y, $z \ldots$, the second is defined by variables x', y', $z' \ldots$

```
┌──────────state──────────      ┌──────────state'──────────
│  x  :  Z                       │  x'  :  Z
│ ────────────────────           │ ────────────────────
│  x > 4                         │  x' > 4
└────────────────────────        └────────────────────────
```

(Actually we don't need to explicitly write $state'$.) In order to relate two successive states, we naturally make use of the schema composition notation introduced above:

```
┌──────────an_operation──────────────
│   state
│   state'
│ ───────────────────────────────
│   P(x, x')
└─────────────────────────────────────
```

We see that the predicate we have in the second part of a state schema represents the invariant of the system we are describing: it will be implicitly respected by all operations which act upon the system.

We can still use the schema calculus: here a complex operation can be decomposed into several simpler cases (using a disjunction of schema); or, it can result from the conjunction of several constraints on before-and-after relations on the state of the system.

6.1.3 Example

Let us try to formalize the search for an element in a table. We need a predefined set which contains all elements that are, or could be, present in the table. We call this set U. Formally, we declare it using square brackets:

$[U]$.

The current state of the table is a subset of U, we represent it by a variable T which is a member of $\mathcal{P}(U)$.

Let us now consider the predicate P. In Z, a natural thing to do is to consider *Ptrue*, the set of elements verifying P, with $Ptrue \in \mathcal{P}(U)$. This predicate is not necessarily defined everywhere, hence we introduce the set *Pdef*, which contains *Ptrue* and represents the domain where P is defined. In other words, we agree that

- $P(x)$ is true if $x \in Ptrue$,
- $P(x)$ is false if $x \in Pdef$ and $x \notin Ptrue$,
- $P(x)$ is undefined if $x \notin Pdef$.

The system state is represented by the following schema.

$$
\begin{array}{|l}
\hline
\quad\qquad\qquad Table \underline{\qquad\qquad\qquad} \\
\; Ptrue, Pdef \; : \; \mathcal{P}(U) \\
\; T \; : \; \mathcal{P}(U) \\
\hline
\; Ptrue \subset Pdef \\
\; T \subset Pdef \\
\hline
\end{array}
$$

However, we have to ensure that *Ptrue* and *Pdef* are kept constant. Then we consider only composed operations built up from the following.

$$
\begin{array}{|l}
\hline
\quad\qquad\qquad Allowed\_\,op \underline{\qquad\qquad\qquad} \\
\; Table \\
\; Table' \\
\hline
\; Ptrue' = Ptrue \\
\; Pdef' = Pdef \\
\hline
\end{array}
$$

The operation we aim at returns an element of T verifying P if there is one. In order to take failure into account, we use a variable b as indicated on page 20. Its domain is $\{true, false\}$, and it is declared as follows:

$bool ::= true \mid false$.

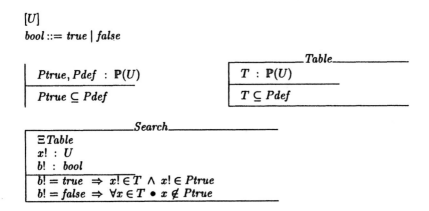

$$[U]$$
$$bool ::= true \mid false$$

$$Ptrue, Pdef \ : \ \mathbb{P}(U)$$
$$Ptrue \subseteq Pdef$$

_____Table_____
$$T \ : \ \mathbb{P}(U)$$
$$T \subseteq Pdef$$

_____Search_____
$\Xi Table$
$x! \ : \ U$
$b! \ : \ bool$
$b! = true \ \Rightarrow \ x! \in T \ \wedge \ x! \in Ptrue$
$b! = false \ \Rightarrow \ \forall x \in T \bullet x \notin Ptrue$

Figure 6.1: Z specification of a table search

The specification of the search operation indicates the expected values of x and b, and it states that T does not change.

_____Search_____
$Allowed\_op$
$x \ : \ U$
$b \ : \ bool$
$T' = T$
$b = true \ \Rightarrow \ x \in T \ \wedge \ x \in Ptrue$
$b = false \ \Rightarrow \ (\forall x \in T) \, x \notin Ptrue$

Remarks. In some cases, the actual notation in Z slightly differs from set theory. Here $\mathcal{P}(U)$ and $(\forall x \in A) \, P(x)$ should be written $\mathbb{P}(U)$ and $\forall x \in A \bullet P(x)$. In Z, the symbol $\subset$ represents strict inclusion, whereas we should use $\subseteq$. Moreover, lexicographic rules of Z allow identifiers to end with a question mark or an exclamation mark. In Z, it is understood that they represent input and output arguments of an operation, respectively. For consistency then, here we should replace x and b with $x!$ and $b!$.

It is also possible, in Z, to define constants with axioms. It is then better to introduce *Ptrue* and *Pdef* in this way and to remove *Allowed_ op*. Finally, the abbreviation $\Xi Table$ can be used for operations that do not modify the table. This is equivalent to declaring *Table*, *Table'* and to stating that nothing changes (that is, $T' = T$). A complete specification of the search for an element in a table using this notation is given in Figure 6.1.

6.1.4 Relations and Functions

In Z, as in set theory, the concept of relation is more primitive than the concept of a function. Let us see what happens with an assertion as simple as $f(x) = y$. Recall that, in first-order logic, f would be interpreted as a *total* function, thus the expression $f(x)$ would make sense. In Z, one often manipulates partial functions or even relations instead of total functions.

In fact the Z type system leads one to consider f from A to B as an element of $\mathcal{P}(A \times B)$, i.e. a relation from A to B. The notation $f(x)$ is then questionable and a number of authors prefer to avoid it. For instance, the assertion $f(x) = y$ may be represented by another one which states that x is in the inverse image by f of the singleton $\{y\}$ – the function from $\mathcal{P}(B)$ to $\mathcal{P}(A)$ that maps any subset Y of B to the set of elements a of A such that $f(a) \in Y$, denoted by $f^{-1}[\_]$, is always total function:

$$x \in f^{-1}[\{y\}] \ .$$

The price to pay is that notations become heavy in many situations where it is straightforward to use functions. A specification style using relational combinators (operators for constructing complex relations from simpler ones) helps to avoid this problem. But the notation becomes more difficult to understand.

6.1.5 Typing

Z semantics are based on the Zermelo–Fraenkel system, without the axiom of choice (which is not used here) and without the replacement schema (cf. § 7.2) [Spi88, CGR93a]. Within ZF, the latter restriction ensures the existence of a class of compartmentalized sets, thus providing a notion of type.

First, we have built-in sets like $\mathbb{Z}$ (positive, null or negative integers) and other sets which are application-specific. We denote these sets by $B_1, B_2, ...$ in what follows, and we consider that they are disjoint. Z includes appropriate restrictions on the use of $\cup$ that prevent us from forming the union of B_i and B_j with $i \neq j$ or constructing a set made of elements taken in different built-in sets. Then the type of a simple element x can be taken as the set B_i of which it is a member. The type of x is also the maximal set S such that $x \in S$.

The property of compartmentalization is preserved when we introduce sets of subsets and Cartesian products of previously formed maximal sets. Then B_i can be regarded as *base types* from which we can form composed types $\mathcal{P}(B_i)$, $P_i \times P_j$, $\mathcal{P}(P_i \times P_j)$, etc. The type of a compound element is again the maximal set of which it is a member. Thus it is not too difficult to check that a Z specification is well typed. Criticisms of this type system will be addressed in § 10.2.10.

In § 2.3.4, we showed that it is important to be able to construct sum types. This concept is available in Z and is referred to as a **free type**. We have already seen a simple example of free type: *bool* ::= *true* | *false*. This statement amounts to the declaration of a set (*bool*), two members in this set (*true* and *false*), and assertions that the latter are distinct elements and are the sole members of *bool*. Let us consider a more significant example: binary trees. This example is also more complex because it is a recursive data structure. We declare it in Z as follows: *tree* ::= *leaf*⟨⟨ℕ⟩⟩ | *bin*⟨⟨*tree* × *tree*⟩⟩. Here, *leaf* and *bin* are injections (respectively from ℕ to *tree* and from *tree* × *tree* to *tree*) that have disjoint ranges and, taken together, cover *tree*. Simple constructors, like *true* and *false*, can be seen as injections from a singleton set. The essential ideas come from algebra and type theory (see Chapters 10 and 12). The point is to guarantee that the axioms induced by a free type are consistent (they don't entail the absurd). To this effect, constructors (i.e. *bin*, *leaf*, *true* and *false* in our examples) must respect a number of rules. Roughly speaking, as constructors are injections, their domain cannot have a larger cardinality than their range, that is, the free type we want to define; for instance their domain cannot be the powerset of the free type. In the Z framework, there is a further technical complication because constructors are basically relations rather than functions.

6.1.6 Refinements

Refining a specification consists of systematically transforming abstract concepts (sets, relations, non-deterministic constructs, etc.) into features available in programming languages: arrays, chained data structures, usual control structures, functions, etc.

Refinement is more difficult in Z than in other formal methods because there is no convenient notation for usual programming constructs such as loops and recursive functions. These concepts are not very easy to handle in Z. However, it is possible to consider data refinement, that is, to relate an abstract data model to a concrete one closer to programming language data structures. For example, in order to represent a set of elements of U by an array (like T in our table example), we can introduce a function t from I to U, where I is an interval of integers.

6.1.7 Usage

Z is above all a notation for writing specification documents. Since its very beginning, its development was oriented towards including richer mathematical notations, e.g. relation combinators. It was not designed with the intention of being supported by software tools. One may quite reasonably guess that this would, in any case, have been beyond the capacities of technologies available in the 1970s. Support tools appeared in the 1990s, mainly for editing and type checking. A recent proof assistant for Z is Z/EVES [Saa97]. On the other hand, many introductory and more avanced books are available (see the bibliographic notes at the end of this chapter) and there is an active user community, especially in Great Britain where a number of industrial projects were developed or re-engineered with Z. Some of these are reported in [CGR93b], [HB95] and [HB99].

Z is mainly used for specifying data and transformations of data. In principle, we can expect to go further, thanks to general set-theoretic concepts included in Z. For instance, can we study interactions between software components running in parallel on different machines? Trajectories of such components can be formalized in Z. However, we are still a long way from the mathematics needed for specialized formalisms such as labeled transitions or process algebra (see Chapter 8). Moreover, the behavior of such systems is very complex and cannot be fully understood without automated support tools.

6.2 VDM

6.2.1 Origins

VDM (Vienna Development Method) was initially a language description method inspired by denotational semantics. Briefly, recall that denotational semantics interprets programs by mathematical functions (cf. § 2.6).

We know that a program may not terminate for certain input data. In the general case, a program is then modeled as a *partial* function — see the concept of partial recursive function in § 3.3.4 and § 3.7.3. On the other hand, total functions are much easier to handle in mathematics. In order to recover total functions, basic sets of values (integers, Booleans and so on) are augmented by an additional value denoted by $\perp$, which represents the undefined.

$\perp$ can be seen as an approximation of all other values in some sense. The rough idea is that $\perp$ represents a value we know nothing about. To formalize the notion of approximation, we consider a relation $<$ such that $\perp < v$ for all "ordinary" v and such that two "ordinary" values are not related by $<$. At the moment we have only two levels of approximation: a very bad one ($\perp$) and a perfect one (the value itself). But for pairs we have more possibilities: either we know nothing ($\langle \perp, \perp \rangle$, which can be considered equivalent to $\perp$), or we know one of the two components ($\langle v_1, \perp \rangle$ or $\langle \perp, v_2 \rangle$), or we know both of them ($\langle v_1, v_2 \rangle$). We have $\perp < \langle v_1, \perp \rangle < \langle v_1, v_2 \rangle$ and $\perp < \langle \perp, v_2 \rangle < \langle v_1, v_2 \rangle$ but $\langle v_1, \perp \rangle$ and $\langle \perp, v_2 \rangle$ are incomparable. In the case of functions defined over an infinite set such as $\mathbb{N}$, the structure of approximations becomes richer and we need concepts of limits coming from topology.

Spaces endowed with a relation $<$ satisfying adequate properties are sometimes called **domains**. Introduced by Dana Scott in 1969 they play a central role in denotational semantics and their theory has been studied in depth. A pedagogical reference is the book of Stoy [Sto77]. In this book we never use the terminology *domain* in the technical meaning mentioned above, but in the ordinary sense of set.

The developers of VDM chose to use the usual set-theoretic concept of function rather than the more complex concept introduced in Scott domains. The notation used in the Vienna method was first called Meta-IV, then VDM-SL (VDM Specification Language). Nowadays, we often use VDM for both the method and the language, and we follow this convention in what follows.

A consequence of the denotational semantics background of VDM is that the concept of (partial) function is more primitive here than the concept of relation. If we need a relation from A to B, we can represent it by a function from $A \times B$ to bool. Z operators for manipulating relations (sequential composition, domain or range restriction, etc.) are still present in VDM but apply to functions.

6.2.2 Typing

Compound objects of VDM are similar to Z schema. Typing is considered from a different perspective, however: in VDM, a piece of logical information declared as the invariant of a compound object is considered as a part of its type, while in Z the type would have been the largest set containing the object. It is thus possible, in VDM, to construct sets having elements of different kinds as members, but type checking is no more decidable: it yields **proof obligations**, that is, assertions that can be automatically stated but that in general can be discharged only with human support.

6.2.3 Operations

Operations describe changes in the object state. They can be specified in an implicit or an explicit manner. The **implicit** manner consists of providing a precondition[1] and a postcondition on objects manipulated by the operation. This is similar to operation descriptions in Z, up to a notational variation: in Z the new state gets a decoration ("$'$") while in VDM it is the previous state ("$\overleftarrow{}$"). For example, incrementing x can be specified by $x = \overleftarrow{x} + 1$. The **explicit** manner for defining operations is closer to refinement than to specification. It consists in describing an algorithm by means of usual constructions (sequence, selection, loop, etc.). In that case, however, the computation steps should be annotated by logical assertions.

6.2.4 Functions

In addition to operations, it is possible to define functions in VDM. In contrast with VDM operations and with functions we find in imperative programming languages, VDM functions do not involve any state change. In fact, we are encouraged in VDM to generously use function definitions in specifications. As for operations, functions can be defined in an implicit manner, by means of a

---

[1]Note that preconditions have a different status in Z and VDM: in VDM they are given in the VDM specification, whereas they are calculated in Z.

precondition on the arguments, and of a postcondition relating the arguments with the result, or in an explicit (algorithmic) manner. Recursive definitions of functions are allowed.

Allowing recursive or even algorithmic definitions of functions at the specification level may seem surprising at first sight. However, a number of functions can hardly be described otherwise: think of the factorial function or calculating income tax.

6.2.5 Three-valued Logic

In VDM, functions are defined and then used in the specification of operations or of other functions, including implicit definitions. In other words, a logical assertion (an invariant, a precondition or a postcondition) can contain occurrences of functions which are defined in another part of the specification. This provides interesting opportunities for structuring VDM specifications. At the same time, this has significant consequences for the underlying logical system. Indeed, functions defined recursively or in an algorithmic way are often partial functions. Then the usual framework of two-valued logic turns out to be too narrow.

Let us consider for example an assertion such as:[2]

$$\forall a, b \quad b > 0 \ \Rightarrow \ \mathtt{div}(a,b) \times b \le a < \mathtt{div}(a,b) \times b + b \ , \tag{6.1}$$

telling us that $\mathtt{div}$ performs an Euclidian division. It is quite easy to find an explicit definition of $\mathtt{div}$ that does not terminate when b is null. In this case $\mathtt{div}(a,b)$ has no value, hence it becomes impossible to give the value *true* or *false* to the logical expression $a \le \mathtt{div}(a,b) \times b < a + b$. However, we feel that (6.1) should be given the value *true*, since the value of $b > 0$ is precisely *false* in the litigious case: we know that the value of $\mathbf{f} \Rightarrow P$ is *true* whichever the value *true* or *false* of P.

In order to deal with such situations, VDM makes use of a three-valued logic. Besides *false* and *true*, we have $\perp$ which denotes the undefined value. We recognize here ideas coming from denotational semantics, which are at the roots of VDM. Truth tables are adequately extended, for example the value of $\mathbf{f} \Rightarrow \perp$ is *true*. However, several three-valued logics are possible. Selecting one of them was a design decision of VDM, and unusual deduction rules could not be avoided (cf. § 5.4).

6.2.6 Usage

A number of VDM applications can be found in language definitions. Despite its name, VDM is more a notation than a method. It is supported by a number of tools. An experimental proof assistant is described in [JJLM91]. Later on, protyping and simulation tools were developed. In the family of VDM, we

---

[2]Of course, $x \le y < z$ is an abbreviation for $x \le y \wedge y < z$.

can cite Raise, which combines the VDM description of data, operations and functions, with CSP, a process algebra for describing message exchanges and synchronization between parallel processes.

6.3 The B Method

The B method can be regarded, to some extent, as a descendant of Z: it was designed by one of the founders of Z, J-R. Abrial, and it maintains the set-theoretic notations used in Z. One of the big differences is that B provides a development process covering specification, refinement, and implementation steps. The way data and operations are presented and structured is also quite different: it is close to imperative programming languages such as Pascal. More precisely, we have the language of guarded commands of Dijkstra (cf. § 4.3.3) enriched with data structures expressed in the set-theoretic notations of Z, providing a uniform framework for specification and development. The main features of B are:

- a specification language (called *abstract machines*);
- a refinement and implementation technique;
- proof obligations associated with each development step;
- structuring mechanisms for decomposing abstract machines;
- tools for supporting and controlling the different tasks.

The B method has been used in industry for several years, notably for railway equipment and signalling [SDM92, BBFM99].

6.3.1 Example

In Figure 6.2 we show a B specification of a variation on the problem of searching for an element in a table. As in § 2.4.4.1 (see the third specification on page 28) we consider here the case of the search for an integer in an interval. The role of U in the previous Z specification is played here by $\mathbb{N}$, denoted[3] by NAT. The role of *Pdef* is played by the interval $[minD..maxD[$ and the role of T by $[min..max[$. The predicate P here is called[4] *Pr* and the operation *Search* returns two results, *bb* and *xx*. Note that in Z, P was represented by the set *Ptrue* whereas here we take a predicate, seen as a mapping from $[min..max[$ to $\mathbb{B}$ (this set is denoted by BOOL in B).

Intuitively, we can imagine that the work space of this machine is an array of Booleans (*Pr*) having *min* and *max* as bounds, which are themselves between *minD* and *maxD*. The latter are fixed once and for all, while *min* and *max*

---

[3]In B, NAT actually represents a *finite* subset of $\mathbb{N}$ that can be written [min..max] (with min < 0 < max), where min and max are fixed parameters depending on the hardware architecture to be used at the implementation level.

[4]Lexicographic detail: identifiers must begin with at least two letters.

MACHINE   table(minD, maxD)

CONSTRAINTS
    $minD \in$ NAT $\wedge$ $maxD \in$ NAT $\wedge$
    $0 < minD$ $\wedge$ $minD \leq maxD$

VARIABLES
    min, max, Pr, bb, xx

INVARIANT
    $min \in$ NAT $\wedge$ $max \in$ NAT $\wedge$
    $minD \leq min$ $\wedge$ $min \leq max$ $\wedge$ $max \leq maxD$ $\wedge$
    $Pr \in min..max - 1 \rightarrow$ BOOL $\wedge$
    $bb \in$ BOOL $\wedge$ $xx \in$ NAT

INITIALIZATION   /* without interest here */

OPERATIONS
    Search $\overset{\mathrm{def}}{=}$
        IF $\exists tt \bullet tt \in min..max - 1 \wedge Pr(tt) =$ true
        THEN
            $bb :=$ true $\|$
            ANY   tt
                WHERE $tt \in min..max - 1 \wedge Pr(tt) =$ true
                THEN $xx := tt$
            END
        ELSE
            $bb :=$ false $\|$
            ANY   tt   WHERE $tt \in$ NAT   THEN $xx := tt$ END
        END

END

Figure 6.2: Table search specification in B

could vary during allocation or disposal operations beyond the scope of this chapter.

Here we chose a fairly low abstraction level for the specification of data structures. But nothing is decided about the search algorithm itself.

6.3.2 Abstract Machines

In B, a specification is structured into units called *abstract machines*. They encapsulate the state of a subsystem as well as operations modifying it or returning a view of it. The idea of encapsulating data and related operations together is well known in computer science, it has most notably been formalized by *abstract data types*. The main components of a B abstract machine are:

– parameters declaration, constants declaration (none in our example) and above all variables declaration — they constitute the internal state of the machine;
– the statement of an invariant, a logical assertion relating the variables, parameters and constants just declared; their type is included in the invariant (the concept of type in B is the same as in Z); the part of the invariant

which relates only parameters and constants is declared separately (in the CONSTRAINTS clause) and there is also a specific clause for constants only;
- the definition of the initial state;
- operations, expressed with *generalized substitutions*, which are a generalization of guarded commands.

Proof obligations are automatically generated in order to ensure that the initial state as well as operations respect the invariant. This is in contrast with Z where, as a simple consequence of the schema calculus, the invariant is naturally included in the postcondition of operations. In some sense B seems less declarative.[5] However, the new state returned by an operation can be specified in a fairly abstract way using logical and set-theoretic notations. Moreover, we can say that B achieves a separation of concerns: we have the opportunity to establish invariant preservation in abstract terms, before going into low level details. It is indeed possible in B to adopt a specification style where the invariant is automatically preserved. But this amounts to delaying the work until later development steps: refinement proof obligations will be more complex. It is far better to work on proof obligations as early as possible. They are an opportunity to check the consistency of the specification and often to correct it, hence the global correctness proof is divided into smaller units.

We see that design decisions for B proof obligations take the whole development cycle (from specification to implementation) into account. Generalized substitutions have been designed with the same concern in mind.

6.3.3 Simple Substitutions and Generalized Substitutions

A **simple substitution** is simply an assignment $x := E$. Indeed we know from § 4.3.3 that the weakest precondition for this transformation to establish the postcondition Q is $[x := E]Q$, that is, the formula Q where E is substituted for all occurrences[6] of x.

Generalized substitutions are combinations of simple substitutions. Among these combinations we have the sequence and the loop, in the language of guarded commands; however, these constructs are allowed only in refinement stages. At the level of specification the following combinators are available:

- parallel composition, corresponding to simultaneous substitutions; for example $x := E \;\|\; y := F$ corresponds to $x, y := E, F$;
- the selection IF C THEN S_1 ELSE S_2 END, which has the expected intuitive meaning; from a logical perspective, it transforms the predicate Q into $C \Rightarrow [S_1]Q \;\wedge\; \neg C \Rightarrow [S_2]Q$;

---

[5] A **declarative** language states *what* should be done, while a **prescriptive** language states *how* it is done. One can consider that we have a specification in the first case and a program in the second case. This distinction was devised in the 1970s in the framework of programming languages, because very high level programming languages like Prolog could be presented as executable specification languages.

[6] Actually, only *free* occurrences, i.e. occurrences which are not in the scope of a quantifier, see Chapter 5.

$$Search \quad \overset{\text{def}}{=}$$
$$\qquad \text{PRE} \quad \exists tt \quad \bullet \quad tt \in min..max - 1 \ \wedge \ Pr(tt) = \text{true}$$
$$\qquad \text{THEN}$$
$$\qquad\qquad \text{ANY} \quad tt$$
$$\qquad\qquad\qquad \text{WHERE} \quad tt \in min..max - 1 \ \wedge \ Pr(tt) = \text{true}$$
$$\qquad\qquad\qquad \text{THEN} \quad xx := tt$$
$$\qquad\qquad \text{END}$$
$$\qquad \text{END}$$

Figure 6.3: Strengthening a precondition in B

- unbounded choice ANY v WHERE $P(v)$ THEN S END, where S depends on the dummy variable[7] v, sometimes shortened in @v $P(v) \to S$. This substitution behaves like S where the choice of v is arbitrary, provided $P(v)$ is true. *Nothing* is said about the intended implementation of this statement: a pseudo-random choice between the different permitted values of v is only one possibility among many others, and in practice it will never be chosen because it is complicated and inefficient! In fact one often refines this construct using a loop, as would be the case in the table search example;
- introduction of a precondition P: PRE P THEN S END, sometimes shortened to $P \mid S$. This substitution is purposely defined only for states verifying P. Its practical use is for stating conditions which guarantee that a given operation can be performed successfully. Ensuring that the operation is called when the aforementioned precondition is true must be done by its user. For example, Figure 6.3 gives a weaker specification of table search, which conforms to the suggestion of § 2.3.5 on page 22.

The construct IF C THEN S_1 ELSE S_2 END is described using two primitive constructs, which are:

- the guard $G \to S$, which behaves like the substitution S from a state where the property G is true;
- the choice between two substitutions $S_1 \ \square \ S_2$.

Their logical definition is si.nple:

$$[G \to S]\, Q \ \overset{\text{def}}{=} \ G \Rightarrow [S]Q \ , \tag{6.2}$$
$$[S_1 \ \square \ S_2]\, Q \ \overset{\text{def}}{=} \ [S_1]Q \wedge [S_2]Q \ . \tag{6.3}$$

Then we take:

$$\text{IF } C \text{ THEN } S_1 \text{ ELSE } S_2 \ \overset{\text{def}}{=} \ C \to S_1 \ \square \ \neg C \to S_2 \ .$$

note that Dijkstra's non-deterministic alternative construct

$$\text{if } B_1 \to S_1 \ \square \ B_2 \to S_2 \text{ fsi} \ ,$$

---

[7]The name of this variable is of concern only inside the block ANY ... END under consideration.

where B_2 is not necessarily the negation of B_1, corresponds here to

$$B_1 \vee B_2 \quad | \quad B_1 \rightarrow S_1 \; \square \; B_2 \rightarrow S_2$$

(see equation (4.16) on page 73).

Unbounded choice ANY ... END is a generalization of $S_1 \; \square \; \cdots S_n$ to an arbitrary number (it can be infinite) of substitutions. Its formal definition is:

$$[@v \, P(v) \rightarrow S] \, Q \stackrel{\text{def}}{=} \forall v \, P(v) \Rightarrow [S]Q \; , \qquad (6.4)$$

which is quite natural if one regards $\forall$ as an infinite conjunction.

6.3.4 The B Refinement Process

At the specification stage, abstract machines use non-deterministic constructs and the whole power of set-theoretic notations, while algorithmic constructs (sequences, loops) are not allowed. During refinement stages, set-theoretic data structures are progressively replaced with data structures closer to programming language data structures, non-determinism is eliminated and generalized substitutions corrresponding to sequences and loops are introduced.

REFINEMENT $table1(minD, maxD)$

REFINES $table$

VARIABLES
    $min1, max1, Pr1, xx1$

INVARIANT
    $min1 = min \quad \wedge \quad max1 = max \quad \wedge$
    $Pr1 = Pr \quad \wedge \quad xx1 = xx \quad \wedge$
    $min1 \leq xx1 \quad \wedge \quad xx1 \leq max1 \quad \wedge$
    $xx1 = max1 \quad \Leftrightarrow \quad bb = \text{false}$

INITIALIZATION   /* Without interest here */

OPERATIONS
    $Search \quad \stackrel{\text{def}}{=}$
        IF $\exists tt \quad \bullet \quad tt \in min1..max1 - 1 \; \wedge \; Pr1(tt) = \text{true}$
        THEN
                ANY tt
                    WHERE $tt \in min1..max1 - 1 \; \wedge \; Pr1(tt) = \text{true}$
                    THEN $xx1 := tt$
                END
        ELSE
                $xx1 := max1$
        END
    END

Figure 6.4: B refinement of table search

Data refinement is illustrated in Figure 6.4 for the example of table search. This refinement step aims essentially at eliminating *bb*. In the refining abstract machine we declare a new space of variables, whose link with original variables is defined by the invariant. In a second stage we could refine the remaining non-deterministic choice by a loop, along the lines indicated on page 31.

Refinement steps are under the control of proof obligations ensuring that invariants are preserved and that a refining machine conforms to the more abstract machine that it refines. Proof obligations are completely defined in the underlying theory of B and they can be automatically generated. The support tools for B include syntax and type checkers, proof obligation generators, code generators and ad hoc automated proof assistants able to deal with propositional logic, first order logic and a huge number of set-theoretic algebraic rules.

The target of code generators is a minimal and simple subset of languages like C, Modula or Ada. Such subsets can reasonably be considered as secure, since only the easiest parts of the compilers are concerned. Indeed, this is made possible because high-level features of programming languages can be considered as redundant here: they are the concern of the specification, whereas the B development cycle starts from truly abstract specifications. At the implementation stage, only low-level data structures and instructions are needed.

6.3.5 Modularity

If we want to develop a whole real-scale system, starting from a huge monolithic specification would be unmanageable. In B it is possible — and recommended! — to decompose a specification into several machines. The big win is that refinement stages are then performed consistently and independently. In particular, proof obligations become smaller, they can be dealt with separately, and maintenance is made easier.

6.4 Notes and Suggestions for Further Reading

Many textbooks present Z and VDM in a manner that is within the reach of every one, for example [PST91, Wor92, WL88] for Z and [Jon90] and [JS90] for VDM. Mike Spivey's reference book on Z is still very useful, though the language has evolved since its publication [Spi89]. The book *Understanding Z* [Spi88] by the same author is not a pedagogical introduction, but gives an early definition of the Z semantics. Free types of Z are described in [Spi89], [Art91] and more recently in [Art98] and [TVD00].

The reference book on B by J.-R. Abrial [Abr96] is both a description of its theoretical foundations and a very detailed definition, illustrated with many examples.

The reader interested in refinement techniques may consult the article by Gardiner [GM91] and the book by de Roever [dRE98].

7. Set Theory

Set theory has a strong influence on formal methods. A straightforward reason for this is that the specification languages considered in the last chapter rely directly upon set theory. More significantly, set theory has strong links with logic:

- as a metalanguage,[1] it provides a semantics for logic via the concept of a model; as an interesting consequence for the use of formal methods, we obtain a means of interpreting logical specifications (cf. § 3.3.1 and § 5.6);
- the axiomatized version(s) of set theory is (are) a first order theory that can be studied as a formal system; for instance, one can try to show that it is consistent (without contradiction). Even more important for us, formalization techniques used in the development of a number of important concepts from set-theoretic primitive concepts can be adapted to the practice of specification methods.

We concentrate here on the Zermelo–Fraenkel axiomatization of set theory. This will be a good opportunity to present a typical technique for enriching a language. Other techniques, e.g. for handling functions, are similar to the ones used in Z and in B. We also comment on how we may deal with inductive or impredicative definitions (corresponding to so-called recursive definitions of programs or data structures).

7.1 Typical Features

7.1.1 An Untyped Theory

A number of set-theoretic operations, such as intersection and union, take arguments sets that, intuitively, have elements of the same kind as members. In contrast, the Cartesian product can be constructed on sets of different kinds and it returns a set having yet another kind. The powerset of a set is not of the same kind as the set itself. Distinguishing the kinds — or what we call the *types* — of sets or elements provides an excellent protection mechanism against many mistakes and errors. But this would excessively hamper the development of set theory. Just think of the way natural numbers are represented in set theory (we will revisit it in § 7.3.1). Moreover, what type should be given to the

---

[1]See page 152.

empty set? Or to the identity function? The answers to these simple questions are not all that simple.[2] *Thus set theory is essentially an untyped theory.* The development of the theory illustrates that it is actually harmful to decompose the universe into elements on the one side and sets on the other. Any item can occur on the left and on the right of the $\in$ symbol. Hence, it is simpler to decide that all items are sets, jumbled together.

7.1.2 Functions in Set Theory

Recall that functions are not a primitive concept in set theory. A function from E to F is a particular relation, that is, an element of $\mathcal{P}(E \times F)$, satisfying a number of properties (uniqueness of the result, and with a domain equal to E if the function is total). To be rigorous, it raises a notational issue: if f and x are two symbols (i.e. two sets), $f(x)$ makes sense only if we have proved beforehand that f satisfies the necessary properties and that x is in the domain of f.

The development of set theory involves a mechanism of *theory extension*, that allows one to enrich the language step by step with new function symbols or new predicate symbols. There is a similar process well known amongst computer scientists, viz. enriching a programming language with user-defined procedures.

7.1.3 Set-theoretic Operations

A very convenient feature of set theory is the collection of operations provided for constructing complex sets from simple sets. Moreover, union, intersection, set difference (symmetric or otherwise), and Cartesian product satisfy many interesting algebraic properties: $\cup$, $\cap$ and $\setminus$ are commutative and associative; $\cup$ and $\cap$ are idempotent; $\varnothing$ is an identity element of both $\cup$ and of $\setminus$, and an absorbing element of $\cap$. One can also identify $(X \times Y) \times Z$ with $X \times (Y \times Z)$ by means of a natural bijection, and $X \times \{\varnothing\}$ and $\{\varnothing\} \times X$ with X, which amounts to saying that $\times$ is associative as well as admitting $\{\varnothing\}$ as an identity element.

These identifications can be seen as abuses of notation, but they are justified from the viewpoint of category theory: intuitively, a product is considered as an object of the theory — a set here — endowed with projections allowing one to retrieve the components of a tuple.

Let p_1 and p_2 denote the two projections in the case of 2-uples (couples), t_1, t_2 and t_3 the three projections in the case of 3-uples (triples); representing $X \times Y \times Z$ by $(X \times Y) \times Z$ amounts to taking $t_1 \stackrel{\text{def}}{=} p_1 \circ p_1$, $t_2 \stackrel{\text{def}}{=} p_2 \circ p_1$ and $t_3 \stackrel{\text{def}}{=} p_2$; choosing $X \times (Y \times Z)$ amounts to taking $t_1 \stackrel{\text{def}}{=} p_1$, $t_2 \stackrel{\text{def}}{=} p_1 \circ p_2$ and $t_3 \stackrel{\text{def}}{=} p_2 \circ p_2$. The chosen representation itself matters little because triples are manipulated only through t_1, t_2 and t_3. We now actually have a kind of *abstract data type*.

---

[2]Typing is good because it prevents us from expressing meaningless things. The problem is that it could equally well prevent us from expressing perfectly good and meaningful things. Designing a good type system is then a significant issue. We revisit this question in Chapter 11.

But if we want to deal with these operations on the same footing as with usual algebraic operations, we come up against an obstacle. Our operations take sets as arguments and return a set. The role of the reference set would then be played by the set of all sets, an inconsistent notion (see Russell's paradox in § 3.1.3).

This leads set theorists to distinguish two kinds of collections, sets and classes. Thus the universe $\mathcal{U}$ of all sets is not a set but a class. Operations can then be defined over members of a class such as $\mathcal{U}$. This works, but the distinction between class and set can be considered to be somewhat artificial.

7.2 Zermelo–Fraenkel Axiomatic System

The are quite a few Zermelo–Fraenkel axioms (ZF in the following). They are defined over a very simple language, without symbols for the union, the intersection, nor the Cartesian product of sets. The latter can be defined by means of clever encodings. Apart from equality, *the only primitive concept is membership*. In summary, Zermelo–Fraenkel set theory is a *first-order* theory defined over an equational language having basically only one predicate symbol ($\in$) apart from =, and no function symbol.

All items are taken from the same grouping. If one looks for a model of set theory, in the sense of § 5.1.3, this grouping or jumble is interpreted as the domain, that is a set, but at the metalanguage level. Items in turn are interpreted as sets, as intuitively intended, only at this second level of interpretation. This is the so-called standard interpretation, but there is nothing to prevent us from imagining other interpretations. We even know, by an application of Löwenheim's theorem (§ 5.6.2), that a denumerable model of ZF exists.

We now briefly present the system of Zermelo and Fraenkel, as described in [Sho77]. This material can be compared, for example, with the underlying theories of Z and B, which are close to, but not exactly, ZF.

7.2.1 Axioms

First, recall that from an axiomatic viewpoint, "set" is nothing but a word, just like "point" or "line" in the axiomatic presentation of geometry. Explaining the meaning of manipulated objects is beyond the scope of an axiomatic theory; its only aim is to let us know the consequences of formulas taken as axioms. The relevance of an axiomatic theory to the real world is a matter of experience and not of formal logic. Here, it is crucial to be able to express in a convenient way that we can form a set y with the elements x satisfying a given property P. This is not always the case, as evidenced by Russell's paradox. The axioms aim precisely at defining when this is the case. A formula expressing this fact is

$$\exists y\, \forall x\, (x \in y \Leftrightarrow P)$$

and we will use the following abbreviation:

$$\mathbf{Set}\{x \mid P\} \ .$$

Here is the list of axioms.

Extensionality: *two sets x and y are equal if they have the same elements:*

$$\forall x \forall y \quad \forall z\, (z \in x \Leftrightarrow z \in y) \Rightarrow x = y \ .$$

An important consequence of this axiom is the following: if there exists a y such that $\forall x\, (x \in y \Leftrightarrow P)$, then y is unique. Thus, as soon as a property $\mathbf{Set}\{x \mid P\}$ is proved, a set is defined. We say that this set is defined **by comprehension**, and it is denoted by $\{x \mid P\}$. Most of the remaining axioms determine the possible forms of P for which we admit that $\{x \mid P\}$ exists.

Powerset: *the set of subsets of x is a set denoted by $\mathcal{P}(x)$:*

$$\forall x \quad \mathbf{Set}\{y \mid \forall z\, (z \in y \Rightarrow z \in x)\} \ .$$

Union: *the union of elements of x is a set denoted by $\bigcup(x)$ —the notation $\bigcup_{y \in x} y$ would be closer to usual conventions:*

$$\forall x \quad \mathbf{Set}\{z \mid \exists y\, (y \in x \wedge z \in y)\} \ .$$

Schema of separation: *extracting from a given set x the elements y satisfying a property $\varphi(y)$ yields a set:*

$$\forall x \quad (\forall y\, \varphi(y) \Rightarrow y \in x) \Rightarrow \mathbf{Set}\{y \mid \varphi(y)\} \ .$$

Schema of replacement: *applying an operation F to the elements of a set x yields a set:*

$$\forall x \quad \mathbf{Set}\{z \mid \exists y\, y \in x \wedge z = F(y)\} \ .$$

In order to define an operation F, one has to extend the language in the following way. One must first take a formula $\phi(u,v)$ such that for all y, there is a unique z such that $\phi(y,z)$. (Formally, one proves $\forall y \exists z\, \phi(y,z)$ and $\forall y\, \forall z \forall z'\, (\phi(y,z) \wedge \phi(y,z') \Rightarrow z = z')$.) Then one introduces a new symbol F and adds the axiom $\forall y\, \phi(y, F(y))$. A formula containing $F(u)$, say $P(F(u))$, is handled as an abbreviation for $\forall v\, \phi(u,v) \Rightarrow P(v)$.

The two last axioms are schemas: any instance of the formula φ (respectively, of the operation F) provides a corresponding separation (respectively, replacement) axiom. The separation schema can be deduced from the replacement schema but is very important in its own right.

Infinity: *there exists a set x which has the empty set as a member and such that for all y which are members of x, there is another member z of x containing the members of y and y itself:*[3]

$$\exists x \ \left(\emptyset \in x\right) \wedge \left(\forall y \ y \in x \Rightarrow y \cup \{y\} \in x\right) \ .$$

This statement is easier to understand if $\emptyset$ is seen as a representation of 0 (zero) and $y \cup \{y\}$ as a representation of the successor of y. We will come back to this later.

Regularity (or foundation): *a non-empty set x contains an element y which is disjoint from x:*

$$\forall x \ \left(\exists y \ y \in x\right) \Rightarrow \left(\exists y \ y \in x \wedge \forall z \ z \in y \Rightarrow \neg(z \in x)\right) \ .$$

This is equivalent (given previous axioms) to stating that the relation $\in$ is well founded. This prevents the construction of infinite chains $x_0 \ldots x_n \ldots$ with $x_{i+1} \in x_i$ for all i. In particular there is no set x such that $x \in x$. But it would be mistaken to think that this axiom aims at avoiding paradoxes: later in this chapter we will mention another axiomatization of set theory without the regularity axiom, and which is just as consistent as ZF.

The system composed of the previous axioms is called ZF. It allows one to recover usual concepts of set theory. In mathematics, a further axiom, the axiom of choice, due to Zermelo, is needed. The ZF system together with the axiom of choice is called ZFC. We state here an informal version of this axiom, which first necessitates the introduction of the concept of a function.

Axiom of choice: *for all families x of non-empty sets, there exists a total function from x to $\bigcup(x)$ mapping every element y of x to an element of y.*
More simply, given a (finite or infinite) family of sets, this axiom allows one to choose an element in each of them. This axiom played a key role in our justification of the principle of well-founded induction in § 3.5.3.

7.2.2 Reconstruction of Usual Set-theoretic Concepts

We see that the empty set, singletons, pairs, intersection, and Cartesian product are not primitive concepts of ZF. Even binary union is not primitive — we have "only" the generalized union. It can of course be recovered, as can the other concepts. Recall that the Cartesian product is needed in order to define relations and functions.

One proceeds step-by-step in a systematic manner: one shows the existence and the uniqueness of an appropriate set, then one introduces a corresponding symbol (this is another application of language extension, previously described in the replacement schema). Uniqueness is shown using the axiom of extensionality. For existence, one almost always uses the schema of separation, which allows us to define a set by comprehension *provided we have already found one*

---

[3]The following formalization uses the abbreviations $\emptyset$ and $\cup$ described below.

in which it is included. This is the key for closing the door on Russell's paradox: we come back to this in § 7.3.3.

Let us illustrate the process of finding the intersection of x and y. We can separate (select) the elements of x which happen to be members of y, because we have:

$$\forall z\, (z \in x \land z \in y) \Rightarrow z \in x \ .$$

The schema of separation allows us to infer:

$$\mathbf{Set}\{z \mid z \in x \land z \in y\} \ .$$

Then we are entitled to define:

$$x \cap y \stackrel{\text{def}}{=} \{z \mid z \in x \land z \in y\} \ .$$

The difference can be defined in the same way, but the union cannot. Here are the main steps, without going into the details:

– The empty set $\varnothing$ is constructed through the separation of elements satisfying **f** in an arbitrary existing set; then one can sequentially form $\mathcal{P}(\varnothing)$ and $\mathcal{P}(\mathcal{P}(\varnothing))$ which is a 2-element set;
– given x and y, one can then form the pair $\{x, y\}$ using the schema of replacement on $\mathcal{P}(\mathcal{P}(\varnothing))$ where the operation F satisfies $F(\varnothing) = x$ and $F(u) = y$ if $\neg(u = \varnothing)$;
– the union of x and y is defined by $x \cup y = \bigcup(\{x, y\})$; it is only at this stage that we have $\mathbf{Set}\{z \mid z \in x \lor z \in y\}$,
  with $x \cup y = \{z \mid z \in x \lor z \in y\}$;
– other set operations (intersection, difference, etc.) are defined directly by separation;
– the concept of an ordered pair is represented by an encoding:

$$\langle x, y \rangle = \{\{x\}, \{x, y\}\} \ ;$$

the Cartesian product $a \times b$ is obtained by separating elements of the form $\langle x, y \rangle$ in $\mathcal{P}(\mathcal{P}(a \cup b))$, with $x \in a$ and $y \in b$.

7.2.3 The Original System of Zermelo

The first system proposed by Zermelo included all previous axioms, with one notable exception: the schema of replacement. The construction of $\{x, y\}$ was directly postulated by the axiom of the pair.

Pair: *the pair made of two sets x and y is a set $\{x, y\}$:*

$$\forall x \forall y \quad \mathbf{Set}\{z \mid z = x \lor z = y\} \ .$$

But a number of set-theoretic developments (e.g. about ordinal and cardinal numbers) could not be recovered in the original system of Zermelo.

7.3 Induction

7.3.1 Reconstruction of Arithmetic

Peano arithmetic can be encoded in ZF. The number 0 is represented by $\varnothing$, the successor operation is represented by $S(x) = x \cup \{x\}$. Then one can prove Peano axioms. The axiom of regularity can be used to show that $S(x) = S(y) \Rightarrow x = y$ for arbitrary x and y (not only for sets representing natural numbers).

The case of the schema of induction is very interesting. Let us first define $\mathbb{N}$. To this effect we consider the predicate **supnat** defined as follows:

$$\mathbf{supnat}(e) \stackrel{\mathrm{def}}{=} \quad \varnothing \in e \ \wedge \ \forall x \ x \in e \Rightarrow S(x) \in e \ .$$

That is, we have $\mathbf{supnat}(e)$ if and only if e contains 0, $S(0)$, ...; intuitively, this means that e is a superset of $\mathbb{N}$. The axiom of infinity precisely states that such an e exists; let us call it $\mathbb{N}'$. In order to define $\mathbb{N}$, we still have to separate the appropriate elements of $\mathbb{N}'$. This amounts to finding a predicate **nat** which characterizes natural integers. We observe that the set $\mathbb{N}$ we want will be the smallest (in the sense of set inclusion) e such that $\mathbf{supnat}(e)$. The predicate **nat** turns out to be "be a member of all e such that $\mathbf{supnat}(e)$":

$$\mathbf{nat}(n) \stackrel{\mathrm{def}}{=} \forall e \ \mathbf{supnat}(e) \Rightarrow n \in e \ .$$

Taking $x = \mathbb{N}'$ in the schema of separation, we can define:

$$\mathbb{N} \stackrel{\mathrm{def}}{=} \left\{ n \mid \mathbf{nat}(n) \right\} \ .$$

The left member of the schema of induction is similar to the definition of **supnat**:

$$P(0) \wedge [\forall x \ P(x) \Rightarrow P(S(x))] \ .$$

Separating in $\mathbb{N}$ the elements x such that $P(x) \wedge x \in \mathbb{N}$, we get a set e satisfying $\mathbf{supnat}(e)$, that is, which both includes $\mathbb{N}$ and is included in $\mathbb{N}$, providing a justification for proofs by induction. In some respect, the definition of $\mathbb{N}$ via **supnat** contains the schema of induction, while the ultimate justification comes from the schema of separation.

In what follows, we use the notations $1, 2, 3$, etc. for $S(0), S(S(0)), S(S(S(0)))$, etc.

Remarks on Typing. Because of the absence of typing, one can write formulas such as $2 = \langle 0, 0 \rangle$ or $3 = 1 \cup \langle 0, 1 \rangle$ without blinking an eye... they are even theorems! It is not difficult to find variants of the previous encodings[4] that do not satisfy these equations (but satisfy other meaningless ones).

---

[4]To be more precise, we can work with variants of the encoding of ordered pairs, of 0, of S, and in general of constructors. Note, however, that the axiom of infinity is formulated with a specific encoding of integers in mind.

7.3.2 Other Inductive Definitions

We can attempt to reuse the same process for defining "recursive" data structures of computer science — here we prefer to use the term "inductive":[5] lists, trees, context-free languages, etc.

Let us illustrate the idea with integer binary trees. We consider a version of binary trees where only leaves are labelled with integers. Here is the corresponding inductive definition:

$$A \;=\; \{n\} \mid \langle A, A \rangle \;.$$

Informally,

- if n is an integer, $\{n\}$ is a tree;
- if a_1 and a_2 are two trees, $\langle a_1, a_2 \rangle$ is a tree;
- all trees can be constructed by application of the two previous clauses.

We represent the two first clauses by a predicate $\texttt{suptree}(e)$, claiming that the set e contains all trees:

$$\texttt{suptree}(e) \quad \overset{\text{def}}{=} \quad \begin{aligned} &[\forall n \; n \in \mathbb{N} \Rightarrow \{n\} \in e] \\ \wedge \; &[\forall a_1 \forall a_2 \; (a_1 \in e \wedge a_2 \in e) \;\Rightarrow\; \langle a_1, a_2 \rangle \in e] \;. \end{aligned}$$

With the goal of formalizing the third clause, let us introduce the predicate saying "to be in all the sets containing all trees":

$$\texttt{tree}(A) \;\overset{\text{def}}{=}\; \forall e \; \texttt{suptree}(e) \Rightarrow A \in e \;,$$

and we would like to define:

$$A \;\overset{\text{def}}{=}\; \{a \mid \texttt{tree}(a)\} \;.$$

Then we come up against an obstacle: the previous version of the axiom of infinity at our disposal does not directly provide a set A' that contains all trees. In fact, such a set can certainly be constructed, by completing the union of $\mathcal{P}(\mathbb{N})$, $\mathcal{P}(\mathbb{N}) \times \mathcal{P}(\mathbb{N})$, $(\mathcal{P}(\mathbb{N}) \times \mathcal{P}(\mathbb{N})) \times \mathcal{P}(\mathbb{N})$, etc. Constructing A' turns out to be complex, probably much more complex than A. On the other hand, $\texttt{tree}$, the characteristic predicate of A, could be defined without significant problems. This may be an argument in favor of working with predicates rather than with sets or models. We consider below another representation of trees.

The previous problem is not raised if we consider the inductive definition of a subset of $\mathbb{N}$ (for example $\{2^n \mid n \in \mathbb{N}\}$), or the inductive definition of a function from $\mathbb{N}$ to $\mathbb{N}$, because it can be separated from $\mathbb{N} \times \mathbb{N}$. Let us take the example of the sequence of Fibonacci, seen as a set of ordered pairs $\langle n, \texttt{fibo}(n) \rangle$ with $n \in \mathbb{N}$. All supersets e of this set satisfy $\texttt{supfibo}(e)$ with:

---

[5] Recall that the meaning of "recursive" in computer science differs from its meaning in logic.

$$\text{supfibo}(e) \quad \overset{\text{def}}{=} \quad \begin{aligned} &\langle 0, 1\rangle \in e \\ \wedge\ &\langle 1, 1\rangle \in e \\ \wedge\ &[\forall n \forall x \forall y\, (\langle n, x\rangle \in e \wedge \langle n+1, y\rangle \in e) \\ &\qquad \Rightarrow \langle n+2, x+y\rangle \in e]\ . \end{aligned}$$

By the axiom of separation we can define:

$$\text{fibo} \overset{\text{def}}{=} \{c \mid \forall e\ \text{supfibo}(e) \Rightarrow c \in e\}\ .$$

7.3.3 The Axiom of Separation

Observe in previous examples that for a set E to admit an inductive definition, we make use of a quantification on a collection of sets *of which E is a member.* Such a formulation is said to be **impredicative**. One may see that a kind of vicious circle exists, and one must be very careful to ensure that no paradox is generated. However, this construction process turns out to be very useful, so useful indeed, that it is not clear we could do without it (see, for example, the introduction to [Lei91]).

Formally, an impredicative definition follows the schema:

$$E \overset{\text{def}}{=} \{x \mid \forall e\ \varphi(e) \Rightarrow \psi(x, e)\}\ , \qquad \text{where } \varphi(E) \text{ is true.}$$

Intuitively, if $\psi(x, e)$ is $x \in e$, E is the smallest set satisfying φ. In the previous examples, the role of φ was played by supnat or supfibo.

We can define a finite set in an impredicative way. Here is a trivial example:

$$\{x \mid \forall e\ (1 \in e \wedge 3 \in e) \Rightarrow x \in e\}\ ,$$

which is a pedantic definition of $\{1, 3\}$.

The application condition of the axiom of separation plays a key role for avoiding paradoxes. An impredicative definition like the one given above for E is admitted only if a set F containing all e such that $\varphi(e)$ has been exhibited beforehand. Otherwise paradoxes like Russell's can be reproduced, taking t for $\varphi(e)$ and $e \notin e \Rightarrow x \in e$ for $\psi(x, e)$. Similarly, there is no set of all sets in ZFC. If such a set U could exist, we could take $e \in U$ for $\varphi(e)$ and $\psi(x, e)$ as before.

However, the application condition of the axiom of separation implies that, except in the special case of natural numbers, much additional work is needed in the construction of inductive data structures.

7.3.4 Separation of a Fixed Point

Fixed points are a traditional device in computer science for explaining inductive definitions. Let us illustrate the idea in the case of $\mathbb{N}$. Intuitively, the inductive definition $n = 0 \mid S(n)$ can be represented by:

$$\mathbb{N} = \{0\} \cup \overline{\mathsf{S}}(\mathbb{N})\ , \tag{7.1}$$

where $\overline{S}(X)$ is the set resulting from the application of S to all elements of X. Of course replacing $=$ with $\overset{\text{def}}{=}$ in (7.1) would make no sense, since the object to be defined occurs on the right-hand side. Hence (7.1) must be regarded as an equation of the form $x = f(x)$ where x is the unknown. In this situation x is called a fixed point of f (see § 3.6). In our example $\mathbb{N}$ is the smallest solution for:

$$X = F(X) , \tag{7.2}$$

with $F(X) \overset{\text{def}}{=} \{0\} \cup \overline{S}(X)$.

In order to state and solve this equation, we need a reference set R where X varies and we have to check that F is monotone, that is

$$X \subset Y \Rightarrow F(X) \subset F(Y) .$$

The technique introduced in § 3.6 consists of showing that the set of post-fixed points of F (the X of R satisfying $F(X) \subset X$) is non-empty, then that the intersection of all post-fixed points is the smallest fixed point of F, which is the solution of (7.2) we are looking for. The reference set we can take here is $\mathcal{P}(\mathbb{N}')$, where, as before, $\mathbb{N}'$ is provided by the axiom of infinity. This axiom actually stipulates that $\mathbb{N}'$ is a post-fixed point, which ensures that the set of post-fixed points is non-empty. The set $\mathbb{N}$ we look for is then the smallest X of $\mathcal{P}(\mathbb{N}')$ such that $F(X) \subset X$. Though it does not explicitly appear, such a definition is in fact impredicative, because we have to state the following when details are worked out (using the axiom of separation):

$$\mathbf{Ens}\{X \mid \quad X \in \mathcal{P}(\mathbb{N}') \wedge F(X) \subset X \wedge$$
$$\forall Y \, (Y \in \mathcal{P}(\mathbb{N}') \wedge F(Y) \subset Y) \Rightarrow X \subset Y\} .$$

This set is actually a singleton, which is precisely defined to be $\{\mathbb{N}\}$. The axiom of separation is used here in a somewhat more involved way than before, because it acts on $\mathcal{P}(\mathbb{N}')$ instead of $\mathbb{N}'$.

7.3.5 Ordinals

The construction of ordinal sets was sketched in § 3.5. They play a key role in set theory and it was absolutely necessary that the axiomatic version of set theory should be able to recover them. Let us just add here that the replacement schema turns out to be essential in this respect (whereas it is scarcely used in regular mathematics, at least not directly). Let us also mention that the axiom of infinity stated above is rich enough: combining $\mathbb{N}$, the schema of replacement and the axiom of union give all the necessary ingredients needed for constructing ordered sets much "larger" than $\mathbb{N}$.

7.4 Sets, Abstract Data Types and Polymorphism

7.4.1 Trees, Again

A model of trees more economical than the one given in § 7.3.2 can be constructed. Instead of an infinite union of Cartesian products, we use, intuitively, an address space that assigns the integer $\overline{1}$ (written with binary digits) to the root, the integer $\overline{10}$ to the first left subtree, the integer $\overline{11}$ to the first right subtree, and so on. We define the ordering relation $\prec$ over $\mathbb{N}$ by $\forall n\ (n \prec 2n) \wedge (n \prec 2n + 1)$. A *branch* B is a subset of $\mathbb{N}$ that contains 1 and that also contains, for all x of B, a unique y satisfying $x \prec y$ (for example a branch can start with 1, 2, 5, 10, 20). A set of leaf addresses is a set L of integers that contains a unique element in every branch. In order to construct a tree of integers from a set of leaf addresses L, we map each member of L to an integer (called its label). The set $\mathbb{B}$ of branches and the set $\mathbb{L}$ of L have to be provided by an appropriate use of the axiom of separation in $\mathcal{P}(\mathcal{P}(\mathbb{N}))$. The set of trees is then $\mathbb{L} \to \mathbb{N}$, the set of total functions from $\mathbb{L}$ to $\mathbb{N}$.

7.4.2 Algebraic Approach

The previous model of trees is quite similar to an encoding that would be used in a software implementation. But one needs some convincing that it corresponds to the expected concept of tree. Of course, no formal proof can be given for such a subjective proposition. But, admittedly, our first (attempted) model based on suptree is much more natural.

We consider an abstract data type[6] tree. This type has two construction operations leaf and bin; leaf constructs an elementary tree which is just a leaf labelled by an integer, bin constructs a new tree from two existing trees. This yields the signatures:

leaf: $\mathbb{N} \to$ tree
bin: tree $\times$ tree $\to$ tree .

In addition, we have axioms stating that all trees are produced by repeated application of leaf and bin, and that two trees are equal if and only if they are constructed by application of the same constructors (on the same arguments). Here is one of these axioms:

$$\forall m \forall n \quad \mathtt{leaf}(m) = \mathtt{leaf}(n) \iff m = n .$$

Clearly, the representation we gave in § 7.3.2 is a model of that abstract data type, where leaf and bin are respectively interpreted by the functions $n \mapsto \{n\}$ and $\langle a_1, a_2 \rangle \mapsto \langle a_1, a_2 \rangle$.

---

[6]In the remainder of this section we employ the terminology introduced in § 10.3.1.

In contrast, the second representation requires more work. The constructor leaf is simply interpreted by the function $n \mapsto \{\langle 1, n \rangle\}$. In order to interpret bin, we need two functions g and d from $\mathbb{N}$ to $\mathbb{N}$, that map the addresses of a tree to addresses of a tree having the same shape, which is the left (or right) subtree of a new tree. We know that every integer can be written in a unique way, either $2n$ or $2n + 1$ depending on its parity. We can then inductively define g and d by

$$\begin{cases} g(1) = 2 \\ g(2n) = 2g(n) \\ g(2n + 1) = 2g(n) + 1 \end{cases} \quad \text{and} \quad \begin{cases} d(1) = 3 \\ d(2n) = 2d(n) \\ d(2n + 1) = 2d(n) + 1 \ ; \end{cases}$$

bin is then interpreted by the function from $\mathcal{P}(\mathbb{N} \times \mathbb{N}) \times \mathcal{P}(\mathbb{N} \times \mathbb{N})$ to $\mathcal{P}(\mathbb{N} \times \mathbb{N})$ that, given two trees a_1 and a_2 returns the tree

$$\text{bin}^{\mathcal{I}_2}(a_1, a_2) = \{\langle g(x), y \rangle \mid \langle x, y \rangle \in a_1\} \cup \{\langle d(x), y \rangle \mid \langle x, y \rangle \in a_2\} \ .$$

We still have to show that, on the one hand, we recover the same interpretation as before (in terms of $\mathbb{B}$ and $\mathbb{L}$) and, on the other hand, the axioms of leaf and bin are satisfied. This is left as an exercise for the reader.

7.4.3 Polymorphism (or Genericity)

The concept of address we use is generic, in the sense that we say nothing about the kind of leaves (more precisely: leaf labels). A soon as $\mathbb{L}$ is constructed, it can be used for building trees that are labelled by elements of any given set X, including a set of trees. For instance, the set of trees of trees of integers is $\mathbb{L} \to (\mathbb{L} \to \mathbb{N})$.

The importance of genericity — also called parametric polymorphism — has been acknowledged for a long time. To define a generic concept of tree, one would like to consider a function tree that maps every set X to $\mathbb{L} \to X$. But tree would then be a member of $\mathcal{U} \to \mathcal{U}$, where $\mathcal{U}$ is the class of all sets. Then it is not a set. We previously had a similar remark about the operations $\cap$, $\cup$, etc. One could look for a more astute process allowing one to interpret types by sets, including polymorphic types. This indeed seemed almost possible to a number of researchers, but then J. Reynolds demonstrated that the answer is negative [Rey85]. We go back to this point in Chapter 11.

7.4.4 The Abstract Type of Set Operations

Just as for trees, one can define an abstract type for sets. This is a well-known example, generally described using two basic constructors: the construction of the empty set and the insertion of an element into a set. Two axioms are introduced in order to stipulate that inserting an element that is already contained in the set has no effect, and the insertion order of elements is irrelevant.[7] The membership predicate and operations such as ∪ and ∩ are then specified using additional axioms — intuitively, they are written with a "recursive" exploration of their arguments in mind. One can then easily prove the algebraic properties one expects on these operations (associativity, etc.).

The same ideas arise when programming with sets. However, let us point out that only *finite* sets are dealt with in this way. Moreover, it is usually accepted that elements of such sets are typed and have the same type. Then a notion of polymorphism is needed if we want to handle Cartesian products or powersets in a natural manner.

7.5 Properties of ZF and ZFC

From a technical viewpoint ZFC is without doubt a great success, because it provides all of the kinds of sets, numbers and structures needed in mathematics. Clearly, limitations coming from the incompleteness theorems of Gödel cannot be avoided. Thus, the consistency of ZF cannot be proven. But there are other results, called **relative consistency** results. In particular, the axiom of choice, which is very non-constructive, was initially the cause of many disputes. In 1938, Gödel showed that if ZF is consistent, then ZFC is consistent as well. In 1963, Cohen proved that the negation of the axiom of choice does not introduce contradictions in ZF as well. This amounts to saying that the axiom of choice, or its negation, cannot be deduced from axioms of ZF.

Another important conjecture about the cardinality of ℝ, called the continuum hypothesis (see page 57), was also proved to be independent from ZF at the same time. Hence one might think of set theory as somewhat arbitrary. In contrast with ℕ, set theory does not have a well-understood concept of a "standard model". For instance, the syntactic model of set theory is certainly not the intended one, because it is denumerable.

7.6 Summary

What is the impact of set theory on formal specification and programming techniques? The most obvious is the universal use of the language of sets. Informal reasoning is sometimes efficiently guided by Euler–Venn's diagrams.[8] There are several opinions regarding set theory itself. Advocates of Z may highlight that ZF has been thoroughly tested as a foundation for mathematics, and hence is a firm basis for designing a specification language. Other researchers prefer to avoid the systematic use of sets, because unexpected complications spoil the initial simplicity of basic concepts (some of them were illustrated above), or because of the intrinsic lack of typing in this theory.

---

[7]We proceeded this way in § 10.5 for representing a table.

[8]The idea of representing what we nowadays call sets by circles goes back to Euler.

Axiomatic set theory is sufficiently powerful to allow one to represent any idea that is needed, for example the data structures of computer science. However, in many cases we end up with quite an arbitrary encoding; then axiomatic set theory may seem closer to an assembly language than to a high level language.

7.7 Notes and Suggestions for Further Reading

So-called "naive" set theory is developed in the book of Halmos [Hal60]. Another well-known reference is Enderton [End77]. The axioms of Zermelo–Fraenkel are presented and discussed in a chapter of the *Handbook of Logic of Mathematical Logic* [Bar77] written by Shoenfield [Sho77].

Further developments are described in Devlin [Dev93], specifically the arithmetic of ordinals and cardinals. At the end of the book there is also a presentation of non-well-founded sets, a variant of ZF without the axiom of regularity. Non-well-founded set theory is used in computer science as a basis for bisimulation and co-induction, which are reasoning techniques relevant to infinite processes and circular data structures. On this topic one may consult the very concise and readable article by Milner and Tofte [MT91].

8. Behavioral Specifications

The table example that we used in previous chapters can be qualified as functional: looking from the outside, we can view it as a function that returns an answer when it is called. We don't have any concerns or get distracted by its internal computation and internal workings. In contrast, we can hardly understand systems which constantly react to their environments if we don't study the series of actions they perform. This is the case for communication protocols, operating systems or command-and-control equipment. For protocols for instance, we have to consider synchronization, to prevent deadlocks, undesired arrival of messages, etc. The complexity of such protocols is by and large, concentrated in these aspects.

Appropriate techniques consist of modeling such systems by, essentially, a graph whose vertices and edges respectively represent possible states and transitions between states, and then characterizing expected behaviors by safety and liveness properties that are expressed over this graph — this is the realm of temporal logic — and finally, verifying that these properties are satisfied.

The following presentation is centered on the general formalism of (labeled or otherwise) transition systems, which will be the semantic pivot between languages such as Unity, CCS or TLA and different variants of temporal logic, including the μ-calculus. At the end of the chapter we mention appropriate verification techniques, specifically *model checking*.

8.1 Unity

Unity [CM89] was first designed in order to elaborate programs that could take advantage of parallel computations that are available on non-von Neuman machines. Such opportunities vary to a considerable extent from one architecture to the other, and so it is better to make no assumptions about control. A Unity program is essentially defined by:

- a state space;
- an initial state, or a set of initial states;
- a set of transitions between states.

Transitions are defined by simultaneous assignments, sometimes with an additional condition which is true by default. Assignments are generally separated by the symbol ⟦. In the original definition of Unity, the state is given by means of declarations similar to those in Pascal, but this is not essential: we could consider variations including data types such as lists, or using more abstract constructs, e.g. set operators or higher-order types. Here we use the term **field** for components of the state rather than **variable**, in order to avoid any confusion with the logical concept of a variable.

8.1.1 Execution of a Unity program

Executing a Unity program consists of choosing *in an undetermined way*[1] one of the assignments, then, if the corresponding condition is true, applying it to the current state and repeating the same cycle again and again. The main idea is that each assignment may contribute to the final result and eventually has the opportunity to be applied. Unity stands for *Unbounded Nondeterministic Iterative Transformations*. The freedom underlying the execution of Unity programs gives them a specification status, all the more so since we will not refrain from using arbitrary mathematical means for defining the space state. We will, however, continue to call them programs.

Let us observe that a Unity program could easily be represented by a B machine, each assignment being encoded by an operation of the form: IF *condition* THEN *multiple substitution* ELSE **skip**. As in B, the weakest precondition calculus introduced in § 4.3 plays an important technical role. The big difference is that in B operations are macroscopic and are executed on external calls, whereas in Unity assignments are rudimentary and they execute spontaneously and perpetually.

$$
\begin{aligned}
&\textbf{Program } T \\
&\quad \textbf{constant } p,q : \text{integer} \\
&\quad \textbf{declare } x,y : \text{integer}; \ t,r : \text{boolean} \\
&\quad \textbf{initially } x{=}p \ \wedge \ y{=}q \ \wedge \ t{=}\text{false} \\
&\quad \textbf{assign} \\
&\qquad\quad r,t := P(x), \text{true if } x \neq y \ \wedge \ \neg t \\
&\qquad \| \quad y,t := x, \text{false if } r \wedge t \\
&\qquad \| \quad x,t := x+1, \text{false if } \neg r \wedge t \\
&\textbf{end}
\end{aligned}
$$

Figure 8.1: Table search in Unity

8.1.2 The Table Example

Recovering strict control over the evaluation order is not very difficult: we just have to take a field as the program counter. We proceed in this way with t in Figure 8.1, where the table search program already presented on page 31 is written in Unity. The field t can even be given a logical interpretation: "r

---

[1] We will see later, however, that this choice must respect a fairness condition.

contains the result of the evaluation of $P(x)$". When $x = y$ the execution reaches a **stable state**: further assignments leave the state unchanged.

But this approach is far from optimal in the spirit of Unity. Figure 8.2 proposes a solution with much more opportunities to take advantage of parallelism. The idea is to have a control flow (or thread) for every possible value of the index. These threads are modeled by the array t. As soon as the result is found, the field f is set to **true**, with the intention of stopping other executing calculations — a more correct phrasing would be: in order to make the state stable.

```
Program P
    constant p, q : integer
    declare x, n : integer; b, f : boolean;
            r : array of boolean; t : array of 0..2;
    initially n, f = p, false ∧ ∀i : p ≤ i < q ⇒ t[i]=0
    assign
    ⟨‖ ∀i : p ≤ i < q ::
                r[i], t[i] := P(i), 1  if  t[i]=0 ∧ ¬f
        ‖   x, f, b, t[i] := i, true, true, 2  if  r[i] ∧ t[i]=1 ∧ ¬f
        ‖   n, t[i] := n+1, 2  if  ¬r[i] ∧ t[i]=1 ∧ ¬f
    ⟩
    ‖   f, b := true, false  if  n=q
end
```

Figure 8.2: Parallel table search in Unity

A good method for designing such programs and reasoning about them consists of considering a state that changes in such a way that it progressively satisfies the desired properties, whatever happens. From a methodological perspective, one distinguishes **safety** properties, which guarantee that every reachable state is acceptable (in other words, nothing bad can happen) from **liveness** properties, which state that a desired state is eventually reached (something good will happen).

In the example in Figure 8.2, safety properties are similar to the invariants I_1, I_2 and I_3 given on page 31: we introduce the subset C of integers i in $[p, q[$ yielding a negative answer (the value of $r[i]$ is **false** and the value of $t[i]$ is 2). The invariant says that the cardinality of C is $n - p$. The main liveness property we expect here is $f = $ **true**. This is also called a **progress property**. However, we must not forget that, at a given time, the chosen assignment may well leave the state unchanged. Assume in our example that $p < q$; an execution continuously choosing the last assignment ($f, b := \ldots$ **if** $n = q$) would never progress. In order to avoid such a situation, Unity imposes a **fairness** condition: each assignment is chosen infinitely many times during an execution.

Other categories of properties have been identified for qualifying system behaviors, such as to be **deadlock free**, or **reachability**. The latter expresses that the system always has the chance of reaching a given situation, for instance

Program H_1
  constant Δ : integer
  declare l_1, d_1, v_1, v_2 : integer; p_1, p_2 : boolean
  initially $l_1 = 0 \,\wedge\, d_1 = \Delta \,\wedge\, p_1 = $false$\,\wedge\, p_2 = $false
  assign
    $l_1 := l_1 + 1$ **if** $l_1 < d_1$
  $\|$ $v_2, p_2 := l_1, $**true**
  $\|$ $d_1, p_1 := \max(d_1, v_1 + \Delta), $**false if** p_1
  $\|$ $d_1 := \max(d_1, v_1 + \Delta)$ **if** p_1
  $\|$ $p_1 := $**false**
 end

Figure 8.3: A synchronization protocol in Unity

to return to the initial state. Note that being deadlock free does not make much sense in Unity, since executions are infinite by construction.

8.1.3 A Protocol Example

Let us consider another program that doesn't aim at computing a result, but at providing a service. Figure 8.3 represents a small clock synchronization protocol.[2] Two stations endowed with a local clock l_i, $i \in \{1, 2\}$ send their own current time through an unreliable medium (messages can be lost, duplicated and their order is not preserved) from time to time. The protocol ensures that the distance between the values of l_1 and l_2 is never greater than the strictly positive constant Δ. Figure 8.3 contains the assignments of the program running in station 1. The medium is represented here by two Booleans p_i, $i \in \{1, 2\}$ telling us whether or not a message for station i is present and by the integer v_i which holds the value of the message if there is one. The capacity of the medium we are considering is then just one message in each direction. Assignments represent, respectively, incrementing the local clock, sending the current time, receiving the time from the distant clock, duplicating, and losing the arriving message. We get the complete system by a **parallel composition** of program P_1 with program P_2 (written $P_1 \| P_2$), where P_2 is identical to P_1 up to an exchange of indices 1 and 2. To put it another way, the state of $P_1 \| P_2$ is made up of the fields of P_1 and of the integers l_2 and d_2; its assignments are the assignments of P_1 and the symmetrical assignments we get by exchanging 1 and 2; finally its set of initial states is the conjunction of the two clauses introduced by the keyword **initially**.

If we take a version of Unity where bags are allowed, we can easily model a medium which does not preserve message order (Figure 8.4). This program can be composed with the program in Figure 8.5 (and its symmetrical counterparts): we get a medium with message losses and duplications. We expect that the protocol satisfies the following properties:

---

[2]The author of this protocol is Gérard Roucairol.

Program HM_1
    constant Δ : integer
    declare l_1, d_1 : integer; c_1, c_2 : **bag of** integer
    initially $l_1 = 0 \wedge d_1 = \Delta \wedge c_1 = \varnothing \wedge c_2 = \varnothing$
    assign
$$l_1 := l_1 + 1 \;\; \textbf{if} \;\; l_1 < d_1$$
$$\| \;\;\;\;\;\;\;\;\; c_2 := c_2 \cup \{l_1\}$$
$$\| \forall v_1 \;\;\; d_1, c_1 := \max(d_1, v_1 + \Delta), c_1 - \{v_1\} \;\; \textbf{if} \;\; v_1 \in c_1$$
end

Figure 8.4: Synchronization protocol using an unbounded channel

Program C_1
    declare c_1 : **bag of** integer
    initially $c_1 = \varnothing$
    assign
$$\| \forall v_1 \;\;\; c_1 := c_1 \cup \{v_1\} \;\; \textbf{if} \;\; v_1 \in c_1$$
$$\| \forall v_1 \;\;\; c_1 := c_1 - \{v_1\} \;\; \textbf{if} \;\; v_1 \in c_1$$
end

Figure 8.5: Unbounded channel with losses and duplications

– safety: $|l_1 - l_2| \leq \Delta$ is always true;
– progress: clocks increase to arbitrary large values.

In this case, progress does not express that executions get closer to a desired situation, but that there are no deadlocks: no state is stable (Δ is non-null). We can also verify a reachability property: from any state (derived from the initial state) one can reach a state where $l_1 = l_2$. An interesting consequence is that we can hope to augment the previous protocol with additional fields and transitions that would model the arrival of an external request and then constrain assignment choices in such a way that l_1 and l_2 would converge to the same value. This is left as an exercise for the reader.

    We will see in § 8.5 how to formalize all these properties in temporal logic. We first present an elementary but very general model for describing behaviors.

8.2 Transition Systems

The systems we model are always presented, to a greater or lesser degree, in the form of a state which changes under the effects of various actions. The state can be thought of at different abstraction levels. It can be the colour (or combination of colours) of a traffic light; the memory space of a real machine; the memory spaces of several machines, with the contents of communication channels of the network that links them together; the tuple of values taken by the fields declared in a program — which may be written in Unity — or as a B specification, an algebraic term, etc.

State changes can be continuous, for analog systems, or discrete, for the systems we consider here: we call them transitions. Most methods adopt a purely observational standpoint, that is, no importance is attached to the internal or external cause that determines the choice between transitions. However, transitions are sometimes associated with events that we want to remember, e.g. a printing request or a message. Then we give them a name, also called a label or an action. We use the term "Kripke model" when transitions are not labeled — we can equivalently consider that all transitions have the same label — and "labeled transition system" or simply "transition system" in the general case.

8.2.1 Definitions and Notations

A **Kripke model** is an ordered pair $\langle S, \mathcal{R} \rangle$ where S is a (finite or infinite) set, called the set of **states** and $\mathcal{R}$ is a binary relation on S, called the **transition relation**.

A **transition system** (or **labeled transition system**) $\mathcal{T}$ is a triple of the form $\langle S, A, (\mathcal{R}_a)_{a \in A} \rangle$ where S is a (finite or infinite) set of states, A is a (finite or infinite) set, called the set of **actions** and each $\mathcal{R}_a$ is a binary relation on S. Equivalently, the family $(\mathcal{R}_a)_{a \in A}$ can be presented as a subset $\mathcal{R}$ of $A \times S \times S$. The reader may like to check that we recover the concept of a Kripke model if A is a singleton set.

We also use the term **automaton** or **state machine** for a transition system, especially when S and A are finite.

For the transition relation of a Kripke model $\mathcal{K}$ one often uses an infix notation such as $\xrightarrow{\mathcal{K}}$ or more simply $\rightarrow$ when there is no ambiguity about $\mathcal{K}$. Similarly the transition relation labeled by a of a transition system $\mathcal{T}$ is denoted by the infix symbol $\xrightarrow[\mathcal{T}]{a}$ or $\xrightarrow{a}$ when the context is clear. Thus $s \xrightarrow{a} t \xrightarrow{b} u$ simply expresses that execution goes from state s to state t using transition a, then to state u using transition b.

8.2.2 Examples

When the number of states of the system is small, we conveniently represent it in a graphical form. Figure 8.6 represents a transition system for a simplified drink vending machine.

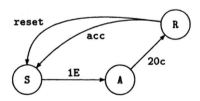

Figure 8.6: A very simple drink vending machine

The standard behavior consists of going from state S (start) to state A (again) when a one euro coin is inserted into the machine (label 1E), then to state R (ready) when a 20 cents coin is inserted (label 20c), then back to the start state when the "accept" button is pressed (label acc) — and a drink is delivered. The customer can also, from state R, push the reset button — then the inserted coins are returned to him/her.

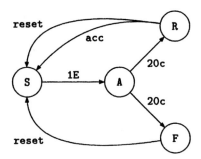

Figure 8.7: A filtering drink vending machine

The vending machine modeled in Figure 8.7 has an additional feature: bad 20 cents coins are rejected. A transition 20c can then lead to state F (failure), where the customer has no other choice but to reset. Formally, this system is defined by the state set $S = \{S, A, R, F\}$, the action set $A = \{1E, 20c, acc, reset\}$, and transition relations $\xrightarrow{1E} = \{\langle S, A\rangle\}$, $\xrightarrow{20c} = \{\langle A, R\rangle, \langle A, F\rangle\}$, $\xrightarrow{acc} = \{\langle R, S\rangle\}$ and $\xrightarrow{reset} = \{\langle R, S\rangle, \langle F, S\rangle\}$.

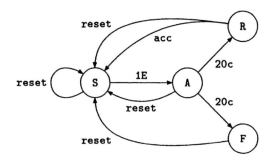

Figure 8.8: A more complete vending machine

The previous model can be augmented by further stipulating that one can push the reset button in every state. This is easy to state in a formal way, by writing $\xrightarrow{reset} = \{\langle s, S\rangle \mid s \in S\}$. Note in Figure 8.8 that the graphical representation becomes fairly complicated. The reader is invited to invent variants of the above system, where, among possible suggestions, one can insert coins in an arbitrary order, or push the accept button in any state (of course, the machine should only perform a state change from state R).

In the model described above, labels happen to correspond to actions initiated by the external environment. This is not necessary. Actually, if our machine delivers a drink as soon as one euro and 20 cents are inserted, without waiting for a confirmation, the label acc is interpreted by a spontaneous action. *But this does not matter to the transition system*: as already indicated, the latter just describes possible sequences of actions without *a priori* interpretation of their meaning.

8.2.3 Behavior of a Transition System

Given a transition system $\mathcal{T}$, a trajectory on $\mathcal{T}$ represents a possible behavior of $\mathcal{T}$. One can imagine that $\mathcal{T}$ defines a state space and motion rules; as in mechanics, a trajectory is a function of time that returns the state of the system at each instant. Since our transitions are discrete, time will be represented by N. Formally, a **trajectory** on $\mathcal{T}$ is a pair of two sequences $(s_n)_{n \in \mathbb{N}}$ and $(a_n)_{n \in \mathbb{N}}$ where:

1. For all integers n, s_n is a state and a_n is an action.
2. For all integers n, $s_n \xrightarrow{a_n} s_{n+1}$.

The component $(a_n)_{n \in \mathbb{N}}$ is called the **trace**. In the case of Kripke models, it is of course superfluous. In the literature trajectories are also referred to as **scenarios**, **executions** or **paths**. We agree that the prefix of a trajectory $\langle (s_n)_{n \in \mathbb{N}}, (a_n)_{n \in \mathbb{N}} \rangle$ will be represented in the form $s_0 \xrightarrow{a_0} s_1 \xrightarrow{a_1} s_2 \ldots$

A trajectory example for the transition system of Figure 8.7 starts with $S \xrightarrow{1E} A \xrightarrow{20c} F \xrightarrow{reset} S \xrightarrow{1E} A \xrightarrow{20c} R \xrightarrow{acc} S$.

The systems modeled in Figures 8.6 and 8.7 have the same set of traces, but have different behaviors: in the former, acc is always allowed after 20c whereas this is not the case of the latter. Traces simply do not provide the relevant pieces of information that would enable us to distinguish between them.

When the system includes deadlocks (also called blocking states, i.e. states s such that for any action a, $\{s' \in S \mid s \xrightarrow{a} s'\} = \varnothing$), the definition of trajectories must be made more general. Trajectories are maximal sequences satisfying the above conditions: either they are infinite, or their last state is a blocking state.

8.2.4 Synchronized Product of Transition Systems

Formalizing more complicated examples using flat transition systems quickly turns out to be quite laborious. It is better to specify the production of a transition system by indirect means, notably:

- the composition of (smaller) systems, as will be considered here;
- the use of higher level concepts or languages, for example Unity or CCS; we go back to this approach in § 8.2.6 and in § 8.3; then transition systems provide an operational semantics for those languages.

Consider n transition systems that we put together: $\mathcal{T}_1 = \langle S_1, A_1, \underset{\mathcal{T}_1}{\longrightarrow} \rangle$, $\mathcal{T}_2 = \langle S_2, A_2, \underset{\mathcal{T}_2}{\longrightarrow} \rangle$ ... $\mathcal{T}_n = \langle S_n, A_n, \underset{\mathcal{T}_n}{\longrightarrow} \rangle$. A complete state of the system is obtained by the **synchronized product** [AN82], which contains a component ranging over S_1, a component ranging over S_2 ... a component ranging over S_n. The state space of the synchronized product of $\mathcal{T}_1, \mathcal{T}_2$ ... $\mathcal{T}_n$ is then the Cartesian product $S_1 \times S_2 ... \times S_n$.

Saying that the whole system goes from state $\langle s_1, s_2, ... s_n \rangle$ to state $\langle s_1', s_2', ... s_n' \rangle$ amounts to saying that each component goes from state s_i to state s_i' using an action taken in $\underset{\mathcal{T}_i}{\longrightarrow}$. This corresponds to the fact that a transition is passed simultaneously in each subsystem, what we call a synchronization. In general we want to restrict the set of possible synchronizations. Typically, an action *send a* of a system will only be synchronized with an action *receive a* of another system. The action set of the synchronized product will then be given by a *subset* Y of $A_1 \times A_2 ... \times A_n$, whose elements are called **synchronization vectors**.

In the general case, we write the synchronized product in the form $\langle \mathcal{T}_1 \parallel \mathcal{T}_2 ... \parallel \mathcal{T}_n; Y \rangle$. It corresponds to the transition system whose state set is $S_1 \times S_2 ... \times S_n$, whose action set is Y and where transition relations are defined by:

$$\langle s_1, ... s_n \rangle \xrightarrow[\langle \mathcal{T}_1 ... \parallel \mathcal{T}_n; Y \rangle]{\langle a_1, ... a_n \rangle} \langle s_1', ... s_n' \rangle \quad \text{iff} \quad s_i \xrightarrow[\mathcal{T}_i]{a_i} s_i' \text{ for all } i \text{ in}[1, n].$$

8.2.5 Stuttering Transitions

In order to represent so-called asynchronous systems that advance in an independent manner, it is convenient to assume that each of them possesses a waiting action e such that executing e leaves the state unchanged. Leslie Lamport uses the term **stuttering transitions**:

$$\xrightarrow{\text{e}} = \{\langle s, s \rangle \mid s \in S_i\}.$$

For example, a synchronization vector $\langle a_1, \text{e}, ... \text{e} \rangle$ allows $\mathcal{T}_1$ to execute action a_1 whereas other systems do nothing. A synchronization vector $\langle a_1, a_2, \text{e}, ... \text{e} \rangle$ allows $\mathcal{T}_1$ and $\mathcal{T}_2$ to synchronize without being disturbed.

Note that introducing stuttering transitions in each state from the outset is good from the modularity viewpoint. A system specified in this way can be embedded in an environment while keeping its own behavior. However, this approach leads one to make a fairness hypothesis on allowed trajectories, in order to disallow trajectories where a system remains indefinitely in the same state even though a change is possible.

8.2.6 Transition Systems for Unity

The **declare** clause of a Unity program U defines its state set S_U (the fields declared are projections of S_U in their respective domain; for example in Figure 8.4, l_1 is a projection of S_U into $\mathbb{N}$). The **initially** clause defines a subset $\mathcal{I}$ of S_U. In order to construct the associated transition system, we give a name a_1, a_2 ... to every assignment introduced after the **assign** clause. Recall that each of them is in the form $s := f_i(s)$ **if** $c_i(s)$ and reads: "if s verifies condition c_i, then the next state is $f_i(s)$ else the next state is still s". Then we define

$$\xrightarrow{a_i} \ = \ \{\langle s, f_i(s)\rangle \mid s \in S_U \wedge c_i(s)\} \ \cup \ \{\langle s, s\rangle \mid s \in S_U \wedge \neg c_i(s)\}.$$

The semantics of the **initially** clause is given by an action i, a preinitial state $*$ which is not in S with $\xrightarrow{i} \ = \ \{\langle *, s\rangle \mid s \in \mathcal{I}\}$. Finally we consider $A_U = \{i, a_1, a_2, ...\}$, the system transition associated with U is then $\mathcal{T}_U = \langle S_U, A_U, (\xrightarrow{a})_{a \in A_U}\rangle$.

One can follow a slightly different point of view where, when the condition c_i evaluates to *false*, the corresponding label is replaced with the stuttering action e, (see § 8.2.5). In this version,

$$\xrightarrow{a_i} \ = \ \{\langle s, f_i(s)\rangle \mid s \in S_U \wedge c_i(s)\} \quad \text{and} \quad A_U = \{i, e, a_1, a_2, ...\}.$$

8.3 CCS, a Calculus of Communicating Systems

In the Unity model, entities cooperate by sharing a common memory. In contrast, approaches based on process algebras put the emphasis on communication. CCS (Calculus of Communicating Systems), due to Robin Milner, is one of the most elegant [Mil89]. We are given a set of actions $A = \{\tau, a, \bar{a}, b, \bar{b}, ...\}$. Processes are constructed as follows: **0** is the process that can do nothing (it is in a deadlock state and cannot communicate); if P and Q are processes and if α is an action, then $\alpha.P$, $P \mid Q$, $P + Q$ and $P\backslash L$ are processes. ".", "|" and "+" are respectively the **prefix** operator, the **parallel composition** operator and the **choice** operator.

Intuitively, τ is the silent action; if α is an action different from τ, α can synchronize with $\bar{\alpha}$ (and reciprocally, considering that $\bar{\bar{\alpha}} = \alpha$). The process $\alpha.P$ performs the action α and then behaves like P. Thus the process $a.b.(a.0 + c.b.0)$ corresponds to the transition system:

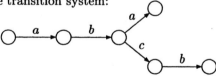

One also can write mutually recursive process definitions in the form $P_1 \stackrel{\text{def}}{=} E_1$, $P_2 \stackrel{\text{def}}{=} E_2$, ... where E_1, E_2, ... represent process expressions in which P_1, P_2, ... can occur. Thus the process $P \stackrel{\text{def}}{=} a.b.(a.P + c.b.0)$ corresponds to the transition system

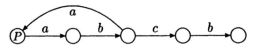

and the systems described in Figures 8.6 and 8.7 can be expressed in CCS respectively by:

$S \stackrel{\text{def}}{=} 1E.20c.(\text{acc}.S + \text{reset}.S)$ and by

$S \stackrel{\text{def}}{=} 1E.(20c.(\text{acc}.S + \text{reset}.S) + 20c.\text{reset}.S)$.

Formally, the state set of the transition system defined by CCS processes P_1, P_2, ... is made up of algebraic subexpressions of P_1, P_2, ..., its action set is A and we get its transition relations by application of the following rules:

– prefix: $\dfrac{}{\alpha.P \xrightarrow{\alpha} P}$;

– choice: $\dfrac{P \xrightarrow{\alpha} P'}{P+Q \xrightarrow{\alpha} P'}$ and $\dfrac{Q \xrightarrow{\alpha} Q'}{P+Q \xrightarrow{\alpha} Q'}$;

– parallel composition without communication :

$\dfrac{P \xrightarrow{\alpha} P'}{P\,|\,Q \xrightarrow{\alpha} P'\,|\,Q}$ and $\dfrac{Q \xrightarrow{\alpha} Q'}{P\,|\,Q \xrightarrow{\alpha} P\,|\,Q'}$;

– communication: $\dfrac{P \xrightarrow{\alpha} P' \quad Q \xrightarrow{\bar{\alpha}} Q'}{P\,|\,Q \xrightarrow{\tau} P'\,|\,Q'}$;

– definition: $\dfrac{A \xrightarrow{\alpha} E'}{P \xrightarrow{\alpha} P'}$ for every definition $P \stackrel{\text{def}}{=} A$.

Note that the parallel composition operator is asynchronous: each component evolves regardless of the other so long as they are not involved in a common action. Stuttering transitions indicated in § 8.2.5 are no more essential in this approach, where modularity is dealt with in a different way (using explicit communication). In the transition systems considered in previous sections, states were explicitly defined and were considered as always being observable through the concept of trajectory. Properties of behaviors considered in § 8.5 are expressed over trajectories and over states. In CCS only transitions are considered as observable; a CCS term (process) can be seen as representing an implicit state, but only its capacity to propose transitions and to continue is important.

CCS also includes the **restriction** operator "\": if P is a process and L is a set of actions different from τ, then $P\backslash L$ is the process that behaves like P but where actions of L are disallowed; P can progress on a branch starting with an action α of L only if this action can be synchronized with the complementary action $\bar{\alpha}$ of a parallel branch of P.

Choice generalizes to an infinite number of processes. If $(P_x)_{x\in\mathbb{N}}$ and $(a_x)_{x\in\mathbb{N}}$ are respectively a family of processes and of actions and if Q is a process, the process $((\Sigma_{x\in\mathbb{N}}\, a_x.P_x)\,|\,\bar{a}_5.Q)\backslash\{a_x \mid x \in \mathbb{N}\}$ evolves necessarily to

$P_5 \mid Q$: this specifies that the second component communicates the value 5 to the first.

A language quite close to CCS called CSP (Communicating Sequential Processes) was proposed by C.A.R. Hoare [Hoa85]. The design of LOTOS, a standardized language for telecommunication protocols, was inspired by CCS and CSP [isob]. However, SDL turned out to be more successful from an industrial perspective, partly because it is founded on more familiar concepts (automata communicating via asynchronous messages transmitted on queued channels) and partly because it benefits from well-developed tool support.

8.4 The Synchronous Approach on Reactive Systems

When a system is composed of several subsystems evolving in an asynchronous manner, possible interleavings of events yield a combinatory explosion of the number of situations to be taken into account. Thus understanding phenomena becomes more complicated, as well as modeling tasks and, of course, verification. However, under a number of conditions, one can follow the so-called synchronous approach, which is well illustrated by the Esterel language [BG92].

The main idea is to consider infinitely fast systems, so that outputs are synchronous with the inputs that cause them. This hypothesis is quite audacious, but it can be interpreted in two ways:

- if one considers a reactive system, that is, a system reacting to stimuli from its environment, it amounts to assuming that the reaction time of the system is smaller than the duration separating two stimuli; it is then essential to be able to bound the reaction time, and control structures of Esterel have been designed accordingly (it is an imperative language with sequences, loops and interrupt mechanisms);
- if one considers subsystems of a synchronous system which has been decomposed in a modular way, it means that the response time of a subsystem with respect to a stimulus provided by another subsystem is null or can safely be considered as null; the big difference with the previous case is that modules and interactions between them are known — sophisticated compilation techniques can be used — whereas the system may have little or no control over its environment.

Another important synchronous language is Lustre [CPHP87]. It is a data-flow language: each synchronization point is represented by the sequence of values successively present at that point and the system is defined by equations relating such sequences. For instance, in the simple case of an **or** logical gate, we can state the equation $s_n = e_n \vee f_n$ in order to express that at each time n, the output s_n is the disjunction of inputs e_n and f_n (this is the idea; the syntax of Lustre avoids the use of indices). Note that here again, outputs are synchronous with inputs. The case of a looping circuit (e.g. a latch) is more interesting: the output at time n also depends on the output at time $n-1$, so we have an equation in the form $s_n = ...s_{n-1}...$

The synchronous approach is particularly suited to embedded applications subject to hard and non-trivial temporal constraints.

8.5 Temporal Logic

Intuitively, temporal logic handles propositions whose truth value evolves over the course of time. Using it for qualifying program behaviors goes back at least to Pnueli [Pnu77]. The idea is quite natural: the state of a system changes during the execution of a program; as a consequence, properties of the state change as well. This is easy to represent in regular logic: if visited states are successively s_0, s_1, ... a proposition p which is successively false, true, ... is represented by a predicate $\hat{p}$ over $\mathbb{N}$ verifying $\hat{p}(0) = false$, $\hat{p}(1) = true$ ... However, the additional argument introduced everywhere turns out to be cumbersome. Moreover it is not sufficient, because the integer 0, 1, ... makes sense only with respect to a given sequence of states. Temporal logic encapsulates the maneuvers we need thanks to a small number of operators.

 Temporal logic is about discrete time. Durations measured with real numbers are beyond its scope.

8.5.1 Temporal Logic and Regular Logic

Most presentations of temporal logic are based on model theory[3] (see § 3.3.1): the meaning of temporal logic formulas is directly defined on models. However, the concept of model used here is somewhat different from the concept used in § 5.1.3 and in § 5.6.1. More precisely, we are given a transition system and each state is mapped to a model in the classical sense. Thus, a proposition or a formula P may be true in some states and false in other states. A formal way to do that consists of introducing a set of elementary propositions $\mathcal{P} = \{P, Q, \ldots\}$ and in mapping each state s to the subset of $\mathcal{P}$ of propositions which are true at s. Equivalently, we can see P, Q, ... as denoting state predicates. We take here the latter standpoint. We will also need trajectory predicates φ, ψ, ... (We can even consider that P, φ, ... are formulas constructed in a first-order or a higher-order language, rather than just a propositional language.)

If we look at syntax, temporal logic formulas combine such predicates as propositions. For example, $P \Rightarrow AF\partial Q$ expresses that if P is true in the current state, then Q is eventually true on every trajectory starting from the current state. Note that state s does not occur in the above formula and that we do not form $P(s)$. This is done only at the level of semantics, recalling that the truth of "propositions" is relative to a state: it makes it explicit that predicates are hidden behind a propositional notation (more generally, that $n{+}1$-ary predicates are hidden behind n-ary predicate notation).

---

[3]A notable exception is the temporal logic of Unity, which is axiomatically defined in [CM89]. See also § 8.5.5.

In order to simplify the exposition, we limit ourselves to Kripke models without blocking states: a trajectory is then a sequence of states $(s_n)_{n \in \mathbb{N}}$ such that for all n, $s_n \rightarrow s_{n+1}$. In the following, we fix a given Kripke model $\mathcal{K}$: all states and trajectories are implicitly about $\mathcal{K}$. Moreover, s and σ always represent a state and a trajectory, respectively.

The semantics of the state predicate P (respectively the trajectory predicate φ) is denoted by $\mathcal{K}, s \Vdash P$ or by abuse of notation, since $\mathcal{K}$ is fixed, $s \Vdash P$ (respectively $\sigma \Vdash \varphi$).

8.5.1.1 Elementary Formulas. We are given atomic formulas P; their truth value $P(s)$ depends *a priori* on the state s. We do not say more about the language defining such formulas. What matters is that our ability to determine $P(s)$ when s is known. We have, not surprisingly:

$$- \ s \Vdash P \ \stackrel{\text{def}}{=} \ P(s) \ \text{ for } P \text{ atomic.}$$

For example, in the initial state s_{init} of the transition system corresponding to the protocol described in § 8.1.3, we have $l_1 = 0$ and $l_2 = 0$, which allows us to state $s_{\text{init}} \Vdash |l_1 - l_2| \leq \Delta$.

The simplest trajectory predicates are constructed by application of the start operator ∂ on a state predicate. In the following $\sigma(i)$ refers to the ith element of σ.

$$- \ \sigma \Vdash \partial P \ \stackrel{\text{def}}{=} \ \sigma(0) \Vdash P \ : P \text{ is true at the beginning of trajectory } \sigma.$$

Coming back to the example of § 8.1.3, we have $\sigma \Vdash \partial(|l_1 - l_2| \leq \Delta)$ for all trajectories σ beginning with s_{init}.

8.5.1.2 Logical Connectors. Temporal logical connectors $\wedge$, $\vee$, etc., are not applied to propositions, but to state predicates (such as P and Q), or to trajectory predicates (such as φ and ψ). Their semantics are defined using corresponding connectors of regular logic, and we proceed similarly for quantifiers:

$$- \ s \Vdash P \wedge Q \ \stackrel{\text{def}}{=} \ s \Vdash P \ \wedge \ s \Vdash Q \quad \text{(similarly for } \vee \text{, etc.),}$$
$$- \ s \Vdash \forall x \ P \ \stackrel{\text{def}}{=} \ \forall x \ s \Vdash P \quad \text{(similarly for } \exists \text{),}$$
$$- \ \sigma \Vdash \varphi \wedge \psi \ \stackrel{\text{def}}{=} \ \sigma \Vdash \varphi \ \wedge \ \sigma \Vdash \psi \quad \text{(similarly for } \vee \text{, etc.),}$$
$$- \ \sigma \Vdash \forall x \ \varphi \ \stackrel{\text{def}}{=} \ \forall x \ \sigma \Vdash \varphi \quad \text{(similarly for } \exists \text{).}$$

The meaning of "$\wedge$", "$\vee$", etc., is not the same on the left and on the right of "$\stackrel{\text{def}}{=}$". On the right, connectors link propositions whereas they link predicates on the left: in the latter case we have (monadic) second-order logic as seen in § 5.5. Trajectory quantifiers introduced in § 8.5.2.2 for translating branching operators are also second-order.

8.5.2 CTL*

Besides "$\wedge$", "$\vee$", etc. we have specific operators. They can be divided in two groups in the temporal logic we consider now, called CTL*.

8.5.2.1 Temporal Operators. The first group includes temporal operators X (*next*), F (*future* or "eventually"), G (*globally*), W (*weak until*) and U (*until*) which build a trajectory predicate from one or two trajectory predicates. In order to define them we need the suffix of σ obtained by removing the k first elements of σ: $\sigma^k \stackrel{\text{def}}{=} (\sigma(k+n))_{n \in \mathbb{N}}$.

- $\sigma \Vdash X\varphi \stackrel{\text{def}}{=} \sigma^1 \Vdash \varphi$: φ will be true on the next step of σ.
- $\sigma \Vdash F\varphi \stackrel{\text{def}}{=} \exists n \ \sigma^n \Vdash \varphi$: φ will eventually be true on σ.
- $\sigma \Vdash G\varphi \stackrel{\text{def}}{=} \forall n \ \sigma^n \Vdash \varphi$: φ will always be true on σ.
- $\sigma \Vdash \varphi W\psi \stackrel{\text{def}}{=} \forall n \ (\forall i \leq n \ \sigma^i \Vdash \neg\psi) \Rightarrow \sigma^n \Vdash \varphi$:
  φ is true on σ while ψ is not true.
- $\sigma \Vdash \varphi U\psi \stackrel{\text{def}}{=} \exists n \ (\sigma^n \Vdash \psi) \wedge (\forall i < n \ \sigma^i \Vdash \varphi)$:
  ψ will eventually be true on σ and until then φ will be satisfied.

Operators W and U are strictly more general than F and G, for example $G\varphi$ is equivalent to $\varphi W\text{f}$. Moreover, $\varphi U\psi$ is equivalent to $\varphi W\psi \wedge F\psi$.

Let us point out that temporal operators are applied to trajectory predicates and not to state predicates. It is therefore possible to combine them, for example in $GF\varphi$ (φ will be infinitely often true) or in $FG\varphi$ (eventually, φ will be continuously true). However, one often needs to apply them to state predicates as well. To this end we use[4] the start operator ∂.

8.5.2.2 Branching Operators. Operators of the second group, E (*exists*) and A (*all*), build a state predicate by quantifying a trajectory property over trajectories starting from the considered state:

- $s \Vdash E\varphi \stackrel{\text{def}}{=} \exists \sigma \ \sigma(0) = s \wedge \sigma \Vdash \varphi$: there exists a trajectory starting from s which verifies φ;
- $s \Vdash A\varphi \stackrel{\text{def}}{=} \forall \sigma \ \sigma(0) = s \Rightarrow \sigma \Vdash \varphi$: every trajectory starting from s verifies φ.

8.5.2.3 True Formulas Everywhere. In order to say that a state predicate P is true in all states of the system, we employ the notation $\mathbb{V}P$. Symbol $\mathbb{V}$ can be seen as an operator which builds a *proposition* from a state predicate. *It is not part of* CTL*: recall that logical connectors of CTL* do not link propositions.

8.5.2.4 Examples. Invariance properties are expressed by formulas of the form $AG\partial P$, which is true in state s if and only if: $\forall \sigma \ \sigma(0) = s \Rightarrow \forall n \ P(\sigma(n))$. Thus, in our first Unity program in Figure 8.1, the formula stating that x is less than y forever is $AG\partial(x < y)$. However, $x < y$ is true only for states that can be reached from an initial state. Initial states are characterized by a predicate I, which is the conjunction of formulas declared after the **initially** clause. Then we should consider the formula $I \Rightarrow AG\partial(x < y)$. In order to state that this formula is true in all states, we write $\mathbb{V}(I \Rightarrow AG\partial(x < y))$.

---

[4]Experienced users will prefer a lightened notation where ∂ is omitted, considering that we have implicit conversions in that case. Indeed, the ∂ operator is absent from most presentations. It is made explicit here for a better understanding of the underlying mathematical model.

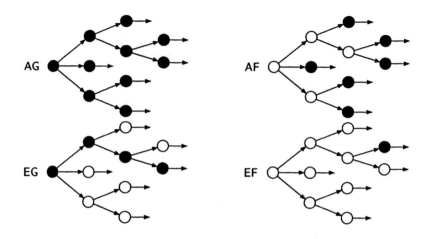

Figure 8.9: The operators of CTL

Similarly, the safety property we expect from the synchronization protocol described in § 8.1.3 is $\mathbb{V}(I \Rightarrow AG\partial(|l_1 - l_2| \leq \Delta))$.

Liveness properties are expressed by operator F, generally just after A. Thus, in the system given in Figure 8.7, we have $AF\partial(s = S)$, and in the program given in Figure 8.2, we have $\mathbb{V}(I \Rightarrow AF\partial(f = \mathtt{true}))$.

The progress property on clocks of the protocol given in § 8.1.3 is more complicated. For example, $I \Rightarrow AF\partial(l_1 \geq 10^{100})$ states only that l_1 will be *very* large. In order to get *arbitrarily* large, the natural statement is: $\forall n \in$ nat $I \Rightarrow AF\partial(l_1 \geq n)$. This formula is allowed in a version of CTL* which includes arithmetic. In the usual version, based on propositional logic, we have to encode progress in a different way, from the idea: "l_1 will always become larger". Assume we have a state predicate *incr* at our disposal; we arrange things in a way such that this predicate is true if and only if during the last fired transition, l_1 was incremented. To this effect, one can insert appropriate fields in the state and update them adequately, without disturbing the rest of the program. (This simple exercise is left to the reader.) Now the progress property says that along every trajectory, *incr* is true infinitely often: $\mathbb{V}(I \Rightarrow AG\,F\partial incr)$.

Reachability properties are obtained by combining E with F: if one controls execution — the choice among competing transitions at each step — a state satisfying the desired property will be reached. Let us consider again the synchronization protocol of § 8.1.3: the reachability of a state where the two local clocks are equal is expressed by $EF\partial(l_1 = l_2)$. The statement $\mathbb{V}(I \Rightarrow AG\partial EF\partial(l_1 = l_2))$ tells us that this equality is reachable from every state of every trajectory starting from the initial state.

8.5.3 CTL

CTL (computation tree logic) is the fragment of CTL* obtained when every occurrence of a temporal operator (X, F, G, W or U) is immediately preceded by a branching operator (A or E). All allowed compound operators (AX, EF, etc.) build state predicates. They are then necessarily applied to sub-formulas systematically headed by the start operator ∂. In practice this operator is implicit. Thus, one can say that in CTL formulas are obtained by repeated application of AX, EF, etc. on *state formulas*. This simplification makes automated verification via *model checking* much easier [BBF$^+$01].

Safety, liveness and reachability properties like the ones described in § 8.5.2.4 are of this kind, but not the progress property $I \Rightarrow$ AG F$\partial incr$. Fairness properties, in the form AGF∂P are excluded as well. In general one cannot express properties about events which are along the same trajectory.

Typical combinations AG ∂P, AF∂P, EG ∂P and EF∂P are illustrated on diagrams in Figure 8.9, where the tree-like character of CTL properties is easy to see. A filled circle represents a state where P is true.

8.5.4 LTL and PLTL

LTL (linear temporal logic) is the fragment of CTL* where only trajectory predicates are considered, that is, predicates built using temporal operators. The idea is that formulas obtained in this way should be verified on all trajectories. Formally, it amounts to putting a unique (and, in practice, implicit) universal quantification A at the beginning of the formula. Thus LTL does not provide a means for considering the existence of different possible behaviors starting from a given state. This is why this logic is called **linear**. For instance, the reachability property expressed by AG∂EFφ has no equivalent formulation in LTL. More generally, this logic does not allow one to distinguish between two transition systems having the same trajectories.

Automated verification research is concentrated on PLTL (propositional LTL), which is the fragment of LTL where non-temporal connectors are those of propositional calculus (first-order quantifiers are forbidden).

Let us mention in passing a traditional notation coming from the modal logic S4, which uses □ for G (forever) and ◇ for F (eventually). This notation is used in TLA, as we will see in § 8.6.

8.5.5 The Temporal Logic of Unity

The very design of Unity involves two ingredients: the programming language presented in § 8.1 and a linear temporal logic endowed with the following particulars: *its operators take state predicates as arguments and they return a proposition*: they are then weaker than LTL operators (which build a trajectory predicate from trajectory predicates); in particular they cannot be embedded.

In contrast, propositions obtained in this way can be combined with logic connectors $\wedge$, $\vee$, etc. The latter can then have "classical" occurrences (as in § 5.1) as well as "temporal" occurrences (as in § 8.5.1.2) in the same formula.

The two basic operators of Unity are **co** and **leadsto** (denoted here by $\rightsquigarrow$). The proposition P **co** Q (for P constrains Q) is defined by $\mathbf{V}(P \Rightarrow \mathrm{AX}\partial Q)$: Q comes immediately after a state verifying P. The original version of Unity used a kind of *weak until*: P **unless** Q, which is defined by $(P \wedge \neg Q)$ **co** $(P \vee Q)$ and turns out to be equivalent to $\mathbf{V}\mathrm{AG}(\partial P \mathrel{\mathsf{W}} \partial Q)$. The proposition $P \rightsquigarrow Q$ expresses that every trajectory where P is initially true eventually reaches a state verifying Q; it is equivalent to $\mathbf{V}\mathrm{AG}(\partial P \Rightarrow \mathrm{F}\partial Q)$.

For instance, a way to formalize the progress property of l_1 in the synchronization protocol of § 8.1.3 is:

$$\forall n \in \mathbf{nat} \ (l_1 = n) \rightsquigarrow (l_1 = n + 1) \ . \tag{8.1}$$

The logic of Unity is originally defined in an axiomatic way by deduction rules. Other rules can be derived, such as the following. *It is the set of such rules that makes* Unity *of practical interest as a verification technique.*

$$\frac{P \rightsquigarrow Q \vee B \qquad B \rightsquigarrow R}{P \rightsquigarrow Q \vee R} \ .$$

8.5.6 Hennessy–Milner Modalities

It is sometimes useful to state properties which refer to transition labels.[5] It is even essential if we work with a language like CCS. To this end, one can use the modalities $[\alpha]$ and $\langle\alpha\rangle$ where α is a label, as in Hennessy–Milner logic [HM85]. They apply to a state predicate P and give new state predicates $[\alpha]P$ and $\langle\alpha\rangle P$. The latter is true in every state s from which a state satisfying P can be reached through a transition labeled by α: $\langle\alpha\rangle P(s)$ if and only if $\exists s' \ s \xrightarrow{\alpha} s' \wedge P(s')$.

Equally, $[\alpha]P$ is true in every state s from which every transition labeled by α leads to a state satisfying P: $[\alpha]P(s)$ if and only if $\forall s' \ s \xrightarrow{\alpha} s' \Rightarrow P(s')$.

For example, in the system given in Figure 8.7 page 131 we have:

- R $\Vdash$ $\langle\mathsf{acc}\rangle\mathsf{t} \wedge \langle\mathsf{reset}\rangle\mathsf{t}$: from the state R one has the option of getting a drink and the option of asking for reimbursement;
- A $\Vdash$ $[20c]\langle\mathsf{reset}\rangle\mathsf{t}$: from the state A, after paying 20 cents, one can still ask for a reimbursement;
- A $\Vdash$ $\langle 20c\rangle(\langle\mathsf{acc}\rangle\mathsf{t} \wedge \langle\mathsf{reset}\rangle\mathsf{t})$: from the state A one can pay 20 cents and then choose between getting a drink or asking for a reimbursement;
- A $\Vdash$ $\langle 20c\rangle(\neg\langle\mathsf{acc}\rangle\mathsf{t} \wedge \langle\mathsf{reset}\rangle\mathsf{t})$: from the state A one can pay 20 cents and then be in position to ask for a reimbursement without having the option of getting a drink.

---

[5]In § 8.5.6 and also in § 8.5.7, the model, that properties are about, is a transition system rather than a Kripke system.

Given that only processed actions are taken into account, one might think that Hennessy–Milner logic is limited to expressing properties of traces, as LTL is limited to properties of trajectories. This would be a mistake. The last property stated above is not true of the first vending machine described in Figure 8.6 whereas both systems have the same traces, as seen previously. In fact modalities $[\alpha]$ and $\langle\alpha\rangle$ are close to branchings expressed by AX and EX.

It is easy to extend the previous modalities by replacing α by a set of actions: $s \xrightarrow{K} s'$ can then be considered as an abbreviation for $\exists \alpha \in K \; s \xrightarrow{\alpha} s'$. In this context, we agree that "$-$" denotes the set of all actions of the system. Thus $[-]P$ is true in any state from which all transitions lead to a state satisfying P. In our example, we have $A \Vdash [-]\langle 1E\rangle t$.

8.5.7 Mu-calculus

The properties just mentioned would also be satisfied by a vending machine that stops working after delivering its first drink or paying money back. The μ-calculus based on Hennessy–Milner logic allows one to specify complex iterative behaviors thanks to the introduction of least fixed points $\mu X.\Phi(X)$ and of greatest fixed points $\nu X.\Phi(X)$, where $\Phi(X)$ represents a state predicate in which the state predicate variable X can occur.

For example, let us consider the formula $P \stackrel{\text{def}}{=} \nu X.\langle reset\rangle \wedge [-][-][-]X$. In a first approximation it can read: P is true if reset can be fired and if, after firing three transitions, reset can again be fired and if, after firing three transitions again, reset can again be fired, and so on. Here, P describes a cyclical behavior with period 3. In the system of Figure 8.7, P is true in states P and F.

More precisely, P is the greatest solution of the fixed-point equation $X = \langle reset\rangle \wedge [-][-][-]X$, that is, the greatest predicate X satisfying $X \Rightarrow \langle reset\rangle \wedge [-][-][-]X$. According to § 3.6, this solution is obtained by successive iterations $P_0 = t$, $P_1 = \langle reset\rangle \wedge [-][-][-]P_0$, $P_2 = \langle reset\rangle \wedge [-][-][-]P_1$, etc. but we have already $P_2 \Leftrightarrow P_1$. To see that, let $|Z|$ denote the set of states satisfying Z, we have $|P_0| = \{S, A, R, F\}$, $|P_1| = \{R, F\}$ and $|P_2| = \{R, F\}$.

This definition by fixed points makes use of the theorem of Knaster–Tarski which asks for a monotony condition. In the μ-calculus, the latter is ensured using a syntactic device: fixed-point variables (like X above) must occur only under an even number of negations.

Terms in the form $\nu X.f(X)$ express properties about full trajectories and then are related to safety. Dually, least fixed points $\mu X.f(X)$ are related to liveness properties.

Fixed points provide a convenient means for defining the semantics of CTL. For example $E(\partial P \cup \partial Q)$ is true in state s if Q is true in s, or if P is true and there exists a next state in which $E(\partial P \cup \partial Q)$ is true. More precisely, $|E(\partial P \cup \partial Q)|$ is the smallest set of states X containing $|Q|$ and containing states s such that $P(s)$ and $s \longrightarrow s'$ with $s' \in X$. This idea is represented in a synthetic way in the formula $\mu X.Q \vee (P \wedge \langle - \rangle X)$, and is the basis of the first verification algorithms for CTL by model checking.

Much more complex properties can be formulated by alternating μs and νs. For instance $\nu X.(\mu Y. P \vee \langle - \rangle Y) \wedge \langle - \rangle X$ represents the CTL formula EG∂EF∂P (there exists a trajectory along which one always has the option, branching off if necessary, of reaching a state satisfying P), whereas $\nu X.\mu Y.(P \vee \langle - \rangle Y \wedge \langle - \rangle X)$ represents the CTL* formula EGF∂P (there exists a trajectory along which P is infinitely often true), which is beyond what CTL and LTL can express. The first formula can be analyzed as follows: $\mu Y.P \vee \langle - \rangle Y$, which is equivalent to EF∂P, is embedded in $\nu X.Q \wedge \langle - \rangle X$, which is equivalent to EG∂Q. The second formula is more subtle: it contains a "true" alternation of fixed-point operators — the two variables X and Y are within the scope of the second fixed-point operator(μ). Still more complex (and delicate) properties can be stated, using additional fixed-point operators alternations, so that we can go beyond the expressive power of CTL*. The interested reader may consult the literature cited at the end of this chapter.

8.6 TLA

With TLA (temporal logic of actions), Leslie Lamport proposed to specify both the expected properties of the behavior of a system and the system itself, all within the framework of a linear temporal logic. To this end, temporal operators are applied to transitions. The latter are described by a binary relation between the current state and the next state using the same convention as in Z: for example, incrementing l_1 is described by a relation A_1 which can be defined by $l_1' = l_1 + 1$ or by $l_1' - l_1 = 1$. Such a relation in TLA is called an **action**. A system that perpetually increments l_1 is specified as follows: $A_1 \stackrel{\text{def}}{=} \square(l_1' - l_1 = 1)$. To specify the initial state, we just need a state formula, for example $Init_1 \stackrel{\text{def}}{=} l_1 = 0$. The conjunction $Init_1 \wedge \square A_1$ makes up our first TLA system.

From a mathematical perspective, one can consider that a TLA formula $\square A$ defines a Kripke model $\langle S, \mathcal{R} \rangle$, by stating a constraint on S and on $\mathcal{R}$. As a first approximation, S is defined by the vocabulary employed in A, which is just l_1 in our example.[6] Each word of the vocabulary denotes a field of S, that we translate to a projection as in § 8.2.6 for Unity. In the case of A_1, at the moment we have $S = \mathbb{N}$ while l_1 boils down to the identity function.[7] The formula A then defines the transition relation $\mathcal{R}$. This yields in example A_1:

$$\begin{aligned} \mathcal{R}_1 &= \{\langle s, s' \rangle \in S \times S \mid l_1(s) - l_1(s') = 1\} \\ &= \{\langle 0, 1 \rangle, \langle 1, 2 \rangle, \langle 2, 3 \rangle, \ldots\} \ . \end{aligned}$$

---

[6] l_1' must be considered as a term obtained by application of the postfix operator $'$ to l_1; this operator is similar to X introduced in § 8.5.2, but it is applied to a term instead of a formula.

[7] Let us mention that according to Lamport, the domain of fields should not be specified. This point is not essential here.

However, as a TLA specification is the conjunction of several formulas having, in general, different vocabularies, but which can overlap — as in Unity, cooperation is modeled by field sharing — one agrees that S is only partially specified by the vocabulary of A. In our example we would have $S = \ldots \times \mathbb{N} \times \ldots$ and l_1 would be the appropriate projection. The transition relation $\mathcal{R}$ is defined as before by formula A, but with the extended interpretation of S. Thus, $\mathcal{R}_1$ becomes, assuming that l_1 is the first projection and that we have another Boolean field:

$$\mathcal{R}_1 = \{\langle\langle 0, f\rangle, \langle 1, f\rangle\rangle, \langle\langle 0, f\rangle, \langle 1, v\rangle\rangle, \langle\langle 0, v\rangle, \langle 1, f\rangle\rangle, \langle\langle 0, v\rangle, \langle 1, v\rangle\rangle,$$
$$\langle\langle 1, f\rangle, \langle 2, f\rangle\rangle, \langle\langle 1, f\rangle, \langle 2, v\rangle\rangle, \langle\langle 1, v\rangle, \langle 2, f\rangle\rangle, \langle\langle 1, v\rangle, \langle 2, v\rangle\rangle,$$
$$\langle\langle 2, f\rangle, \langle 3, f\rangle\rangle, \langle\langle 2, f\rangle, \langle 3, v\rangle\rangle, \langle\langle 2, v\rangle, \langle 3, f\rangle\rangle, \langle\langle 2, v\rangle, \langle 3, v\rangle\rangle,$$
$$\ldots\}$$

Now, if we augment the previous specification with a second formula $A_2 \stackrel{\text{def}}{=}$ $(l_1 \leq 1 \wedge b_2' = f) \vee (l_1 > 1 \wedge b_2' = v)$, the conjunction $A_1 \wedge A_2$ yields the transition relation:

$$\mathcal{R}_1 \cap \mathcal{R}_2 = \{\langle\langle 0, -\rangle, \langle 1, f\rangle\rangle, \langle\langle 1, -\rangle, \langle 2, f\rangle\rangle, \langle\langle 2, -\rangle, \langle 3, v\rangle\rangle, \ldots\}$$

where the joker "$-$" represents the two values f and v. The important point to remember is that the transition relation we get by composition is no longer the Cartesian product of transition relations, but their intersection.

It is possible to present the composition using a more general construct called the fibered product. The product and the intersection are two special cases of fibered products. We will not expand this remark here.

The terminology of TLA is different from that employed for transition systems in § 8.2.1: in the former case an action is a subset of $S \times S$, in the latter an action is a label (associated to a subset of $S \times S$).

As with the product, composition by conjunction entails a synchronization of transitions of all components. If we want A_1 to evolve as well as another system that does not mention l_1, the remedy is the same as in § 8.2.5: offering a choice between modification and stuttering. In order to simplify the writing, in TLA we have the notation $[R]_{\langle z\rangle}$ for $R \vee (z' = z)$. One would then write: $\square[l_1' - l_1 = 1]_{\langle l_1\rangle}$.

One of the main points of TLA is that behaviors are specified by stuttering invariant formulas: formulas such that, if they are satisfied by a trajectory σ, they are also satisfied by any trajectory we get by inserting or removing state repetitions in σ. For this, formulas are essentially in the form $\square[R]_{\langle z\rangle}$.

Let us illustrate the idea on Roucairol's protocol written in Unity in Figure 8.4. The first component performs three actions at will:

- $N_1 \stackrel{\text{def}}{=} l_1 < d_1 \wedge l_1' - l_1 = 1$ incrementation,
- $E_1 \stackrel{\text{def}}{=} c_2' = c_2 \cup \{l_1\}$ sending,
- $R_1 \stackrel{\text{def}}{=} \ldots$ receiving, not detailed here.

Note that the firing condition of N_1 is represented by a conjunction with $l_1 < d_1$. One would define actions N_2, E_2 and R_2 in a symmetrical manner. The desired behavior Ψ is then specified as follows (be warned that actions relative to channel c_i are put together; their behavior could be augmented by losses and duplications):

$$C_1 \stackrel{\text{def}}{=} R_1 \vee E_2 \qquad\qquad\qquad C_2 \stackrel{\text{def}}{=} R_2 \vee E_1$$
$$HM_1 \stackrel{\text{def}}{=} N_1 \vee C_1 \vee C_2 \qquad\qquad HM_2 \stackrel{\text{def}}{=} N_2 \vee C_2 \vee C_1$$
$$\Psi \stackrel{\text{def}}{=} \Box[HM_1]_{\langle l_1, d_1, c_1, c_2 \rangle} \wedge \Box[HM_2]_{\langle l_2, d_2, c_2, c_1 \rangle}$$

We still have to ensure progress and fairness of the behavior, using a conjunction with a suitable formula, and without introducing any parasitic safety property. To this end one uses particular formulas noted $\text{WF}_f(A)$ or $\text{SF}_f(A)$. They are defined by means of $\Diamond$ and $\Box$, and they express that action A (HM_i in our example) is fairly fired and modifies fields listed in f. They represent stuttering invariant properties.

Reasoning is performed in TLA using about fifteen deduction rules. Some of them are as simple as $\frac{P \Rightarrow Q}{\Box F \Rightarrow \Box Q}$, but rules on fairness properties are more complex. The reader may consult [Lam94] and [Aba90].

8.7 Verification Tools

Previous sections presented different approaches for accurately specifying systems composed of several entities evolving at the same time, as well as the properties we expect them to satisfy.

8.7.1 Deductive Approach

For verifying these systems, one can proceed by decomposition and formal deductions, notably in the framework of Unity or TLA. The user then has to properly organize his or her understanding of the phenomena under consideration and to master formal reasoning to a good extent. Proof environments can then provide valuable assistance. A good specialized tool is STeP [BBC+95]. Libraries on top of general proof assistants such as LP, Isabelle and Coq, are also available or in development, see for example the work of Crégut and Heyd [HC96]. The strong point of this approach is that the user may use powerful mathematical devices for structuring specifications and proofs, and for explaining when and why the system works.

8.7.2 Verification by Model Checking

Considering that the number of possible scenarios for a system composed of several entangled subsystems increases very quickly, including for small systems, a different approach was invented in the 1980s and transpired to be quite efficient and effective: building the Kripke model of the whole system (labels are generally ignored), that is, the graph of all global states and possible transitions between them, then computing the truth value of expected propositions on each state — hence the name **model checking**. This is possible provided that the graph is finite (hence extendable data structures like unbounded queues or trees are not allowed), and that properties are expressed in a *propositional* temporal logic.

Without going into detail, the verification of a CTL formula is based on fixed-point computations (see § 8.5.7). It is linear in the size of the graph and in the size of the formula. However, the number of states is itself essentially exponential: introducing a one byte variable in the two protocol entities is enough to make the number of global states explode by 32,000. This approach has actually proven to be really successful since the introduction of techniques for representing graphs and formulas in a compact way where common parts are shared, in particular thanks to the use of BDDs (binary decision diagrams).

The automated verification of a PLTL formula Φ is more complex because it is expressed about a path instead of a state. One translates its negation $\neg\Phi$ into an observer automaton and then computes the synchronized product of the latter with the system to be verified. The property is satisfied if and only if the language recognized by the product is empty. The verification remains linear in the size of the graph but becomes exponential in the worst case in the size of the formula.

Moreover, note that in a number of environments, the property to be verified has to be directly expressed in the form of an automaton, without using temporal logic. A similar idea is used in proofs by **bisimulation**, though in a technically very different way. The basic principle there is to check that a given automaton (e.g. a CCS process) has the same observable behavior (in terms of labels) as a second automaton, the latter being considered as an abstract view of the former.

From the user perspective, model checking can relieve him or her of an exhaustive amount of reasoning on a huge number of specialized situations. Another valuable aspect of this approach is that if a property is not satisfied, model checking algorithms produce a counter-example scenario. The main difficulties are in the modeling steps of the system and of expected properties. Automated verification is made possible by adequate limitations in the languages (propositional logic, bounded data structures). Remaining means of expression have to be used very cleverly.

8.8 Notes and Suggestions for Further Reading

The general formalism of transition systems is described in [Arn94]. The languages Unity and TLA are respectively defined in [CM89] and [Lam94]. The process algebras CCS and CSP are dealt with in [Mil89] and [Hoa85], respectively.

The synchronous approach is described in [Hal93]. Interested readers may also consult papers on Esterel [BG92], Lustre [CPHP87] and Signal [BLJ91].

Excellent syntheses on temporal logic can be found in [Eme90] and [Sti92]; however, Unity and TLA are not covered. Let us also mention that some temporal logics include modalities about the past, they are for instance exploited in Lustre for safety properties. The reference manual for the STeP environment is [BBC$^+$95]. The μ-calculus is studied in [Bra92] and in [AN01].

Reference books on verification techniques through model checking are also available. McMillan's book is still very valuable [McM93], while [CGP99] is centered on underlying theory and implementation technologies. [BBF$^+$01] is more synthetic and provides useful practical advice, as well as an overview of the main software tools available. Two of the most prominent are SMV [Berct], which is based on CTL, and SPIN, based on LTL [Hol97]. Original papers on model checking are [QS82] and [CES83]. The relative merits of branching as opposed to linear time temporal logics have been a matter of debate since the early 1980s. For a recent paper on this issue, the reader is referred to [Var01].

9. Deduction Systems

In the propositional case, a formula P has only a finite number of interpretations: there are exactly 2^n of them, where n is the number of atomic propositions used in P. The truth table method makes it easy to determine whether P is satisfied, is a tautology, or is a logical consequence of a finite set of propositions. This is a *semantic* technique: it is based on a study of models of P.

In contrast, the topic of **proof theory** is to know the consequences of a set of axioms by *purely syntactic* means. The central concept is then the deductive consequence relation, denoted by $\vdash$. This relation is *a priori* different from the semantic relation $\models$. It is defined by so-called logical axioms (for example, $(P \wedge Q) \Rightarrow P$) and rules called inference rules or deduction rules. Recall that in model theory, the symbol $\models$ can also be used to state that a formula P is valid ($\models P$). In a similar way, $\vdash P$ denotes that P is a theorem.

We will need to express syntactical manipulations on the deductive consequence relation itself. To this effect we will introduce ordered pairs $\Gamma \vdash \Delta$ called sequents. Then we will have proof trees made of sequents and having a sequent as their conclusion. This yields a more general concept of a theorem and leads us to use different symbols for stating theorems and for representing deductive consequences, so that we could write $\vdash \quad \Gamma \vdash \Delta$. According to [Gal93], the symbol $\vdash$ comes from Girard.

Logical axioms are always true. They should not be confused with axioms which are proper to a given theory and define the latter. A well-known example are Peano axioms, which define arithmetic. Such axioms are called proper axioms, or non-logical axioms.

There are three main approaches for defining $\vdash$: Hilbert's approaches, which uses many axioms and very few inference rules and two approaches due to Gentzen (natural deduction and sequent calculus) which have the converse features: few axioms and many deduction rules. We start with these three methods.

We will also sketch two other techniques for calculating consequences. The first was developed by Dijkstra and Scholten in the framework of their calculational approach to programming. The second is rewriting systems, which provide efficient tools for equational reasoning — prosaically: replacing equals with equals.

The chapter ends with the relationship between truth and provability. One would expect that provable formulas are true and conversely. This is correct for first-order logic, but arithmetic makes the situation more complicated. The main results come from works originally motivated by the foundations of mathematics, considered as the typical place for studying formal reasoning. The practice of formal methods is mainly concerned by intrinsic limitations related to *fully automatic* proof search techniques.

9.1 Hilbert Systems

The impact of Hilbert systems seems less important than the other approaches for computer science. Therefore we limit ourselves to propositional logic.

The notation $\vdash P$ states that the proposition (or the formula) P is proved. In particular, axioms will be noted in this way.

In the framework of propositional logic, a Hilbert system has only one deduction rule, called **modus ponens**. It can read, P and Q being arbitrary propositions: if $P \Rightarrow Q$ is proved and if P is proved, then Q is a theorem as well:

$$\frac{P \Rightarrow Q \quad P}{Q} \ .$$

First-order logic includes a second inference rule, the **generalization rule** which reads: if P is a theorem, then $\forall x P$ is a theorem as well. For example, from $x > 0 \Rightarrow 2.x > 0$ we deduce $\forall x \ x > 0 \Rightarrow 2.x > 0$:

$$\frac{P}{\forall x \, P} \ .$$

The construction of proofs is quite simple. The intuitive idea is to present a proof in the form of a tree[1] where nodes are labeled by an instance of an inference rule and where leaves are labeled by an axiom. The proved theorem is on the root. The precise definition of a **proof** is as follows:

– if $\vdash A$ is an axiom,

$$\frac{}{A} \ \textbf{ax}$$

is a proof with conclusion A; $\frac{}{A} \ \textbf{ax}$ may be regarded as a rule without premise;
– if $\mathcal{D}_1$ and $\mathcal{D}_2$ are two proofs with respective conclusions C_1 and $C_1 \Rightarrow C_2$, then the tree

$$\frac{\begin{array}{cc} \mathcal{D}_2 & \mathcal{D}_1 \\ C_1 \Rightarrow C_2 & C_1 \end{array}}{C_2} \ \textbf{mp}$$

with root *modus ponens*, where P and Q are respectively instantiated by C_1 and C_2, with immediate left subtree $\mathcal{D}_2$, and with immediate right subtree $\mathcal{D}_1$, is a proof with conclusion C_2.

---

[1] The concept of a tree can itself be formalized, see page 83.

A formula P is a **theorem** if there exists a proof tree with conclusion P. Note that we will sometimes indicate explicitly, on proof trees, the name of the rules we use near the corresponding fraction lines.

The main part of information is actually contained in the axioms. These axioms, on which we agree independently from any theory (in the sense defined in § 5.6.1), are called **logical axioms**. They are chosen in a way such that all valid formulas can be deduced. Many axiom systems satisfy this condition. All these systems are equivalent (that is, the axioms of one system are theorems of any other axiom system). For illustration purposes, here are some axioms of a well-known system due to Hilbert and Ackermann [HA28]:

$$\vdash P \Rightarrow (Q \Rightarrow P) \ ,$$
$$\vdash (P \Rightarrow (P \Rightarrow Q)) \Rightarrow (P \Rightarrow Q) \ ,$$
$$\vdash (P \Rightarrow Q) \Rightarrow ((Q \Rightarrow R) \Rightarrow (P \Rightarrow R)) \ .$$

This system includes twelve additional axioms about $\wedge$, $\vee$, $\Leftrightarrow$ and $\neg$ connectors [GG90, p. 112]. They are actually axiom **schemas**: real axioms are obtained if we substitute any proposition of the considered language for the symbols P, Q, R. For example, from the schema $\vdash P \Rightarrow (Q \Rightarrow P)$, we get, in a language including the proposition symbol P:

$$\vdash P \wedge P \Rightarrow (\neg P \Rightarrow P \wedge P) \ .$$

Now we can provide a proof of $P \Rightarrow P$, using the two first axioms:

$$\frac{\dfrac{}{(P \Rightarrow (P \Rightarrow P)) \Rightarrow (P \Rightarrow P)} \ \textbf{ax} \qquad \dfrac{}{P \Rightarrow (P \Rightarrow P)} \ \textbf{ax}}{P \Rightarrow P} \ \textbf{mp} \ .$$

The presentation of this tree can be simplified, because we know that the formulas displayed at the level of leaves are necessarily axioms and that *modus ponens* is used in all other places:

$$\frac{(P \Rightarrow (P \Rightarrow P)) \Rightarrow (P \Rightarrow P) \qquad P \Rightarrow (P \Rightarrow P)}{P \Rightarrow P} \ .$$

We will see below more varied proof trees, where it is better to keep explicitly the name of the rules which are used.

It is regrettable that Hilbertian axiomatic systems are somewhat contrived. Axioms are sometimes complicated. It is a shame that the proof of $P \Rightarrow P$ is not trivial. It is hard to claim that this formalization of logic represents *usual* logical reasoning. The situation gets even worse in Hilbert systems invented with the purpose of minimalizing the number of axioms. *In practice, no proof assistant is based on this approach.*

In mathematics (group theory, topology, geometry, etc.) a system of (proper) axioms plays an important role; for a given theory; there is few room for alternative systems. In contrast, logic already offers a large number of possible systems, this is already the case for propositional logic. This actually suggests that no one tautology is more fundamental than the others.

However Hilbert systems are quite convenient for the mathematical study of logic, in order to know whether every provable proposition is true and conversely — these properties are respectively called soundness and completeness. In this respect a particular relation turns out to be important: the deducibility relation.

The axioms of a theory (in the sense given in § 5.6.1) are called **proper axioms**, or **non-logical axioms**. Let Γ be a set of closed formulas. A closed formula P is a **deductive consequence** of Γ (we also simply say that P is deducible from Γ), which we note $\Gamma \vdash P$, if P can be proved using *modus ponens* — and the generalization rule in the case of first-order logic — from logical axioms and formulas of Γ.

Let Γ be an axiom system allowing one to prove $P \Rightarrow Q$; if we insert the hypothesis P in Γ, we observe (thanks to *modus ponens*) that $\Gamma, P \vdash Q$. The converse property seems natural and can actually be proved, but more work is required.

Theorem 9.1 (deduction)
If $\Gamma, P \vdash Q$ then $\Gamma \vdash P \Rightarrow Q$.

This theorem is proved by induction on (the length of) the proof trees corresponding to $\Gamma, P \vdash Q$, by inspecting the different possible cases. Warning: it is important to distinguish the proof of the previous theorem and the objects it talks about, which are themselves proofs and theorems. There are two language levels, and the first is the **metalanguage** of the second.

The metalanguage is the language we use for defining, commenting or explaining another language. It is a natural language in most cases, such as English. In the present case the metalanguage involves basic mathematical concepts in order to explain the syntax of logic as well as concepts related to deduction. In this respect, Theorem 9.1 is a **metatheorem**.

We will see that natural deduction and sequent calculus take the opposite view with relation to Hilbert systems: the meaning of implication will be based on the two last (and symmetric) properties formalized by *modus ponens* and the deduction theorem. Other connectors will also be systematically treated in a symmetric way.

9.2 Natural Deduction

With **natural deduction**, Gentzen introduced a formalization more faithful to regular reasoning.

9.2.1 Informal Presentation

Let us start with a simple example. We want to show that the square of an even number is even, given that the product of an even number by an arbitrary number is even. The formula to be proved is:

$$[\forall x \;\text{even}(x) \Rightarrow \forall y \;\text{even}(x.y)] \Rightarrow [\forall a \;\text{even}(a) \Rightarrow \text{even}(a.a)] \;. \tag{9.1}$$

This example has no mathematical interest, but it allows us to illustrate the meaning of quantifiers and implication. As in usual reasoning, our first step is to prove that, from the hypotheses $\forall x \;\text{even}(x) \Rightarrow \forall y \;\text{even}(x.y)$ and $\text{even}(a)$, we can deduce $\text{even}(a.a)$.

We then assume $\forall x \;\text{even}(x) \Rightarrow \forall y \;\text{even}(x.y)$ and we consider for x an arbitrary a. We then have $\text{even}(a) \Rightarrow \forall y \;\text{even}(a.y)$. Let us now assume that a is even. We can deduce that for all y, $a.y$ is even, then that $a.a$ is even. Hence we have $\text{even}(a) \Rightarrow \text{even}(a.a)$, for any a. We deduce $\forall a \;\text{even}(a) \Rightarrow \text{even}(a.a)$. This formula was proved under the hypothesis $\forall x \;\text{even}(x) \Rightarrow \forall y \;\text{even}(x.y)$, hence we conclude (9.1).

Let us split up this reasoning into its components. First we prove that, from the hypotheses:

$$\forall x \;\text{even}(x) \Rightarrow \forall y \;\text{even}(x.y) \qquad \text{and} \tag{9.2}$$

$$\text{even}(a) \tag{9.3}$$

we can deduce:

$$\text{even}(a.a) \;. \tag{9.4}$$

Let us take an arbitrary a for x in (9.2). We have then:

$$\text{even}(a) \Rightarrow \forall y \;\text{even}(a.y) \;. \tag{9.5}$$

Let us now consider the hypothesis (9.3): a is even. From (9.5) and (9.3) we get that $a.y$ is even for all y:

$$\forall y \;\text{even}(a.y) \;, \tag{9.6}$$

then (9.4) if we take a for y. This conclusion depends on the hypothesis (9.3), so we have:

$$\text{even}(a) \Rightarrow \text{even}(a.a) \;, \tag{9.7}$$

and this for an arbitrary a, that is for an a on which we don't have any hypothesis. We deduce:

$$\forall a \;\text{even}(a) \Rightarrow \text{even}(a.a) \;, \tag{9.8}$$

which was proved under the hypothesis (9.2), hence (9.1).

The inferences used in the previous example have one of the following shapes:

– if from P we can prove Q, we have a proof of $P \Rightarrow Q$, more precisely a proof of $P \Rightarrow Q$ without the hypothesis P; to put it otherwise: in order to prove $P \Rightarrow Q$ it is enough to prove Q under the hypothesis P; thus we got (9.6) from (9.3), hence (9.7);

- if we proved $P \Rightarrow Q$ on the one hand and proved P on the other, then we have a proof of Q; for instance we deduced (9.6) from (9.5) and (9.3);
- if we proved P (P may contain free occurrences of x) *without any hypothesis on the variable x* then we have a proof of $\forall x P$; in our example, see how we deduced (9.8) from (9.7); however, we could not deduce $\forall a$ **even**$(a.a)$ from (9.4), because a hypothesis on a was still present!
- if we proved $\forall x P$ then we have a proof of $[x := t]P$ where t is an arbitrary term; for example, we deduced (9.4) from (9.6).

We observe that each connector $\star$ is defined by an **introduction rule** and an **elimination rule**. An introduction rule determines how we can get a formula having $\star$ as its main connector, while an elimination rule shows how, from such a formula, we can get one of its immediate subformulas. This corresponds to a general thought line: in the framework of natural deduction, the behavior of each connector is defined by introduction and elimination rules. Here are the rules for conjunction and disjunction:

- if we proved P and we proved Q, then we have a proof of $P \wedge Q$;
- if we proved $P \wedge Q$, then we have a proof of P (similarly, we also have a proof of Q);
- if we proved P (similarly, if we proved Q), then I have a proof of $P \vee Q$;
- if we proved $P \vee Q$, and if in each case we can prove R, then we have a proof of R.

Natural deduction includes no logical axiom; but one manipulates deductions *under hypotheses*. The typical way to discharge these hypotheses is to use introduction rules for $\Rightarrow$. A **proof** is a special case of deduction in which no hypothesis is left; finally, as in Hilbert systems, a **theorem** is a formula for which there exists a proof.

The example of even numbers illustrated this process. One of the simplest examples is the proof of $P \Rightarrow P$. First we put the hypothesis P, and we have a trivial deduction of P under this hypothesis. Using the introduction rule for $\Rightarrow$, we immediately get a deduction of $P \Rightarrow P$ without hypothesis.

9.2.2 Formal Rules

The formalization of natural deduction inference rules takes the shape of fractions, as in Hilbert systems. Each rule is identified by a name such as $\wedge_i$ (introduction of $\wedge$), $\wedge_{e1}$ or $\wedge_{e2}$ (respectively left and right elimination of $\wedge$). The rules of the system NJ of Gentzen are given in Figure 9.1. We comment on them now.

In a proof, hypotheses are identified by a number between parentheses. When a hypothesis is discharged, its number is recalled on the corresponding inference rule (introduction of $\Rightarrow$, elimination of $\vee$ or of $\exists$). When one of these three rules is applied, it is possible to discharge one, several (see the example of Figure 9.2) or zero occurrences of the same hypothesis; *all* occurrences marked by the appropriate number are discharged. A given formula may be used several

$$\frac{P \qquad Q}{P \wedge Q}\wedge_i \qquad\qquad \frac{P \wedge Q}{P}\wedge_{e1} \qquad \frac{P \wedge Q}{Q}\wedge_{e2}$$

$$\frac{\begin{array}{c}\overbrace{P}^{(n)}\\ \vdots \\ Q\end{array}}{P \Rightarrow Q}\Rightarrow_{i(n)} \qquad\qquad \frac{P \Rightarrow Q \qquad P}{Q}\Rightarrow_e$$

$$\frac{\bot}{P}\bot_e$$

$$\frac{P}{\forall x\, P}\forall_i \qquad\qquad \frac{\forall x P}{[x := t]P}\forall_e$$

$$\frac{P}{P \vee Q}\vee_{i1} \qquad \frac{Q}{P \vee Q}\vee_{i2} \qquad\qquad \frac{P \vee Q \qquad \begin{array}{c}\overbrace{P}^{(m)}\\ \vdots \\ R\end{array} \qquad \begin{array}{c}\overbrace{Q}^{(n)}\\ \vdots \\ R\end{array}}{R}\vee_{e(m,n)}$$

$$\frac{[x := t]P}{\exists x\, P}\exists_i \qquad\qquad \frac{\exists x\, P \qquad \begin{array}{c}\overbrace{P}^{(n)}\\ \vdots \\ Q\end{array}}{Q}\exists_{e(n)}$$

Figure 9.1: The system NJ of Gentzen

times as a hypothesis and then have several occurrences. These occurrences may be marked by the same number or by different numbers. In the latter case, they will be discharged on different logical steps.

In order to apply the rule $\forall_i$, it is necessary that no hypothesis where x is free is left, as we have seen above: such a hypothesis would constrain x, while we want x to be arbitrary! By a similar reasoning, a side condition for applying the rule $\exists_e$ is that in all hypotheses except P, x cannot occur free.

The symbol $\bot$ denotes here the *absurd*, like **f** in § 5.1.2, and not the undefined value we introduced on page 88 for 3-valued logics. Of course, there is no introduction rule for $\bot$. In order to use this constant, we can consider it as a hypothesis. For example, Figure 9.3 contains a proof of $P \Rightarrow ((P \Rightarrow \bot) \Rightarrow \bot)$.

The negation $\neg P$ is not a primitive concept in natural deduction, it is considered as an abbreviation for $P \Rightarrow \bot$. For example, we have a proof of $P \Rightarrow \neg\neg P$ in Figure 9.3. Similarly, $P \Leftrightarrow Q$ is considered as an abbreviation for $(P \Rightarrow Q) \wedge (Q \Rightarrow P)$.

$$\frac{\dfrac{\overbrace{P \wedge Q}^{(1)}}{Q} \wedge_{e2} \quad \dfrac{\overbrace{P \wedge Q}^{(1)}}{P} \wedge_{e1}}{\dfrac{Q \wedge P}{P \wedge Q \Rightarrow Q \wedge P} \Rightarrow_{i(1)}} \wedge_i$$

Figure 9.2: Commutativity of conjunction

$$\frac{\dfrac{\overbrace{P}^{(1)} \quad \overbrace{P \Rightarrow \bot}^{(2)}}{\bot} \Rightarrow_e}{\dfrac{(P \Rightarrow \bot) \Rightarrow \bot}{P \Rightarrow ((P \Rightarrow \bot) \Rightarrow \bot)} \Rightarrow_{i(1)}} \Rightarrow_{i(2)} \qquad \frac{\dfrac{\overbrace{P}^{(1)} \quad \overbrace{\neg P}^{(2)}}{\bot} \Rightarrow_e}{\dfrac{(\neg P) \Rightarrow \bot}{P \Rightarrow \neg\neg P} \Rightarrow_{i(1)}} \Rightarrow_{i(2)}$$

Figure 9.3: Introduction of a double negation

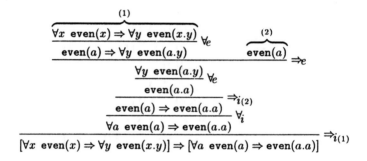

Figure 9.4: Example of even numbers

9.2.2.1 Formalized Examples. Figure 9.2 presents a half of the proof of commutativity of $\wedge$, while Figure 9.3 etablishes $P \Rightarrow \neg\neg P$. The example of even numbers is formalized in Figure 9.4.

These proof trees can be read in two ways. The easiest is from the top to the bottom. Reading a proof in this direction corresponds to the way semi-formal proofs are usually presented. The explanation given above for even numbers is an example of this kind. The reader has just to check that all steps are correct. Figure 9.2 could then read: assume $P \wedge Q$; using $\wedge_e$ twice, we deduce Q on the one hand and P on the other, hence $Q \wedge P$ by $\wedge_i$; we conclude $P \wedge Q \Rightarrow Q \wedge P$ by $\Rightarrow_i$.

In contrast, when we want to *construct* a proof tree, it is generally easier to start from its root. One is then constantly guided by the shape of the current goal. Thus, in order to prove $P \wedge Q \Rightarrow Q \wedge P$ we have to prove $Q \wedge P$, that is, Q and P separately, from the hypothesis $P \wedge Q$. We will see below (§ 9.2.5) how this strategy is supported by software tools.

Let us revisit the total correctness of the linear search algorithm, which was proved on page 25. One of the properties of the loop variant was based on the fact that $x \leq N$ was a loop invariant:

"We still have to show that the property $v \geq 0$ (...) $x \leq N$ [is left invariant]. At the beginning of an iteration step, we have necessarily $\neg P(x)$ which yields $x \neq N$, since N satisfies $P(N)$; hence $x \leq N$ boils down to $x < N$; after the assignment x:=x+1, this yields $x \leq N$ as expected, since N and x are integers."

The property to be proved can be formulated as follows

$$P(N) \wedge \neg P(x) \wedge x \leq N \Rightarrow x+1 \leq N . \tag{9.9}$$

It is necessary to make precise the theory we work with. An option would be to consider a theory of relative integers, but in order to avoid the introduction of additional material, let us keep Peano arithmetic. Thus we just need a constant N, a predicate symbol P and we assume P(N). The expression $x+1$ is represented by $S(x)$; as in § 5.3.2.2, $x \leq y$ is defined as $x < S(y)$. Our goal is then to prove, under the hypothesis P(N):

$$\neg P(x) \wedge x < S(N) \Rightarrow S(x) < S(N) . \tag{9.10}$$

We will use the following axiom for equality:

$$x = N \Rightarrow [P(N) \Rightarrow P(x)] . \tag{9.11}$$

On this example we will construct the proof tree in the bottom-up direction. If we look at the shape of the goal (9.10), a natural strategy is to attempt to prove $S(x) < S(N)$ under the additional hypothesis $\neg P(x) \wedge x < S(N)$. The current goal boils down to $x < N$ if we admit that we have the following lemma:

$$\forall x \forall y \; x < y \Rightarrow S(x) < S(y) . \tag{9.12}$$

This lemma is available on any decent proof tool, however a formal proof is given below. At this stage we have the following partial tree:

$$
\frac{
\dfrac{\dfrac{\dfrac{\forall x \forall y \; x < y \Rightarrow S(x) < S(y)}{\forall y \; x < y \Rightarrow S(x) < S(y)} \;_{\forall e}}{x < N \Rightarrow S(x) < S(N)} \;_{\forall e} \qquad \dfrac{\overbrace{\neg P(x) \wedge x < S(N)}^{(1)} \qquad \left. \begin{array}{c} \vdots \\ \end{array} \right\} \text{to be provided}}{x < N}
}{
\dfrac{S(x) < S(N)}{\neg P(x) \wedge x < S(N) \Rightarrow S(x) < S(N)} \;_{\Rightarrow i(1)}
} \;_{\Rightarrow e}
$$

We still have to show $x < N$ under the hypotheses $P(N)$ and $\neg P(x) \wedge x < S(N)$. As the goal is atomic, we now proceed from the top to the bottom: let us split the second hypothesis into $\neg P(x)$ and $x < S(N)$. Both of them are atomic as well, but a Peano axiom (see § 5.3.2.1 on page 85) happens to state that $x < S(N)$ implies $x < N \lor x = N$. Eliminating $\lor$ allows us to consider the two cases $x < N$ and $x = N$ separately. The branch to be constructed then has the following shape:

$$\cfrac{\dfrac{x < S(N)}{x < N \lor x = N} \Rightarrow_{e<} \qquad \overbrace{x < N}^{(2)} \qquad \overbrace{x = N}^{(3)}}{\begin{array}{ccc} & \vdots & \vdots \\ & x < N & x < N \end{array}} \; \lor_{e(2,3)} \quad \cdot$$

Proving $x < N$ from $x < N$ is trivial. We are left with proving $x < N$ from $x = N$... and the additional hypotheses $\neg P(x)$ and $P(N)$, which yield the absurd thanks to the equality axiom:

$$\cfrac{\neg P(x) \qquad \cfrac{\overbrace{x = N}^{(3)} \quad P(N)}{\cfrac{P(x)}{}}\Rightarrow_{e=}}{\cfrac{\bot}{x < N}\bot_e} \Rightarrow_{e} \quad \cdot$$

The notation $\Rightarrow_{e=}$ is an abbreviation for two consecutive eliminations of $\Rightarrow$ from the equality axiom (9.11). We proceed in a similar manner for the comparison axiom. Note that constructing a proof using additional axioms does not raise any special difficulty. This amounts to working under the hypothesis that these axioms are satisfied. The proof of $x < N$ is then:

$$\cfrac{\cfrac{\overbrace{\neg P(x) \land x < S(N)}^{(1)}}{\dfrac{x < S(N)}{x < N \lor x = N}\Rightarrow_{e<}}\land_{e2} \quad \overbrace{x < N}^{(2)} \quad \cfrac{\cfrac{\overbrace{\neg P(x) \land x < S(N)}^{(1)}}{\neg P(x)}\land_{e1} \quad \cfrac{\overbrace{x = N}^{(3)} \; P(N)}{P(x)}\Rightarrow_{e=}}{\dfrac{\bot}{x < N}\bot_e}\Rightarrow_e}{x < N}\lor_{e(2,3)} \quad \cdot$$

9.2.2.2 An Arithmetical Example. In order to illustrate how we can formalize reasoning by induction, let us prove the property (9.12) we used earlier:

$$\forall x \, \forall y \; x < y \Rightarrow S(x) < S(y) \; . \tag{9.12}$$

This formula is proved by induction on y. It means that we prove:

$$\forall y \; x < y \Rightarrow S(x) < S(y) \tag{9.13}$$

using the axiom of induction (5.7) on page 86 that we recall here:

$$(x<0 \Rightarrow S(x)<S(0)) \quad \wedge$$
$$(\forall y \ (x<y \Rightarrow S(x)<S(y)) \Rightarrow [x<S(y) \Rightarrow S(x)<S(S(y))]) \qquad (9.14)$$
$$\Rightarrow \forall y \ x<y \Rightarrow S(x)<S(y) \ .$$

We get (9.12) from (9.13) by applying the rule $\forall_i$. All hypotheses on x must be removed. The formula (9.13) is a trivial consequence of (9.14) as soon as:

$$x<0 \Rightarrow S(x)<S(0) \qquad \text{and} \qquad (9.15)$$
$$\forall y \ (x<y \Rightarrow S(x)<S(y)) \Rightarrow [x<S(y) \Rightarrow S(x)<S(S(y))] \qquad (9.16)$$

are proved. We entrust the reader with the task of checking it by means of a sufficiently wide sheet of paper (hint: use $\Rightarrow_e$, $\wedge_{e1}$ and $\wedge_{e2}$).

We will need Peano axioms concerning $<$. They were given on page 85, but we recall them here:

$$\forall x \ \neg(x<0) \ , \qquad (9.17)$$
$$\forall x \forall y \ x<S(y) \Leftrightarrow x<y \vee x=y \ . \qquad (9.18)$$

It is easy to prove (9.15) by reducing it to the absurd and using (9.17).

$$\cfrac{\cfrac{\cfrac{\dfrac{\forall x \ \neg(x<0)}{\neg(x<0)}\forall_e \qquad \overbrace{x<0}^{(1)}}{\bot}\Rightarrow_e}{S(x)<S(0)}\bot_e}{x<0 \Rightarrow S(x)<S(0)}\Rightarrow_{i(1)} \qquad .$$

Proving (9.16) boils down to proving $S(x) < S(S(y))$ from $x < y \Rightarrow S(x) < S(y)$ — this is the induction hypothesis — and from $x < S(y)$:

$$\cfrac{\cfrac{\cfrac{\cfrac{\overbrace{x<y \Rightarrow S(x)<S(y)}^{(2)} \qquad \overbrace{x<S(y)}^{(3)}}{\begin{array}{c}\vdots\end{array}\Big\} \text{ to be provided}}{S(x)<S(S(y))}}{x<S(y) \Rightarrow S(x)<S(S(y))}\Rightarrow_{i(3)}}{(x<y \Rightarrow S(x)<S(y)) \Rightarrow [x<S(y) \Rightarrow S(x)<S(S(y))]}\Rightarrow_{i(2)}}{\forall y \ (x<y \Rightarrow S(x)<S(y)) \Rightarrow [x<S(y) \Rightarrow S(x)<S(S(y))]}\forall_i \qquad .$$

The second comparison axiom (9.18) tells us that, in order to prove $S(x) < S(S(y))$, it is enough to prove:

$$S(x)<S(y) \vee S(x)=S(y) \ . \qquad (9.19)$$

On the other hand, the same axiom yields $x<y \vee x=y$ from $x<S(y)$, which allows us to reason on two cases. When $x<y$ the induction hypothesis allows us

to conclude $S(x) < S(y)$. In the second case $(x = y)$ an equality axiom provides $S(x) = S(y)$. In order to simplify the tree we replace (9.19) with $S(x) \leq S(y)$ on two occurrences. This is a harmless presentation trick.

$$
\cfrac{
  \cfrac{
    \overbrace{x < S(y)}^{(3)}
  }{x < y \vee x = y} \Rightarrow_{e<}
  \quad
  \cfrac{
    \cfrac{
      \overbrace{x < y \Rightarrow S(x) < S(y)}^{(2)} \quad \overbrace{x < y}^{(4)}
    }{S(x) < S(y)} \Rightarrow_e
  }{S(x) \leq S(y)} \vee_{i1}
  \quad
  \cfrac{
    \cfrac{
      \overbrace{x = y}^{(5)}
    }{S(x) = S(y)} \Rightarrow_{e=}
  }{S(x) \leq S(y)} \vee_{i2}
}{
  \cfrac{S(x) < S(y) \vee S(x) = S(y)}{S(x) < S(S(y))} \Rightarrow_{e<}
} \vee_{e(4,5)}
$$

9.2.2.3 Some Remarks About Axioms. In Hilbert systems, the meaning of logical connectors is encoded in ad hoc axioms, so that one could get the impression that logic is just a somewhat arbitrary game of symbols [GLT89]. In contrast, natural deduction embeds the meaning of logical connectors in inference rules corresponding to regular reasoning. This makes the latter approach much more satisfactory.

The symmetry introduction-elimination we have for each connector is reminiscent of the relation constructor-destructor of algebraic abstract data types. It turns out to be very important in the development of the theory, especially for its relationship with type systems and λ-calculus. We will revisit it in Chapter 11.

Finally, let us remark that though NJ does not include any axiom on logical connectors, nothing prevents us from introducing axioms about non-logical symbols. In our examples axioms about equality and arithmetic are employed.

9.2.3 Toward Classical Logic

Something is missing in the system NJ: one cannot prove all tautologies in it! To this effect it is necessary to add the law of excluded middle, or, equivalently, an elimination rule for double negations:

$$
\cfrac{}{P \vee \neg P} \text{ EM}
\qquad\qquad
\cfrac{\neg\neg P}{P} \neg\neg e \quad .
$$

The system we get is called NK, it is complete (cf. § 9.8) for first order classical logic. In fact, the system NJ represents exactly *intuitionistic logic*, a logic we already talked about on page 42.

Adding a law such as EM is entirely compatible with usual reasoning. Combined with $\vee_e$, one gets the form "if P entails Q and $\neg P$ entails Q as well, then Q is proved". But EM (as well as $\neg\neg e$) breaks the symmetry and the cohesion of the system, and then complicates the study of NK. Therefore, Gentzen introduced another system which is perfectly symmetrical and is much more satisfactory for classical logic: the sequent calculus.

However, it is important not to confuse this calculus with natural deduction presented with sequents. This way of presenting natural deduction is sometimes the most convenient. In passing, note that a formalism inspired from natural deduction, which aims at defining the semantics of programming languages, and is therefore called natural semantics [Kah87], is usually presented with sequents.

9.2.4 Natural Deduction Presented by Sequents

A **sequent** is an ordered pair composed of a finite sequence of formulas Γ and of a formula P, noted $\Gamma \vdash P$. Such a sequent represents the judgement "P is derivable under the hypotheses of Γ".

$\Gamma \vdash P$ can be seen as a deduction tree of the previous presentation, where we keep only the leaves (non-discharged hypotheses) and the root (the conclusion). Everything goes on as if one takes a snapshot of the simplified deduction tree at each step, and then displays these snapshots along a tree.

The sequence of formulas Γ may include different occurrences of a hypothesis H, so that H can be discharged at different stages. Two contexts Γ_1 and Γ_2 which are identical up to the order of formulas they contain can be considered as equivalent. In other word, contexts can be considered as multisets rather than sequences.

Examples. The simplest proof one can construct in natural deduction is the derivation of P under the hypothesis P. With sequents, we get the judgement $P \vdash P$, which is an axiom in this presentation. More generally, axioms are all sequents having the shape $\Gamma \vdash P$ where P is a member of Γ. (In the framework of NK, one has to add the excluded middle or the elimination of double negations.) Inference rules indicate how we go from a sequent to the next. For instance, Figure 9.5 gives the rules about conjunction and implication. Observe that every formula in the context Γ corresponds to a bundle of hypotheses to be discharged simultaneously; we no longer need to use a mark which links a bundle to the step where it is discharged; this is the main advantage of this presentation of natural deduction. The two styles can be compared in Figure 9.6.

$$\frac{\Gamma \vdash P \quad \Gamma \vdash Q}{\Gamma \vdash P \wedge Q} \wedge_i \qquad \frac{\Gamma \vdash P \wedge Q}{\Gamma \vdash P} \wedge_{e1} \qquad \frac{\Gamma \vdash P \wedge Q}{\Gamma \vdash Q} \wedge_{e2}$$

$$\frac{\Gamma, P \vdash Q}{\Gamma \vdash P \Rightarrow Q} \Rightarrow_i \qquad \frac{\Gamma \vdash P \Rightarrow Q \quad \Gamma \vdash P}{\Gamma \vdash Q} \Rightarrow_e$$

Figure 9.5: Rules of NJ presented with sequents

$$\cfrac{\cfrac{\overbrace{P \wedge Q}^{(1)}}{Q} {\scriptstyle \wedge e2} \qquad \cfrac{\overbrace{P \wedge Q}^{(1)}}{P} {\scriptstyle \wedge e1}}{\cfrac{Q \wedge P}{P \wedge Q \Rightarrow Q \wedge P} {\scriptstyle \Rightarrow i(1)}} {\scriptstyle \wedge i} \qquad\qquad \cfrac{\cfrac{P \wedge Q \vdash P \wedge Q}{P \wedge Q \vdash Q} {\scriptstyle \wedge e2} \qquad \cfrac{P \wedge Q \vdash P \wedge Q}{P \wedge Q \vdash P} {\scriptstyle \wedge e1}}{\cfrac{P \wedge Q \vdash Q \wedge P}{\vdash P \wedge Q \Rightarrow Q \wedge P} {\scriptstyle \Rightarrow i}} {\scriptstyle \wedge i}$$

Figure 9.6: Commutativity of conjunction (2 styles)

9.2.5 Natural Deduction in Practice

Searching a formal proof is much easier in natural deduction than in a Hilbert system. Most deduction steps are guided by the structure of the formula to be proved. But the size of formal proofs remains large. Moreover, when we write everything explicitly, we see that a given subformula has to be written several or many times. Using such techniques by hand quickly becomes tedious — then error prone! — for realistic proofs.

However, natural deduction is well suited to interactive automated proof assistants. For example, it is used in HOL and Coq. At each stage, the current sequent is displayed, then the user calls a deduction rule and a new sequent or set of sequents is displayed. In practice, it is generally better to specify a combination of deduction rules by means of a language of tactics.

Let us see how the example of Figure 9.6 is proved with Coq. We introduce the goal

$$P \wedge Q \Rightarrow Q \wedge P \quad .$$

As this goal has the shape $A \Rightarrow B$, we naturally try the rule $\Rightarrow_i$. This is implemented by the tactic called Intro. A new hypothesis $P \wedge Q$ will be generated and we can provide its name, say h1, as a parameter of Intro. The system then displays the sequent h1:$P \wedge Q \vdash Q \wedge P$. In order to prove the conjunction $Q \wedge P$ we try the rule $\wedge_i$, which is called Split. Two subgoals are generated, the first displayed by Coq is h1:$P \wedge Q \vdash Q$. We then want to use the hypothesis h1 by eliminating its main connector. To this effect we use the tactic Elim, with h1 as a (mandatory) parameter. The second subgoal is solved in the same way.

Several steps can be put together into a sequence of tactics, which is written in our example:

```
Intro h1; Split; Elim h1.
```

Note that fully automatic tactics can also be used for such simple formulas.

Each basic tactic represents a deduction rule, but in the general case, a tactic just states instructions aiming at carrying on the construction of the proof tree. In some cases, an automated tactic can elaborate a full branch. Thus the real proof we obtain is not the visible script of tactics (which is also the thing one edits and keeps in a file), but the internal proof tree, that is built by the system.[2] This object is checked by a very small kernel, which has only one task: inspecting whether, or not, rules are correctly applied. This technology allows one to design and implement proof assistants which are both reliable and open. We go back to this point in § 12.5.

Note in passing that the size of scripts is generally smaller than the size of the corresponding proof trees. To give a rough idea, here is a detailed script (shorter ones can be found, but they use advanced features) for the proof given above for (9.10), under the assumption P(N).

$$\neg P(x) \,\wedge\, x < S(N) \,\Rightarrow\, S(x) < S(N) \ . \tag{9.10}$$

```
Intro h1; Apply succ_monot.
Elim h1; Intros h1l h1r;
Case ax_comp2 with h1r introducing h2 h3.
      Trivial.
      Elim h1l; Rewrite h3; Assumption.
```

The first line introduces $\neg P(x) \wedge x < S(N)$ as a hypothesis named h1 and then applies a lemma named succ_monot, which states that the successor function is monotonic (9.12). This yields the new subgoal $x < N$. In the second line, h1 is split into $\neg P(x)$ and $x < S(N)$, respectively called h1l and h1r. Then we reason on the two cases we get when we apply ax_comp2 (the second comparison axiom (9.18)) to h1r. The fourth line corresponds to the trivial case $x < N$ (h2 is automatically used behind the scene). The last line proceeds by elimination of the conclusion of h1l which is $\bot$; we are left with the subgoal $P(x)$ which boils down to the hypothesis P(N), thanks to the equality h3.

9.3 The Sequent Calculus

In natural deduction, the concept of a theorem becomes of secondary importance with relation to the deducibility relation. This is still more true with sequent calculus. The main difference between the intuitionistic sequent calculus (called LJ by Gentzen) and natural deduction (NJ) is the replacement of elimination rules, governing how the main connector of the conclusion can be eliminated, with **left introduction** rules, governing what can be deduced from a compound hypothesis, given what is deduced from its components. On the other hand, classical logic (LK) is no longer obtained by the introduction of an ad hoc axiom, but by using an entirely symmetric concept of a sequent. A **classical sequent** is a couple of two finite sequences of formulas Γ and Δ, noted $\Gamma \vdash \Delta$. As we did before, we agree that sequences which are the same up to a permutation are considered as identical. Intuitively, the sequent $\Gamma \vdash \Delta$ can read: "the *conjunction* of hypotheses contained in Γ entails the *disjunction* of formulas contained in Δ". For example, $A, B \vdash C, D$ is similar to $A \wedge B \Rightarrow C \vee D$.

---

[2]By the way, it is possible to print the tree in natural language [TBK92].

It is more natural to start the study of sequent calculus with the classical system LK. We get the intuitionistic calculus LJ from LK by confining derivations within the space of **intuitionistic sequents**, which are sequents where the right part has at most one formula.

Another difference between natural deduction and sequent calculus stands in the status of negation, which is no longer an abbreviation built upon $\Rightarrow$ and $\perp$, but a plain connector: it is even the vault key of the symmetry of the system. Indeed, a formula can pass from one sequent side to the other by means of a negation: see the introduction rules for $\neg$ in Figure 9.9.

9.3.1 The Rules of the Sequent Calculus

The rules of LK can be divided into three groups: a group of structural rules (Figure 9.7), a group on identity (Figure 9.8) and a group of logical rules (Figure 9.9). These rules can all be read top down (if the premises are good,[3] so is the conclusion), or bottom up (searching to prove the conclusion reduces to searching a proof of the premises).

The **structural rules** tell us something about the structure of sequents and not about the structure of formulas. They define how the stock of hypotheses and conclusions is handled. Although no logical connector is involved in these rules, essential properties of the logic they formalize are determined by them [GLT89]. **Thinning** (or **weakening**) **rules** allow us to introduce "useless" formulas and to consider as axioms only sequents in the form $P \vdash P$. **Contraction rules**, when read bottom up, allow us to repeat a formula that may be used in several ways. They correspond to the building of packets of hypotheses in natural deduction (when occurrences of several hypotheses are gathered, that is, identified by the same number in our first presentation in § 9.2.2).

$$\frac{\Gamma \vdash \Delta}{P, \Gamma \vdash \Delta} \text{ aff}_l \qquad \frac{\Gamma \vdash \Delta}{\Gamma \vdash \Delta, Q} \text{ aff}_r$$

$$\frac{P, P, \Gamma \vdash \Delta}{P, \Gamma \vdash \Delta} \text{ ctr}_l \qquad \frac{\Gamma \vdash \Delta, Q, Q}{\Gamma \vdash \Delta, Q} \text{ ctr}_r$$

Figure 9.7: Structural rules of LK

The **identity group** consists of two rules: the **axiom** $A \vdash A$, where we can without loss of generality restrict ourselves to the cases where A is atomic, and the **cut rule** which formalizes the usual concept of a lemma. Everybody can intuitively convince themselves that the cut rule is sound when Δ is empty: P

---

[3]Understand: *provable* or *valid*; the first alternative remains valuable in the case of LJ.

plays the role of a lemma derived from Γ, the "consequence" Δ' can then be deduced from Γ and Γ'. The general case where Δ is non-empty boils down to this special case if one considers that formulas of Δ can be freely transferred to the left-hand side, then put back to the right-hand side.

$$\frac{}{A \vdash A}\ \text{ax} \qquad \frac{\Gamma \vdash \Delta, P \qquad P, \Gamma' \vdash \Delta'}{\Gamma, \Gamma' \vdash \Delta, \Delta'}\ \text{cut}$$

Figure 9.8: Identity group of LK

The price to pay for each double transfer is a double negation (see the logical rules) which costs nothing in classical logic. The problem is not raised in intuitionistic logic since Δ is necessarily empty. Moreover the previous reasoning can be made symmetrical for LK: let us make Γ' empty in a similar way, $P \vdash \Delta'$ expresses that Δ' refutes P (this is, as we could say, an anti-lemma) and $\Gamma \vdash P$ expresses that P refutes Γ. Let us also remark that, if we regard Γ and Δ' as formulas, the cut rule states that $\vdash$ is a transitive relation.

Most **logical rules** ($\wedge$, $\vee_{D_1}$, $\vee_{D_2}$, $\Rightarrow$, $\forall$ and $\exists_r$ of Figure 9.9) are constructed by analogy with intuitionistic natural deduction. Rules $\vee_l$ and $\exists_l$ are constructed by duality with $\wedge_r$ and $\forall_r$. As a result, we get a kind of left/right symmetry for each connector on the one hand, and a duality between $\wedge$ (respectively $\forall$) and $\vee$ (respectively $\exists$) on the other.

At first sight, one may wonder that $\Rightarrow_l$ distinguishes two contexts $\Gamma \vdash \Delta$ and $\Gamma' \vdash \Delta'$, and then does not seem to be reducible to a combination of $\vee_l$ and $\neg_l$. It is actually possible to identify $\Gamma' = \Gamma$ and $\Delta' = \Delta$ in LK; this variant is discussed below (Figure 9.11). The version presented here is compatible with the intuitionistic case: as in all rules where Δ comes with an additional formula, it is enough to impose that Δ is empty. Δ' consists of at most one formula.

The self-duality of negation expressed in $\neg_l$ and $\neg_r$ provides an interpretation of sequents in terms of refutation and of proof. Proving a sequent is, depending on one's preference, to refute a formula on the left-hand side or, to prove a formula on the right-hand side, in the context made of the remainder of the sequent. In other words, if we have a sequent $P, \Gamma \vdash \Delta, Q$, we can equally well say that we prove Q in the context $P, \Gamma \vdash \Delta$, or that we refute P in the context $\Gamma \vdash \Delta, Q$.

9.3.2 Examples

In order to illustrate a number of the previous rules, we give in Figure 9.10 the proof of the excluded middle law (note the use of a contraction on the right), and the example of even numbers already presented in natural deduction.

$$\frac{\Gamma \vdash \Delta, P}{\neg P, \Gamma \vdash \Delta} \neg_l \qquad\qquad \frac{P, \Gamma \vdash \Delta}{\Gamma \vdash \Delta, \neg P} \neg_r$$

$$\frac{P, \Gamma \vdash \Delta}{P \wedge Q, \Gamma \vdash \Delta} \wedge_{l_1} \quad \frac{Q, \Gamma \vdash \Delta}{P \wedge Q, \Gamma \vdash \Delta} \wedge_{l_2} \qquad \frac{\Gamma \vdash \Delta, P \quad \Gamma \vdash \Delta, Q}{\Gamma \vdash \Delta, P \wedge Q} \wedge_r$$

$$\frac{P, \Gamma \vdash \Delta \quad Q, \Gamma \vdash \Delta}{P \vee Q, \Gamma \vdash \Delta} \vee_l \qquad \frac{\Gamma \vdash \Delta, P}{\Gamma \vdash \Delta, P \vee Q} \vee_{r_1} \quad \frac{\Gamma \vdash \Delta, Q}{\Gamma \vdash \Delta, P \vee Q} \vee_{r_2}$$

$$\frac{\Gamma \vdash \Delta, P \quad Q, \Gamma' \vdash \Delta'}{P \Rightarrow Q, \Gamma, \Gamma' \vdash \Delta, \Delta'} \Rightarrow_l \qquad \frac{P, \Gamma \vdash \Delta, Q}{\Gamma \vdash \Delta, P \Rightarrow Q} \Rightarrow_r$$

$$\frac{[x := t]P, \Gamma \vdash \Delta}{\forall x P, \Gamma \vdash \Delta} \forall_l \qquad \frac{\Gamma \vdash \Delta, P}{\Gamma \vdash \Delta, \forall x P} \forall_r *$$

$$\frac{P, \Gamma \vdash \Delta}{\exists x P, \Gamma \vdash \Delta} \exists_l * \qquad \frac{\Gamma \vdash \Delta, [x := t]P}{\Gamma \vdash \Delta, \exists x P} \exists_r$$

Rules $\forall_r$ and $\exists_l$ must respect the restriction already discussed in NJ: x cannot possess free occurrences in the context (that is, in Γ or in Δ).

Figure 9.9: Logical rules of LK

9.3.3 Cut Elimination

The major theorem of sequent calculus is:

Theorem 9.2 (Gentzen's Hauptsatz)
Every provable sequent can be proved without the cut rule.

The proof of Gentzen is constructive: it provides an algorithm for eliminating cuts. This theorem is interesting because cut-free proofs enjoy properties which are not satisfied in the general case. One of the most important is the **subformula property**: all formulas which occur in a cut-free proof are subformulas of the formula (or of the sequent) to be proved. This is clear because no rule but the cut rule has a formula (P) in its premises which does not occur in its conclusion ($\Gamma, \Gamma' \vdash \Delta, \Delta'$).

As the cut rule is redundant, what is the point of introducing it? Indeed:

- the cut rule is useful in practice because, combined with contractions, it allows one to factorize inferences; note that contractions are used in an essential way when a given quantified formula is instantiated on several places of the same proof;
- the cut rule turns out to be very convenient in the development of the theory. For example, one may want to inverse logical rules. Consider $\wedge_r$: if $\Gamma \vdash \Delta, P \wedge Q$ is derivable, we would like to infer that $\Gamma \vdash \Delta, P$ is derivable (and similarly for $\Gamma \vdash \Delta, Q$). Indeed, we can easily derive the sequent

$$\dfrac{\overline{\rule{2cm}{0pt}}\ \textbf{ax}}{\dfrac{P \vdash P}{\dfrac{\vdash P, \neg P}{\dfrac{\vdash \neg P \vee P, \neg P}{\dfrac{\vdash \neg P \vee P, \neg P \vee P}{\vdash \neg P \vee P}\ \text{ctr}_r}\ \vee_{r1}}\ \vee_{r2}}\ \neg_r}$$

$$\dfrac{\dfrac{\overline{\rule{3cm}{0pt}}\ \textbf{ax}}{\text{even}(a) \vdash \text{even}(a)} \quad \dfrac{\dfrac{\dfrac{\overline{\rule{4cm}{0pt}}\ \textbf{ax}}{\text{even}(a.a) \vdash \text{even}(a.a)}}{\forall y\ \text{even}(a.y) \vdash \text{even}(a.a)}\ \forall_l}{\dfrac{\text{even}(a) \Rightarrow \forall y\ \text{even}(a.y), \text{even}(a) \vdash \text{even}(a.a)}{\dfrac{\forall x\ \text{even}(x) \Rightarrow \forall y\ \text{even}(x.y), \text{even}(a) \vdash \text{even}(a.a)}{\dfrac{\forall x\ \text{even}(x) \Rightarrow \forall y\ \text{even}(x.y) \vdash \text{even}(a) \Rightarrow \text{even}(a.a)}{\dfrac{\forall x\ \text{even}(x) \Rightarrow \forall y\ \text{even}(x.y) \vdash \forall a\ \text{even}(a) \Rightarrow \text{even}(a.a)}{\vdash [\forall x\ \text{even}(x) \Rightarrow \forall y\ \text{even}(x.y)] \Rightarrow [\forall a\ \text{even}(a) \Rightarrow \text{even}(a.a)]}\ \Rightarrow_r}\ \forall_r}\ \Rightarrow_r}\ \forall_l}\ \Rightarrow_l}}{}$$

Figure 9.10: Proof examples using LK

$P \wedge Q \vdash P$, then we get the desired result using a cut on $P \wedge Q$:

$$\dfrac{\Gamma \vdash \Delta, P \wedge Q \quad \dfrac{P \vdash P}{P \wedge Q \vdash P}\ \wedge_{l1}}{\Gamma \vdash \Delta, P}\ \text{cut} \quad ;$$

– one may also consider proofs making use of proper axioms. For example, the two first axioms of Peano can be represented by the sequents:

$$0 = S(x) \vdash \quad ,$$
$$S(x) = S(y) \vdash x = y \ .$$

Gentzen's theorem is generalized as follows: all cuts can be eliminated except the ones where a proper axiom is used.

The dynamics of the cut elimination process is fairly complex. The idea of the algorithm is to make cuts going upwards to the leaves of the derivation tree. Each lemma in the form $\forall x P$ is potentially usable in an infinite number of instances, so it is *a priori* not obvious that the process terminates.

True eliminations occur in the case of a cut with an axiom. Propagating a cut coming from logical inferences may have the effect that the number of cuts increases, but as a compensation, new cuts are about subformulas. Here is an example in order to illustrate this phenomenon.

$$\frac{\Gamma, A \vdash B}{\Gamma \vdash A \Rightarrow B} \Rightarrow_r \qquad \frac{B \vdash B \qquad \Gamma' \vdash A}{\Gamma', A \Rightarrow B \vdash B} \Rightarrow_l$$
$$\frac{}{\Gamma, \Gamma' \vdash B} \text{cut} \quad .$$

On the next step, the cut on $A \Rightarrow B$ is replaced with two "smaller" cuts, one on A and the other on B:

$$\frac{\dfrac{\Gamma' \vdash A \qquad \Gamma, A \vdash B}{\Gamma, \Gamma' \vdash B} \text{cut} \qquad B \vdash B}{\Gamma, \Gamma' \vdash B} \text{cut} \quad .$$

The second cut is on an axiom, it is immediately eliminated:

$$\frac{\Gamma' \vdash A \qquad \Gamma, A \vdash B}{\Gamma, \Gamma' \vdash B} \text{cut} \quad .$$

The most dangerous cuts are the ones which occur immediately after a contraction, because propagating them entails a duplication without a straightforward counterpart. During cut elimination, the proof size may increase in a hyperexponential way (it may be $4^{4^{\cdot^{\cdot^{4^h}}}}$, where h is the height of the initial proof and where the iteration number of exponentials depends on the size of cut formulas). This measures the complexity of the elimination process: the algorithm is not supposed to be actually performed on real proofs.

This positive result of Gentzen is very important. Among a number of applications in computer science, it will be seen in Chapter 11 that it lies at the root of recent developments of computational paradigms within a logical framework.

9.4 Applications to Automated Theorem Proving

The deduction systems presented above formalize the concept of a proof: they first aim at *recognizing* a proof. On the other hand, constructing a proof tree is much less simple, at least when we go beyond propositional calculus.

 Recall that a closed formula P is not necessarily always true or always false: its truth value generally depends on its atomic components.

Two well-known techniques implemented in automated proof search tools called "tableaux" and resolution, are traditionally presented from a model-theoretic perspective. The sequent calculus provides another viewpoint based on proof theory.

The first method works on arbitrary formulas. We present here the propositional version. Given a proposition P, we will see, thanks to a systematic decomposition procedure of P, how a counter-example for P or a proof of P can be constructed.

When P is no longer a proposition, but a closed first-order formula, we face an additional difficulty: intuitively, we have to construct witnessing values for individual variables. There are systematic search procedures that eventually yield a proof of P if there is one, while the search for a counter-example does not terminate in the general case: the problem is then only semi-decidable. This will be summarized below (§ 9.8.1).

We will not indicate how to adapt the method of semantical tableaux to first-order logic. However, this will be done for the resolution principle, which works with a restricted set of formulas (restrictions are about the use of connectors) but has as its main interest the computation of witnessing values thanks to the *unification* algorithm.

9.4.1 Sequents and Semantical Tableaux

In order to mechanize the search for a proof of a given sequent, it is necessary that the process of applying rules (with a bottom-up reading) terminates.

Logical rules of LK possess a remarkable property: they decompose each formula into its components, so that the number of used logical connectors decreases. As cut rules are not mandatory, only contraction rules are still problematic. However we can still avoid them in classical propositional logic, thanks to the variants of $\wedge_l$, $\vee_r$ and $\Rightarrow_l$ given in Figure 9.11.

$$\frac{P,Q,\Gamma \vdash \Delta}{P \wedge Q, \Gamma \vdash \Delta} \wedge_l \qquad \frac{\Gamma \vdash \Delta, P, Q}{\Gamma \vdash \Delta, P \vee Q} \vee_r \qquad \frac{\Gamma \vdash \Delta, P \qquad Q, \Gamma \vdash \Delta}{P \Rightarrow Q, \Gamma \vdash \Delta} \Rightarrow_l$$

Figure 9.11: A variant of LK

These rules are equivalent to the rules of Figure 9.9 (one pass from a version to the other using weakenings and contractions), but the new ones have an advantage for automated proof search: if the conclusion is provable, the premise (or the premises) is (are) provable as well. Such rules are said to be revertible or invertible: intuitively, no piece of information is lost when we go from the conclusion to premises. Then we can forget contraction rules without loss of completeness. Remaining rules provide an algorithm for verifying tautologies which is quite simple to implement.

Weakening rules are tried as a last resort: when we get a sequent $\Gamma \vdash \Delta$ only made up of atomic propositions, two cases are possible: either Γ and Δ have a common formula A; in this case $\Gamma \vdash \Delta$ is derivable from the axiom $A \vdash A$ using weakening rules (in practice we don't need to perform these steps, computing the intersection is sufficient); or Γ and Δ are disjoint, then there is no way to prove $\Gamma \vdash \Delta$; as this sequent is needed in order to derive the sequent S we search a proof for, (because of reversibility of the rules we use) we conclude that S is not provable.

The same algorithm can be presented — under a different form — from a model-theoretic perspective, so that we are led to the method of **semantical tableaux**. First the concept of a tautology is extended in the obvious way to sequents, with the analogy between the sequent $A_1, \ldots, A_m \vdash B_1, \ldots, B_n$ and the formula $(A_1 \wedge \ldots \wedge A_m) \Rightarrow (B_1 \vee \ldots \vee B_n)$ in mind: a sequent $\Gamma \vdash \Delta$ is tautological if every interpretation where all propositions of Γ are true satisfies at least one proposition of Δ.

Reciprocally, a counter-example to the latter sequent is provided by any interpretation where all propositions of Γ are given the truth value *true* and all propositions of Δ are given the truth value *false*. The rules of the last variant of LK considered above are such that the conclusion admits a counter-example if and only if one of the premises admits this counter-example, which is another way of stating that the rules are sound and invertible. When we reach a sequent made up only of atomic propositions, we have two cases:

- the two sides of the sequent possess a common proposition A; it is then obvious that the sequent is not semantically refutable, since A cannot simultaneously take the values *true* and *false*;
- the two sides are disjoint, so we immediately get a counter-example.

Thus it can be shown that a formula F is a tautology if and only if no branch of the search tree starting from F reaches a counter-example.

Though this presentation rests on providing truth values to propositions, the semantical tableaux method is very different from the truth table method. The latter becomes less efficient as the number of atomic propositions becomes larger. Actually, only the former method can be generalized to infinite sets of propositions and to first-order logic. An example of an automated tool based on semantical tableaux is $3T^AP$ [HBG94].

9.4.2 From the Cut Rule to Resolution

Since the 1960s, a number of researchers, following Gilmore, Davis and Putnam, and Robinson set out to look for a feasible semi-decision procedure, based on the work done by Jacques Herbrand in the 1930s. The programming language Prolog is generally presented as an application of the *resolution principle* due to Robinson [Rob65].

9.4.2.1 Resolution in the Framework of Propositional Logic. The resolution principle is easy to present from sequent calculus. Let us start with the propositional case. It is well known that, using De Morgan laws and replacing $P \Rightarrow Q$ with $\neg P \vee Q$, every proposition can be put in the form of a conjunction of **clauses**, where a clause is a disjunction of **literals**, and a literal is either an atomic proposition, or the negation of an atomic proposition:

$$\neg A_1 \vee \ldots \vee \neg A_m \vee B_1 \vee \ldots \vee B_n \ .$$

In order to prove a proposition P from a conjunction of clauses $C_1 \ldots C_k$, we first put P in clausal form $P_1 \wedge \ldots \wedge P_l$ in turn. Our problem then boils down to separately proving each clause P_j from $C_1 \ldots C_k$. Thus we can without loss of generality restrict ourselves to reason with clauses only.

Which inference rules can we use on clauses? It happens that *only one is enough*: the resolution rule that, from two clauses $\Gamma \vee R$ and $\neg R \vee \Gamma'$, denoting respectively

$$\neg A_1 \vee \ldots \qquad \vee \ldots \vee \neg A_m \vee B_1 \vee \ldots \vee R \vee \ldots \vee B_n \quad \text{and}$$
$$\neg A_1' \vee \ldots \vee \neg R \vee \ldots \vee \neg A_m' \vee B_1' \vee \ldots \qquad \vee \ldots \vee B_n' \quad ,$$

allows us to deduce the clause $\Gamma \vee \Gamma'$ (the disjunction of all literals of Γ and Γ'). As usual, this can be stated by means of a fraction:

$$\frac{\Gamma \vee R \qquad \neg R \vee \Gamma'}{\Gamma \vee \Gamma'} \quad .$$

The soundness of this rule is easy to explain if we agree that the clause

$$\neg A_1 \vee \ldots \vee \neg A_m \vee B_1 \vee \ldots \vee B_n$$

represents the sequent

$$A_1, \ldots, A_m \vdash B_1, \ldots, B_n \quad :$$

the resolution principle simply corresponds to the cut rule.

We can also understand why the resolution rule is sufficient, thanks to Gentzen's Hauptsatz. If we translate clauses into the language of sequents, we need *a priori* structural rules, the identity group and logical rules of LK. However, the latter are of no use here since our sequents are without a logical connector!

The theorem of cut elimination seems to indicate that the resolution rule is useless as well, but beware: here we want to prove a sequent corresponding to a clause P_j from the sequents corresponding to clauses $C_1 \ldots C_k$, so that the latter are interpreted as *proper axioms*. We know that, in contrast to cuts with logical axioms $A \vdash A$, cuts with proper axioms cannot be eliminated. However, the resolution rule can safely be restricted to the cases where at least one of the premises is among $C_1 \ldots C_k$.

These ideas are explained in more detail in [GLT89]. The reader can also find there a justification for the removal of contraction and weakening rules in the case of Prolog. In the purely logical fragment of Prolog, a program is a set of (first-order) clauses which have at most one positive literal. They are called **Horn clauses**, and they correspond exactly to intuitionistic sequents.

9.4.2.2 Resolution in the Framework of First-order Logic.
In order to illustrate the resolution principle when we have first-order variables, consider the two formulas saying that every human being is mortal and that Socrates is a human being:

$$\forall x \; \texttt{human}(x) \Rightarrow \texttt{mortal}(x) \; ,$$
$$\texttt{human}(\texttt{Socrates}) \; .$$

We can put the first formula in clausal form

$$\neg\texttt{human}(x) \lor \texttt{mortal}(x)$$

where it is implicit that x is universally quantified. In the special case where x is Socrates, this yields:

$$\neg\texttt{human}(\texttt{Socrates}) \lor \texttt{mortal}(\texttt{Socrates}) \; .$$

The resolution rule for propositions can then be applied:

$$\frac{\texttt{human}(\texttt{Socrates}) \qquad \neg\texttt{human}(\texttt{Socrates}) \lor \texttt{mortal}(\texttt{Socrates})}{\texttt{mortal}(\texttt{Socrates})} \; .$$

In fact, the full resolution rule performs the substitution and the simplification in one step:

$$\frac{\texttt{human}(\texttt{Socrates}) \qquad \neg\texttt{human}(x) \lor \texttt{mortal}(x)}{\texttt{mortal}(\texttt{Socrates})} \; .$$

In the general case we have to find a substitution for both premises so that, after performing the substitution, they contain two opposite literals. For example, in the following deduction, we substitute 0 for n in the first premise, $S(m)$ for x and $S(p)$ for y in the second:

$$\frac{\neg(m+n=p) \lor S(m)+n=S(p) \qquad \neg(x+0=y) \lor x=y}{\neg(m+0=p) \lor S(m)=S(p)} \; .$$

The procedure for computing the smallest **unifier** of two terms or of two atomic formulas, which is the most general composition of substitutions making these terms (or these formulas) identical, is called **unification**. (There is actually an equivalence class of unifiers identical up to a renaming of variables.) In the previous example, the smallest unifier we chose is

$$[n := 0] \circ [x := S(m)] \circ [y := S(p)] \; .$$

Unification algorithms examine simultaneously the two terms to be unified, according to their syntactical structure, while adding substitutions when, at the same location, one term has a variable v and the other has either a variable, or a term that does not contain v; however, if we have something impossible, e.g. two different constants at the same location, the algorithm stops and returns a failure.

Explaining resolution using sequents is just as easy when we consider predicates instead of propositions. We consider sequents without quantifier but containing free occurrences of variables. A substitution step uses the rule:

$$\frac{\Gamma \vdash \Delta}{[x := t]\Gamma \vdash [x := t]\Delta} \ ,$$

which is easy to derive from the rules of LK. By repeating such substitutions and then applying a cut rule we get the resolution rule:

$$\frac{\Gamma \vdash \Delta, R \qquad \neg R', \Gamma' \vdash \Delta'}{\sigma\Gamma, \sigma\Gamma' \vdash \sigma\Delta, \sigma\Delta'} \ ,$$

where σ is the most general unifier of R and R'.

The previous examples of deductions read more easily with sequents:

$$\frac{\vdash \text{human}(\text{Socrates}) \qquad \text{human}(x) \vdash \text{mortal}(x)}{\vdash \text{mortal}(\text{Socrates})} \ ,$$

$$\frac{m+n=p \vdash S(m)+n=S(p) \qquad x+0=y \vdash x=y}{m+0=p \vdash S(m)=S(p)} \ .$$

In the following, clauses are noted in the form of sequents.

9.4.2.3 Skolemization. As we consider only quantifier-free sequents, this amounts to agreeing that variables are universally quantified on the whole formula we would get, after all literals are placed on the right-hand side of the sequent. For example, the sequent $A(x) \vdash B(x)$ should be understood as equivalent to $\forall x\, A(x) \Rightarrow B(x)$.

At first sight, the expressive power of first-order logic is weakened by this limitation: using both quantifiers should be allowed at any place in a formula. However, it is possible to put any formula in **prenex** form

$$\mathcal{Q}_1 x_1 \ \ldots \ \mathcal{Q}_n x_n \, M \ ,$$

where $\mathcal{Q}_i$ represents $\forall$ or $\exists$, and where M, called the **matrix**, contains no quantifier.

Existential quantifiers can also be removed by introducing new function symbols, called **Skolem functions**. For example, in the formula

$$\exists x \, \forall y \, \exists z \, P(x, y, z) \ ,$$

x depends on nothing while z depends on y; introducing the constant a and the unary function f we get:

$$\forall y \, P(a, y, f(y)) \ .$$

This process of eliminating existential quantifiers is called **skolemization**, and leads to the **Skolem normal form**. What really justifies this transform is the following theorem.

Theorem 9.3
Let $\{F_1, \ldots F_n\}$ be a set of formulas and let $S_1, \ldots S_n$ be their respective Skolem normal forms, $\{F_1, \ldots F_n\}$ is inconsistent[4] if and only if $\{S_1, \ldots S_n\}$ is inconsistent.

---

[4] A set of formulas is defined to be inconsistent if we can infer the absurd, which is formalized here by the empty sequent. We come back to this concept later (§ 9.8.2).

In practice, this means that, in order to prove that P is a consequence of the clauses $C_1, \ldots C_k$, we will reason by reduction to the absurd: proving that $\neg P$ is impossible. To this effect we put $\neg P$ in skolemized clausal form $P_1, \ldots P_l$. Then we try to deduce the empty clause from $C_1, \ldots C_n, P_1, \ldots P_l$. This search is made much easier thanks to the preliminary process of removing connectors and quantifiers.

For example, we want to prove that there exists a mortal being knowing that Socrates is a human being and that every human is mortal. These two hypotheses are modeled by the sequents

$$\vdash \text{human(Socrates)} \quad \text{and} \quad \text{human}(x) \vdash \text{mortal}(x) \ .$$

Now, skolemizing $\exists x\, \text{mortal}(x)$ would lead us to dead end: $\text{mortal}(a)$, where a is a new constant, cannot be deduced from the two previous sequents. In contrast, if we consider the negation $\neg \exists x\, \text{mortal}(x)$, corresponding to the sequent $\text{mortal}(x) \vdash$, we can deduce the empty sequent from the three previous sequents. Note that we use here a top down strategy for constructing the proof tree, with the idea of confronting axioms with the sequent to be refuted (initially $\text{mortal}(x) \vdash$) in mind.

$$
\frac{\vdash \text{human(Socrates)} \quad \dfrac{\text{human}(x) \vdash \text{mortal}(x) \quad \text{mortal}(x) \vdash}{\text{human}(x) \vdash}}{\vdash} \ .
$$

Let us illustrate the use of skolemization, with a proof that, if there exists a common lower bound to all elements:

$$\exists z\, \forall x \quad z \leq x \tag{9.20}$$

then every element has a lower bound:

$$\forall x\, \exists y \quad y \leq x \ . \tag{9.21}$$

We put (9.20) in normal form. We introduce a Skolem constant m for z:

$$\vdash m \leq x \ . \tag{9.22}$$

Then we consider the normal form of the negation of (9.21), which leads us to introducing a Skolem constant — say n — for x this time (we implicitly exploit the dual identities $\neg \forall u\, P \Leftrightarrow \exists u\, \neg P$ and $\neg \exists u\, P \Leftrightarrow \forall u\, \neg P$):

$$y \leq n \vdash \ . \tag{9.23}$$

The proof itself has only one resolution step, using the unifier $[y := m] \circ [x := n]$:

$$\frac{\vdash m \leq x \quad y \leq n \vdash}{\vdash} \ .$$

9.4.2.4 Uses of the Resolution Principle. The resolution rule is just one part of a full proof search procedure. At each step, we still need to choose a pair of clauses on which the rule should be applied. A number of different strategies are possible, some of them are guaranteed to find a derivation of the empty clause if there is one, in theory. For an exposition of the most important, the reader may consult the book of Chang and Lee [CL73]. The resolution principle is actually used in proof tools for first order logic, e.g. Otter [McC94].

9.4.3 Proofs in Temporal Logic

Temporal logic was presented in § 8.5. Proofs for linear temporal logic can be formalized using an axiomatic approach (Figure 9.12) or a sequent calculus based approach (Figure 9.13). These systems are sound and complete (cf. § 9.8) for Kripke semantics on the considered fragments (they do not include U for instance).

$$\Box(A \Rightarrow B) \Rightarrow (\Box A \Rightarrow \Box B)$$
$$\Box A \Rightarrow A$$
$$\Box A \Rightarrow \Box\Box A$$
$$\Diamond A \stackrel{\text{def}}{=} \neg\Box\neg A$$

Figure 9.12: Axioms of temporal logic (system S4)

$$\frac{\Gamma, A \vdash \Delta}{\Gamma, \Box A \vdash \Delta} \qquad \frac{\Gamma \vdash A, \Delta}{\Gamma \vdash \Diamond A, \Delta}$$

$$\frac{\Box\Gamma \vdash A, \Diamond\Delta}{\Box\Gamma \vdash \Box A, \Diamond\Delta} \qquad \frac{\Box\Gamma, A \vdash \Diamond\Delta}{\Box\Gamma, \Diamond A \vdash \Diamond\Delta}$$

Figure 9.13: Rules of sequent calculus for system S4

9.5 Beyond First-order Logic

The deduction systems introduced in the previous sections can be extended to second-order and higher-order logic. We will revisit this point in Chapter 11 with the presentation of system F.

9.6 Dijkstra–Scholten's System

In the previous deduction systems, particularly the systems of Gentzen, logical equivalence is not handled directly: it has to be first translated by a double implication. In contrast, the connector of equivalence plays a pivotal role in the calculus of Dijkstra and his followers, which was designed for favoring the conciseness of proofs. We limit ourselves to the propositional fragment in what follows.

9.6.1 An Algebraic Approach

Deductions are regarded as rewriting of logical expressions. One goes from one line to the next by replacing a subexpression with an equal subexpression. "Deduction rules" are then considered as logical identities which have an algebraic flavor, such as $(a + b)^2 = a^2 + 2ab + b^2$.

The first logical connector one starts with is equivalence $\Leftrightarrow$. One postulates the following properties of this operation:

- it is associative: $(A \Leftrightarrow B) \Leftrightarrow C = A \Leftrightarrow (B \Leftrightarrow C)$;
- it is commutative: $A \Leftrightarrow B = B \Leftrightarrow A$;
- it admits $\mathbf{t}$ as an identity element: $A \Leftrightarrow \mathbf{t} = A$;
- it is the (Leibniz) equality on Boolean values: $A \Leftrightarrow B$ is another way to write $A = B$ when A and B are logical expressions.

Disjunction is then introduced with similar postulates: commutativity, associativity, idempotentness ($A \vee A = A$), distributivity over equivalence. Syntactically, $\vee$ (as $\wedge$ and $\Rightarrow$) takes precedence with relation to $\Leftrightarrow$. The implication $A \Rightarrow B$ and the conjunction $A \wedge B$ are respectively defined by:

$$A \Rightarrow B \stackrel{\text{def}}{=} A \vee B \Leftrightarrow B \qquad \text{and} \qquad A \wedge B \stackrel{\text{def}}{=} A \Leftrightarrow B \Leftrightarrow A \vee B \ .$$

The expression $A \Leftrightarrow B \Leftrightarrow A \vee B$ has to be regarded as a whole, and certainly not as the conjunction of $A \Leftrightarrow B$ and $B \Leftrightarrow A \vee B$. It can be compared with an algebraic expression such as $p + q + p.q$.

The last operator to be introduced in this approach is negation, which is respectively related to equivalence and to disjunction by the following postulates:[5]

$$\neg(A \Leftrightarrow B) \Leftrightarrow \neg A \Leftrightarrow B \qquad \text{and} \qquad \neg A \vee A \ .$$

The constant $\mathbf{f}$ is defined as the negation of $\mathbf{t}$:

$$\mathbf{f} \stackrel{\text{def}}{=} \neg \mathbf{t} \ .$$

---

[5]If we think of the relation between NJ and NK, it is interesting to note that $\neg\neg A \Leftrightarrow A$ can be derived from the first postulate (first prove $A \Leftrightarrow \neg B \Leftrightarrow \neg A \Leftrightarrow B$), whereas it is not the case of the law of excluded middle.

9.6.2 Displaying the Calculations

Proving a formula X amounts to making it equal to t using a sequence of rewriting steps. Calculations are displayed in the following way, in order to provide the justification of each step and make reading easier:

$$X$$
$$= \qquad \{\text{evidence for } X = Y \text{ (or for } X \Leftrightarrow Y)\}$$
$$Y$$
$$= \qquad \{\text{evidence for } Y = Z \text{ (or for } Y \Leftrightarrow Z)\}$$
$$Z$$

etc.

Evidences are more or less explicit, depending on the context. In the examples given below they are quite detailed. First we give a proof of $A \vee t$ which makes use of the equality $(X \Leftrightarrow X) = t$.

$$A \vee t$$
$$= \qquad \{(X \Leftrightarrow X) = t \text{ , with } X := A\}$$
$$A \vee (A \Leftrightarrow A)$$
$$= \qquad \{\text{distributivity of } \vee \text{ over } \Leftrightarrow\}$$
$$(A \vee A) \Leftrightarrow (A \vee A)$$
$$= \qquad \{(X \Leftrightarrow X) = t \text{ , with } X := A \vee A\}$$
$$t \qquad .$$

When the formula X to be proved has the shape $R \Leftrightarrow S$, it is simpler to rewrite R to S (the fact that t is an identity element for $\Leftrightarrow$ ensures that the two processes are equivalent). We will proceed below in this way, for proving that $A \Leftrightarrow B$ is equivalent to $(A \Rightarrow B) \wedge (B \Rightarrow A)$. From a more general perspective, as soon as properties of implication are proved, for example transitivity, one is allowed to use steps such as

$$\vdots$$
$$X$$
$$\Rightarrow \qquad \{\text{evidence for } X \Rightarrow Y\}$$
$$Y$$
$$\vdots$$

in order to prove that the first line entails the last, or

$$\begin{array}{ll} \vdots \\ X \\ = & \{\text{evidence for } Z \Rightarrow (X \Leftrightarrow Y)\} \\ Y \\ \vdots \\ \mathbf{t} \end{array}$$

in order to prove that Z entails the first line.

This way of displaying calculations is also used in the framework of imperative program calculation [Coh90, Kal90] (as introduced in Chapter 4) and of functional programming [Bir95].

9.6.3 The Role of Equivalence

The fact that $\Leftrightarrow$ is an equality plays a very important role:

- as soon as $A \Leftrightarrow B$ is at our disposal, occurrences of A can be replaced with B in an expression (law of Leibniz);
- all previous identities can be written with $\Leftrightarrow$ instead of $=$;
- as $\Leftrightarrow$ is associative and commutative, many identities can be read in several ways.

This leads one to manipulate multiples equivalences without parentheses: $X \Leftrightarrow Y \Leftrightarrow Z \cdots \Leftrightarrow T$. In a sequence such as the latter, one can delete two occurrences of the same formula: $X \Leftrightarrow X$ is $\mathbf{t}$, which is the identity of $\Leftrightarrow$.

One of the most noticeable multiple identities is certainly the *golden rule*, which is, among other things, a definition of $\wedge$:

$$R \wedge S \Leftrightarrow R \Leftrightarrow S \Leftrightarrow R \vee S \ .$$

This rule admits six permutations, and each permutation can be parenthesized in five ways; considering that R and S play symmetrical roles, we still have eleven different uses of the golden rule.

The previous ideas are illustrated in Figure 9.14, where it is proved that double implication (i.e. traditional equivalence) is identical to the notion of an equivalence which is axiomatized here.[6] Note that this theorem needs a fairly longer proof in other frameworks:

- the proof that double implication is associative is an interesting benchmark for automated tautology verification systems; this is one case where the method of truth tables is more efficient than the method of semantical tableaux;
- proving that double implication is a Leibniz equality requires an induction on the structure of formulas.

---

[6]It is quite instructive to prove the same theorem by progressively identifying $A \Leftrightarrow B \Leftrightarrow ((A \Rightarrow B) \wedge (B \Rightarrow A))$ to $\mathbf{t}$, and using the right instance of $X \Rightarrow Y \Leftrightarrow X \Leftrightarrow X \vee Y$.

$$
\begin{array}{ll}
& (A \Rightarrow B) \wedge (B \Rightarrow A) \\
= & \quad \{ \text{ definition of } \Rightarrow; \text{ commutativity of } \vee \} \\
& (A \vee B \Leftrightarrow B) \wedge (A \vee B \Leftrightarrow A) \\
= & \quad \{ \text{ golden rule; associativity of } \Leftrightarrow \} \\
& A \vee B \Leftrightarrow B \Leftrightarrow A \vee B \Leftrightarrow A \Leftrightarrow (A \vee B \Leftrightarrow B) \vee (A \vee B \Leftrightarrow A) \\
= & \quad \{ A \vee B \Leftrightarrow A \vee B \text{ is an identity element; commutativity of } \Leftrightarrow \} \\
& A \Leftrightarrow B \Leftrightarrow (A \vee B \Leftrightarrow B) \vee (A \vee B \Leftrightarrow A) \\
= & \quad \{ \text{ distributivity of } \vee \text{ over } \Leftrightarrow: \text{ factorisation of } A \vee B \} \\
& A \Leftrightarrow B \Leftrightarrow A \vee B \Leftrightarrow B \vee A \\
= & \quad \{ \text{ commutativity of } \vee \} \\
& A \Leftrightarrow B \; .
\end{array}
$$

Figure 9.14: Double implication in Dijkstra's system

This shows that the axioms for equivalence we have seen here contain a lot of information. In practice, they turn out to be sufficient more often than one would expect; it is worth translating an equivalence into a double implication only as a last resort.

9.6.4 Comparison with Other Systems

The approach presented here is clearly an axiomatic one. This said, deductions are not of the same kind as in Hilbert systems: here we have equational reasoning, *modus ponens* is not primitive and is even avoided.

A closer look shows that axioms are chosen in the spirit of an algebraic theory. Each primitive operation ($\Leftrightarrow$, $\vee$ and $\neg$) comes with its own algebraic properties or with algebraic properties related to other operations. So it may be better to consider this calculus as an algebra rather than a logic. In other words, it is a structure defined by non-logical axioms (see page 149). This is consistent with the fact that this approach has nothing to do with foundational issues, in contrast with formal logic as designed at the beginning of the 20th century [DS90].

The set $\mathbb{B}$ of Booleans endowed with conjunction, disjunction and negation admits a number of laws already mentioned on page 47: idempotence, commutativity, associativity, distributivity. It then makes up what is called a **Boolean algebra**. There is a similar algebra in set theory with the operations union, intersection and complementation.

Those algebras can also be presented from the concept of a **Boolean ring**. A unitary ring is a commutative group endowed with a distributive law having an identity element, for example $\langle \mathbb{Z}, +, . \rangle$. A Boolean ring is a unitary ring where every element is idempotent for the second law. The powerset of a set endowed with symmetrical difference $\backslash$ and intersection makes up a Boolean ring, as well as $\langle \mathbb{B}, \oplus, \wedge \rangle$, where $\oplus$ is defined by $A \oplus B \stackrel{\text{def}}{=} \neg(A \Leftrightarrow B)$. One can also interpret $\mathbb{B}$ by $\{0, 1\}$, $\oplus$ by the addition modulo 2 and $\wedge$ by the product.

An important property of Boolean rings is that every expression can be reduced into a form which is unique up to permutations, called its Stone normal form. A number of derivations perfomed in the system presented here amount to computing a Stone normal form[7] in the dual Boolean ring $\langle \mathbb{B}, \Leftrightarrow, \vee \rangle$.

At the same time, $\Leftrightarrow$ plays the role of an equality and then makes it possible to perform rewriting steps. We revisit this original view on deduction below.

To conclude this comparison with the previous systems, note that a number of passages in [DS90, vG90a, Coh90, Kal90] explicitly consider logic as an arbitrary symbol game. This is regrettable, because the systems of Gentzen go beyond this standpoint, which was previously defended by Hilbert. The purely formalist approach to logic was not that much of a success, it was even to some extent refuted by the failure of Hilbert's program [NNGG89, Gir87b]: see below the incompleteness theorems of Gödel. However, let us mention the work of A.J.M. van Gasteren [vG90a], which shows that a careful examination of the formal shape of expressions can provide valuable heuristics for solving some problems.

9.6.5 Choosing Between Predicates and Sets

Most logical connectors correspond to an operation over sets: $\vee$ corresponds to $\cup$, $\wedge$ corresponds to $\cap$, $\neg$ corresponds to complementation in a reference set (which has to be fixed in advance). We don't have a regular notation for the set operations corresponding to $\Leftrightarrow$ and $\Rightarrow$ (recall that $A \subset B$ is not a set but a logical expression), but let us introduce one for the set operation corresponding to $\Leftrightarrow$, say $\odot$, so that we get two similar theories. The set $A \odot B$ is the complement of symmetrical difference $A \setminus B$ in the reference set.

The algebraic properties of $\Leftrightarrow$, $\vee$ and $\wedge$ can immediately be transposed to $\odot$, $\cup$ and $\cap$. The choice between formalizing a given problem using logical operations, or using set operations, may then seem nothing more than a matter of taste.

However, the identity between $\Leftrightarrow$ and Boolean equality has additional specific advantages. Thus, every theorem in the form $X \Leftrightarrow Y \Leftrightarrow Z \ldots$ represents several identities at once, allowing one to replace X with $Y \Leftrightarrow Z \ldots$, or Y with $X \Leftrightarrow Z \ldots$, or $X \Leftrightarrow Y$ with $Z \ldots$, and so on whereas $A \odot B \odot C \ldots$ represents only one set. In particular, the golden rule

$$X \wedge Y \Leftrightarrow X \Leftrightarrow Y \Leftrightarrow X \vee Y$$

contains in a compact way at least five common identities on set, related to intersection, union and symmetrical difference:

$$
\begin{aligned}
A &= B \setminus (A \cap B) \setminus (A \cup B) & (A \cap B) &= A \setminus B \setminus (A \cup B) \\
A \setminus B &= (A \cap B) \setminus (A \cup B) & (A \cup B) &= A \setminus B \setminus (A \cap B) \\
& & A \setminus (A \cup B) &= B \setminus (A \cap B).
\end{aligned}
$$

This remark, together with the fact that set theory is sometimes more complicated than expected, leads Dijkstra to consider that predicate calculus is

---

[7]This remark was communicated to the author by Gerard Huet.

more convenient than set constructions. For example, in his approach to formal specification, the space state of a program is described by a logical formula rather than a set expression, as would be the case in Z or in B.

9.6.6 Uses of Dijkstra–Scholten's System

This system is well suited to pencil and paper manipulations. Dijkstra's school attaches importance to the quality of proofs presentation. Though entirely formal, proofs are always concise and easy to check, even in a number of non-trivial programmation problems. Many calculational steps use the associativity and the commutativity of $\wedge$, $\vee$, $\Leftrightarrow$, notably when we have chains of equivalences. A skilled eye should be able to recognize an interesting pattern in a chain — note that automated reasoning in the presence of associative and commutative operations is not that easy.

Doing formal proofs in this framework turns out to be an art, with its own guiding heuristics [DS90, vG90a, Coh90, Kal90]. The proofs we get using this approach are quite different from the ones provided by traditional proof theory: the latter are easily checked by a program, but non-trivial ones soon become too large for human eyes to spot. So one may consider that the approach to formal proofs presented in this section provides more convincing arguments; however, automated help is needed for realistic scale problems, and techniques based on sequent calculus or on natural deductions seem more apropriate [Rus93].

9.7 A Word About Rewriting Systems

A well-known technique has been developed for automating equational reasoning: rewriting systems. We will provide an example on page 200.

The general situation is as follows. We are given a finite number of equalities $S_i = T_i$, from which we want to prove a goal $A = B$. If the terms A or B contain an instance of S_i (or of T_i), we can replace it with the corresponding instance of T_i (or of S_i). For example, if we take $x + x = 2 * x$ for granted, we can replace the goal

$$(a + b) * (a + b) = a * a + 2 * a * b + b * b$$

with

$$(a + b) * (a + b) = a * a + a * b + a * b + b * b .$$

Equational reasoning consists of iterating such substitutions, until we get an equation where the two sides are syntactically identical. However, in the framework of automated proof search, we have to avoid cyclic sequences of transformations, where $A = B$ would be replaced with $A_1 = B_1$, ... and finally $A_n = B_n$ would be transformed into $A = B$ again. Such a cycle is very easy to get: just use an equality in one direction and then in the reverse direction.

We also have to avoid a potentially infinite sequence of transformations, which may happen e.g. with equations such as $x = e \star x$ (an arbitrary term t may then be replaced with $e \star t$, then with $e \star (e \star t)$, etc.).

A central idea is then to restrict the use of equations given as axioms: they have to be *oriented*, that is, we have to choose one direction, either from the left to the right, or conversely. This choice yields a **rewriting rule**. But of course, one then runs the risk of becoming unable to prove a number of theorems: indeed, in many reasonings one uses a given equality in one direction at one stage, and in the reverse direction at a later stage.

In order to recover a rewriting system having the same consequences as the original equations, new rules stemming from the axioms have to be added. This process, called **completion**, was introduced by Knuth and Bendix in 1970 [KB70].

Let us explain this somewhat more formally. We look for a set of rules $G_i \to D_i$ such that:

1. $G_i = D_i$ is a consequence of the equations given as axiom.
2. Given an arbitrary term t_0, every sequence $t_0, t_1 \ldots t_n \ldots$ (where t_{k+1} stems from t_k by application of a rule $G_i \to D_i$) eventually reaches a unique term which depends on t_0 only, called its **normal form**.
3. Two terms which are equal modulo the axioms possess the same normal form.

Then, in order to know whether $A = B$ is a consequence of the axioms, we just have to compute the normal forms of A and B and then to compare the results.

Bringing this basic idea into actual play, however, raises non-trivial issues. Important research developments came out, as well as interesting support software systems such as REVE [FG84, Les86], RRL [KZ95] and LP [GG89, GG91] for completion and rewriting, and Spike [BR95, Bou94, BKR92] for inductive proof of equations. More recent (and efficient) systems are Maude [CDE+99] and Elan [BKK+98].

In passing let us point out the importance of termination: the normalization process should be guaranteed to terminate, and this is an essentially delicate problem. Theoretical and practical tools were developed in the framework of rewriting systems for proving that a relation is noetherian (see the definition on page 52). This is a technical matter, where ordinals naturally have an important place.

Rewriting systems are strongly related to algebraic specification techniques, since specifications are written using equations in this framework. We go back to it in Chapter 10.

9.8 Results on Completeness and Decidability

If we want to prove theorems in a mechanical way, propositional, first-order and second order logic don't offer the same possibilities. In fact this even depends on the theory we consider. We give here a brief account of some of the main known results. For a number of them (particularly for incompleteness theorems), only an approximate statement is given, because a precise statement would necessitate too many technical preliminaries. A state-of-the-art survey is available in [Rab77] (and [Gri91], in French). Apart from the basics on model theory and proof theory already presented, we rely on the concepts related to calculability introduced in Chapter 3.

Completeness was also introduced and illustrated in Chapter 3. Its intuitive meaning is that everything which is true is provable. But this may be understood in two ways, since, given a set of formulas Γ, one may consider truth either in a *class* of models of Γ, or in *one* special (intended) model of Γ.

9.8.1 Properties of Logics

We first define a number of properties about *logics*. Our framework is classical first-order or higher-order logic. We agree that every first-order language defines a logic — within which several theories can be described. Some results depend on the number and on the arity of the symbols defining the language considered. In what follows P and Γ represent respectively a closed formula and a set of closed formulas.

A logic is **sound** if the deductive consequence relation implies the semantic consequence relation, i.e. if $\Gamma \vdash P$ entails $\Gamma \models P$. A logic is **complete** if the semantic consequence relation implies the deductive consequence relation, i.e. if $\Gamma \models P$ entails $\Gamma \vdash P$. A logic is **decidable** if there exists an algorithm that finds whether an arbitrary formula admits, or does not admit, a proof, using a finite number of steps; a logic is **semi-decidable** if there exists an algorithm that finds a proof of any theorem, using a finite number of steps (it may be the case that the algorithm does not terminate if the input formula is not a theorem); in the other cases the logic is said to be **undecidable**. We have the following results.

Theorem 9.4
Propositional logic and predicate logics of arbitrary high order are sound.

This is simply because logical axioms are valid and deduction rules propagate validity.

Theorem 9.5 (Schröder)
Propositional logic is decidable.

Theorem 9.6 (Post)
Propositional logic is complete.

Theorem 9.7 (completeness, Gödel)
Given any first-order language, the corresponding first-order logic is complete.

To put it otherwise, if a formula φ is true in *all* models of a family of formulas, then φ has a formal proof.

Theorem 9.8
Second-order logics are incomplete (even weak monadic logic).

This theorem is a consequence of the incompleteness of arithmetic (see below) and of the fact that arithmetical truth can be characterized by a finite number of second-order axioms.

Theorem 9.9 (Church)
First-order logic is semi-decidable. More precisely, if the language of a first-order logic contains at least one function symbol or binary predicate symbol, the validity of an arbitrary formula cannot be mechanically decided.

One generally considers recursively axiomatizable theories, and a consequence is that their theorems make up a recursively enumerable set. When the conditions of Church's theorem are satisfied, which is the most frequent case, we get semi-decidables theories. Thus, theorems of predicate calculus can be recursively enumerated (since first-order logic is complete), but not the other formulas.

In order to get positive decidability results beyond first-order, one has to consider very restricted languages. However, the following result remains true for second-order monadic logic.

Theorem 9.10
Equational logic with an arbitrary number of unary relation symbols and at most one unary function symbol is decidable.

9.8.2 Properties of Theories

Now we define properties about *theories*. The concept of completeness we use here for theories is a syntactical concept: a theory T is **(syntactically) complete** if for every closed formula P one has either $T \vdash P$, or $T \vdash \neg P$. It is clear that two models of a complete theory cannot be distinguished, since any closed formula has the same truth value in each of them. In this respect one can say that a complete theory characterizes a unique model.

The simplest example of an incomplete theory is the empty theory: if A is a unary predicate symbol, neither $\vdash \forall x\, A(x)$ nor its negation is a theorem; even more simply, if B is a proposition symbol, neither $\vdash B$ nor $\vdash \neg B$ is a theorem. A more interesting example is group theory, which states nothing about $\forall xy\ (xy = yx)$, since there exist commutative groups as well as non-commutative groups. Hence group theory is not complete, and there is no cause for alarm here.

In contrast, a number of theories are designed with a precise intended model in mind. This is typically the case with natural integers endowed with usual arithmetical operations. In such a case, one is interested in the consequences which are true in one model (the so-called standard model), and not in every model of the axioms. Let us recall the axioms for addition.

$$\forall x \quad x + 0 = x$$
$$\forall x \forall y \quad x + S(y) = S(x + y) \ .$$

One would expect that $\forall x\ 0 + x = x$, which is true in $\mathbb{N}$, is a consequence of the previous axioms. Actually we also need the induction schema: there exist models of the two previous axioms where $\forall x\ 0 + x = x$ is not satisfied.

Note that if a system of axioms is incomplete one may try to complete it by introducing additional axioms.

A theory T is **inconsistent** if one of the three following equivalent conditions is verified:

- $T \vdash \mathbf{f}$,
- there exist a formula P such that $T \vdash P$ and $T \vdash \neg P$,
- for all P, one has $T \vdash P$.

In the opposite case T is said to be **consistent**; note that, this is the first property about a theory one would expect. The complementarity between completeness and consistency should be noted. To coin a phrase, we could say that a complete and consistent theory tells the truth and nothing but the truth.

A **decidable theory** is defined in the same way as a decidable logic.

Theorem 9.11
A first-order theory is consistent if and only if it has a model.

This theorem is actually another formulation of the completeness theorem for first-order logic.

Theorem 9.12 (Herbrand)
A set T of first-order clauses is inconsistent if and only if there exists a finite set of closed instances of clauses of T which is inconsistent as well.

Thanks to this theorem, the search for a proof in predicate calculus can be reduced to the search for a proof in propositional calculus. It plays an essential role in the semi-decision procedures based on the resolution principle, as already mentioned in § 9.4.2.

Theorem 9.13 (Turing)
A recursively and complete axiomatizable theory is decidable.

Theorem 9.14
The arithmetic of Peano is undecidable, as well as any consistent theory that contains it.

For any consistent extension of PA (Peano's arithmetic), it is even possible to exhibit a closed formula, which is neither provable nor refutable, and which is, however, true in the intended model (Rosser). Then there is no first-order characterization of the standard model of arithmetic. This is an essential limitation which cannot be repaired by adding appropriate axioms.

Beware: a formula which is true in *all* models of PA is provable by means of axioms of PA: this is the meaning of the completeness theorem (9.7). The first incompleteness theorem of Gödel states that *the* standard model of PA contains at least one formula which is true but cannot be proved using the axioms of PA. Note that we have already seen that there exist non-standard models of PA (see § 5.3.2.3). The original proof of Gödel shows how to construct such a formula, inspired by the paradox of the liar:[8] using tricky codings he was able to encode arithmetic formulas, then arithmetic proofs, by integers, so that he could write a formula stating its own unprovability. Less artificial theorems have been discovered recently [PH77, KP82].

Here is an example of a statement which is true but beyond the proof power of PA, taken from [KP82]. Let us choose an arbitrary natural number n, for example 266, and a basis b, for example 2, so we write: $266 = 2^8 + 2^3 + 2^1$. Exponents are then represented in the same basis, and so on. In our example this yields $266 = 2^{2^{2+1}} + 2^{2+1} + 2^1$. Now consider the following process: we add 1 to b in this representation, we subtract 1 to the new value, then again with the new values of b and n if n is non-zero, and so on. In our example the second value of n is $3^{3^{3+1}} + 3^{3+1} + 2$, that is about 10^{38}; the first values of $\langle b, n \rangle$ are approximately $\langle 2, 266 \rangle$, $\langle 3, 10^{38} \rangle$, $\langle 4, 10^{616} \rangle$, $\langle 5, 10^{10,000} \rangle$. Though it may seem strange, the incredible growth of n eventually stops — the basis becomes equal to the number. The process then amounts to letting n be decremented by 1 at each step, so that the sequence is finite (the process stops when $n = 0$). However, this cannot be proved in Peano's arithmetic.

The **second theorem of Gödel** is the most celebrated because of its epistemologic consequences. It states that the consistency of arithmetic cannot be proved by simple induction on natural numbers. Later, Gentzen proved the consistency of arithmetic by means of a stronger induction principle.

Note that an important fragment of arithmetic, called **Presburger arithmetic**, is decidable. Its essential difference with Peano arithmetic is that terms cannot include a product $x.y$ where x and y are variables. In other words, terms are linear expressions with integer coefficients.

Theorem 9.15
Presburger arithmetic is decidable.

9.8.3 Impact of These Results

A good knowledge of the previous results is useful when one uses a proof assistant — it is even a must in the design of a such tool. Positive results open possibilities, negative ones bring impassable theoretical barriers to light.

The actual impact of decision or semi-decision results depends on the complexity of the computations they entail. Unfortunately, even in the simple case of the propositional calculus, deciding the satisfiability or the validity of a proposition is up to now believed to need a computation time which is, in the

---

[8]Epimenides says that he is lying; if this is true, i.e. Epimenides *is* a liar, then he is telling the truth, so he is not a liar — a contradiction. If this is false, i.e. Epimenides is telling a lie when he says that he is a liar, then that means he is *not* a liar — a contradiction again.

worst case, an exponential function of the size of the formula.[9] Beyond this, results are truly disastrous: for many decidable problems, theoretical upper bounds lead to computation times which would be larger than the age of the universe. However, on problems we encounter in actual practice, the efficiency of decision procedures is sometimes drastically improved by clever techniques or by appropriate restrictions. This is notably the case for Boolean formulas and Presburger arithmetic — in particular, in the latter framework, it is a good idea to consider formulas without existential quantifiers.

In summary, fully automated proof search can only be carried out in a less expressive logic, so that only very specific classes of problems can be handled in this way. In the general case, the skills and knowledge of the user seems to be the determining factor. Software support tools are of course very useful. As they have to be interactive rather than fully automated in many situations, one may consider that there is no point in restricting oneself to a limited language, say, first-order logic. Indeed, support tools based on higher-order logic (e.g. PVS, Coq or Isabelle) become more widely used nowadays. Of course they are much more user-friendly when "simple" subgoals can be solved by efficient decision procedures.

9.9 Notes and Suggestions for Further Reading

The principles and algorithms used by automated proof tools for first-order logic, notably the resolution principle, are often presented from a model-theoretic perspective, using the fact that a special model, called the model of Herbrand, is sufficiently representative of the general situation: this model is built upon the set of syntactic terms that can be constructed in the language under consideration. The book of Chang and Lee [CL73] provides a good synthesis along these lines.

Logic is presented from a sequent calculus perspective in more recent books, such as the one of Jean Gallier [Gal86], devoted to first-order classical logic, or the book of Girard, Lafont and Taylor [GLT89], which contains a good introduction to natural deduction and second-order logic. Reference books on proof theory include [Tak75], [Sch77] and [Gir87b].

Natural deduction inspired a theory of programming language semantics called natural semantics by Kahn [Kah87], which is close to the structural operational semantics of Plotkin [Plo81]. These two approaches are also explained and compared in [NN92]. Natural semantics has been implemented in Centaur [JRG92], an experimental software tool for prototyping programming languages.

---

[9]To be more precise, determining whether a Boolean formula has a model is the *NP-complete* problem *par excellence*: many combinatory problems (knapsack, optimization, etc.) can be reduced to it. For such problems, algorithms able to find a solution in exponential time (in the worst case) are known, but until now there is no proof that a polynomial time solution does not exist, though it seems highly improbable.

The books [Coh90] and [Kal90] present the approach of Dijkstra and Scholten to logic and its application to the design of correct algorithms. A thorough development of the logical part is given in [DS90].

Reference publications on rewriting include the article of Huet and Oppen [HO80], the chapter of Dershowitz and Jouannaud [DJ90] in [vL90b] and the book of Baader and Nipkow [BN98].

A translation of the original of Gödel on his incompleteness theorems is available in [NNGG89]. The article is preceded by a long explanation of F. Nagel and J-R. Newman, and then followed by an interesting presentation of J-Y. Girard about the program of Hilbert, its epistemological stakes and the consequences of its failure. Many results on decidability and indecidability are given in [Rab77].

10. Abstract Data Types, Algebraic Specification

At first glance, algebraic specification techniques may seem to have less relevance to industrial applications than other methods. They are, however, worth studying because they benefit from extensive theoretical research and have had a great influence on other specification techniques, and more importantly, on computer science in general, notably with the concept of the abstract data type. Typing is a well-known concept in computer science. It is not only a means of protection against a number of mistakes, but also a methodological tool. We start with an informal discussion of the uses of typing and several interpretations of this notion. As a first approximation, a type can be regarded as a set. Unfortunately, one has to be more cautious with this interpretation, than one would expect. We will therefore consider more abstract concepts of a type.

10.1 Types

Adding a Boolean value to a string hardly makes sense. Types are basically used for ensuring that such situations do not occur. To this end, types are assigned to the relevant expressions (terms, formulas, commands, etc.) of the language we are considering. When a given operation, say addition, is applied to its arguments, we can then check that the latter have the expected type. **Type-checking** a given expression consists of verifying that all its components have the expected type. The key to type-checking is a means to determine, given an expression E and a type T, whether or not E has the type T (denoted $E : T$). There are several options for a typing system.

– If one wishes type-checking to be performed statically (at compile-time), this problem has to be decidable; then the quantity of information carried by types tends to be limited.
– A part of type-checking can also be performed at run-time; one of the most well-known examples, previously introduced in Chapter 2, concerns array indices, which must be kept between two bounds. We can no longer guarantee the absence of run-time faults, yet it is still possible to have the program terminate in a graceful manner. Another disadvantage is the additional time required by these verification steps. Unless explicitly stated otherwise, we will consider only static type-checking below.

– If one wishes to have an expressive typing system, free of run-time penalties, the proof that the program is well typed must be carried out with the help of the programmer. Let us also mention that, in a development with the B method, typing information is added to the invariant and the assertions, so that type checking yields proof obligations; however, the typing system of B is not terribly rich, in order that proof obligations corresponding to type checking can be automatically discharged.

The type associated to an object is not necessarily unique. For example, if we consider the set-theoretic interpretation of a type, many sets containing a given item could be seen as a possible type for this item. This yields a possible interpretation of subtyping, a concept related to inheritance in object-oriented languages.

Also, we often want to give several types to a function, but for other reasons. The idea can be illustrated with the simple case of the identity function, which can be considered to have the types int → int, bool → bool, ..., that is, in general, $T \to T$ where T is an arbitrary type. Such types are called *polymorphic* types in the framework of functional languages, or *generic* types in the framework of languages such as Ada or Eiffel.

A number of functions on lists, such as catenation or the computation of the length of a list, are in the same category as the identity function: the *algorithm* used is exactly the same. Note that addition is also a polymorphic operation, because it can be defined over integers, floating numbers, vectors, matrices, etc.; but here the underlying algorithm is different for each case. The former kind of polymorphism is called **parametric** polymorphism, whereas the latter kind is called **ad-hoc** polymorphism. In the following we will limit ourselves to parametric polymorphism.

10.2 Sets as Types

In a typed programming language, a variable v is associated with a type, which is generally seen as the collection of the possible values of v. Typing the variables amounts then to specifying the set of the possible states, or equivalently, a constraint on the execution of the program. We can then say that types provide an invariant. For illustration purposes, let us imagine a programming language having a Pascal-like syntax, where types are sets.

10.2.1 Basic Types

For example, we can interpret the declaration:

```
var x: {a,b,c};
    y: {d,e};
```

as a specification requiring that the state space of the program is a strict subset of $\{a, b, c, d, e\}^2$, which is, specifically, $\{a, b, c\} \times \{d, e\}$. If we add:

z: $\mathbb{N}$;

the state space becomes $\{a, b, c\} \times \{d, e\} \times \mathbb{N}$.

10.2.2 A First Glance at Dependent Types

More advanced languages, such as Cayenne [Aug98], allow more sophisticated type declarations, where the *type* of a component depends on the *value* of another component. Thus:

var x: $\mathbb{N}$;
    y: 0 .. x;

would specify that the state space is $\{(x, y) \in \mathbb{N}^2 \mid y \leq x\}$. As a classical example, we can consider the (Gregorian) calendar. As a first approximation, we can take $\{1 \ldots 31\}$ as the type of the day of the month, but a more accurate typing would be $\{1 \ldots f(m, a)\}$, where m and a represent, respectively, the current month and year, and where f is a well-known function. Such types are referred to as *dependent types*. One may also use a logical formulation for them, for example $1 \leq q \wedge q \leq f(m, a)$. In this framework, type-checking involves coping with logical inferences, which make it more complex. We will revisit dependent types at the end of Chapter 11.

10.2.3 Type of a Function

Functions can also be given a type. For example, a possible type for addition is $\mathbb{N} \times \mathbb{N} \to \mathbb{N}$. In fact, the main purpose of typing is to ensure that applying a function to its arguments does indeed make sense. In our example, we want to reject an expression such as $a + b$ if either a or b is not in $\mathbb{N}$. On the other hand, assuming that $a + b$ is well typed, we know that $a + b$ is a member of $\mathbb{N}$, hence it can be, in turn, one of the arguments of a further addition.

10.2.4 Type Checking

Using the constructs which are available in the language under consideration (arrays, function application, tuples or whatever), one may form an expression E. Saying that E has the type T amounts here to saying that E is a member of T. Type-checking may then be seen as membership checking.

10.2.5 From Sets to Types

In the language imagined so far, types are defined in a set-theoretic notation. Now, we could ask ourselves if any set could actually serve as a type.

Consider the set of even non-negative numbers, denoted here by $2\mathbb{N}$. There are functions or programs that require such numbers as arguments, as we have already seen at the end of § 3.5.2. Let f_e be such a function, and assume we are given two non-negative even numbers p and q. Then we can form $f_e(p)$ and $f_e(q)$. But it is unclear, at this stage, whether or not $f_e(p+q)$ should be accepted. Since $x \in 2\mathbb{N}$ implies $x \in \mathbb{N}$, $p+q$ makes sense, but as the type of $+$ is $\mathbb{N} \times \mathbb{N} \to \mathbb{N}$, we can only conclude that $p + q \in \mathbb{N}$. On the other hand, we have more, i.e., $p + q \in 2\mathbb{N}$, so we could in principle write $f_e(p+q)$. But this cannot be decided with the type of $+$ only, we need additional knowledge from number theory. For a more complex example, consider a function $f_{\bar{c}}$ that takes as input a number which is *not* a cube. Then one could write $f_{\bar{c}}(a^3 + b^3)$ if a and b are positive integers — this is a special case of Fermat's last theorem. Such examples are somewhat artificial, but actually, everyday programming provides arbitrarily difficult situations — and involves data structures which are more complex than $\mathbb{N}$. We cannot afford to embed any amount of mathematics in a static type-checking procedure.

Hence, in the general case, we have to admit that when we apply a function to arguments, the type of the result is provided by the type of the function, and nothing more. In general, this entails a loss of information. It also means that type-checking is not equivalent to membership checking: we may have $E \in T$ whereas $E : T$ does not hold.

10.2.6 Towards Abstract Data Types

In our example, we then give up the idea that $f_e(p+q)$ is well typed. But it is still possible to recover something very close to $f_e(p+q)$. The price to pay is the introduction of new symbols for functions that return even numbers: for example, the addition $+_e$ of two even numbers, which has the type $2\mathbb{N} \times 2\mathbb{N} \to 2\mathbb{N}$, or the multiplication $*_{e_1}$ of an even number by a non-negative integer, which has the type $2\mathbb{N} \times \mathbb{N} \to 2\mathbb{N}$. These operations behave exactly like $+$ and $*$, on their respective domains. Using $+_e$ and $*_{e_1}$ (and if we declare that the constant 2 has the type $2\mathbb{N}$) we can easily construct many well-typed expressions under the form $f_e(E)$, including $f_e(p +_e q)$. Moreover, type-checking can be performed as normal, by a systematic syntactical inspection of the expression.[1]

In fact, we just worked in the spirit of abstract data types, as we will see in § 10.3. The conclusion is that $2\mathbb{N}$ is useful as a type, provided that we are given functions returning a result in $2\mathbb{N}$.

10.2.7 Coercions

If we know that an expression E has the type $2\mathbb{N}$, we certainly would like to be able to use it in situations where something of type $\mathbb{N}$ is expected. In the general case, inferring that $E : U$ from $E : T$ is possible provided we have

---

[1] Of course, the correctness of this approach still relies upon arithmetical facts, which provide evidence that $+_e$ and $*_{e_1}$ return even numbers. This task is independent from the general type-checking procedure. We can then say that the relevant arithmetical facts are now used in a controlled manner, as defined by the occurrences of $+_e$ and $*_{e_1}$ in the expression to be type-checked.

some evidence that T is a subtype of U. In the set-theoretic interpretation of types, this amounts to $T \subset U$. But subset checking is at least as difficult as membership checking. Such inferences can be guided by syntactic means, just as before: introduce an explicit identity function $i_{T,U}$ from T to U that maps any x, considered as a member of T, to the same x, considered as a member of U. Thus, in our example, $i_{2N,N}(p) + i_{2N,N}(q)$ has the type N.

In general, we are also interested in converting integers to real or floating-point numbers, etc. This may involve a change in the internal representation, so that $i_{T,U}$ is no longer a function that maps x to itself but, more generally, an *injection*. Such injections are referred to as **coercions**. They can often be declared once-and-for-all, and left implicit in expressions, in order to keep them simple — type-checking should then be completed in such a way that the coercions are recovered.

10.2.8 A Simpler Approach

One may conclude that allowing arbitrary subsets of N or, more generally, of any given type, to be considered as types, drives us to cumbersome notations. An alternative approach, which is followed in Z and in B, consists of taking a collection of sets for types, in such a way that, for all x, we have a unique type T_x such that $x \in T_x$. (For example, the type of integers will be Z.) As a consequence, the intersection of two different types must be empty.

Let A be a set, and f be an operation defined over S. If we want to be able to assign a type to f, all elements of A must be of the same type, say T, so that f will have the type $T \to U$, for some type U (actually the type of f will be $\mathcal{P}(T \times U)$, since, in set theory, functions are special cases of relations). We see that the only sets A we can work with, are the subsets of types. The type T_x of x is then the greatest set containing x.

As we have seen in Chapter 6, type-checking is not very difficult in this framework. Examples such as the one with even numbers are dealt with using invariants or assertions instead of types.

10.2.9 Unions and Sums

In the previous approach, we cannot build a set with elements of different types. It is problematic because one sometimes needs to handle several things on an equal footing, say, integers and pairs of integers, whereas $Z \cup Z \times Z$ is not allowed.

On the other hand, we know that a direct use of $Z \cup Z \times Z$ is not that useful: given a data item x which is a member of $Z \cup Z \times Z$, one generally wants to eventually perform a computation which depends on the source of x. But it is not obvious that its source can be recovered. In the implementation of many programming languages, we cannot distinguish a 64-bit integer from a pair of two 32-bit integers. Even in set theory, integers and Cartesian products are encoded in such a way that a given element can be interpreted in several ways. As we have seen in § 3.1.2, a better concept is the *sum*. Recall that the sum

$A + B$ of two sets A and B can be defined as $\{\texttt{false}\} \times A \cup \{\texttt{true}\} \times B$. This is a special subset of $\mathbb{B} \times (A \cup B)$. But it would be strange to allow that $\{\texttt{false}\} \times A$ and $\{\texttt{true}\} \times B$ may be mixed, whereas A and B may not.

What we need here is a *structured* or *abstract* view of $A + B$, where $A \cup B$ is hidden. Such an approach is still more relevant if we consider slightly more complicated data structures, such as binary trees: recall that the type of binary trees is similar to an infinite sum $A + (A \times A) + (A \times (A \times A)) + ((A \times A) \times A) + \ldots$. Here again, we will see in § 10.3 that abstract data types are helpful.

Note also that the difficulty pointed out here is overcome in Z due to the introduction of the concept of a free type. Indeed, Z free types provide a notation for the special abstract data types that we need in such situations.

10.2.10 Summary

Interpreting a type as a set has the immediate benefit of simplicity. However, a direct use of typed sets is not that helpful when we want to represent a number of well-typed regular data structures used in computer science. It is certainly not by chance that set-theory is an untyped theory, where $3 = 1 \cup \langle 0, 1 \rangle$ is a perfectly legal equality. Let us end this discussion with two comments.

- On the one hand, set theory is too general: the set of set-theoretic functions from $\mathbb{N} \times \mathbb{N}$ to $\mathbb{N}$ is not countable, whereas only the set of computable functions is relevant for programs, and this is a countable set[2] (cf. § 3.3.4).
- From another perspective, set theory is not general enough, if one needs to describe a polymorphic (also called generic) type [Rey85], as was already mentioned in Chapter 7.

10.3 Abstract Data Types

The general idea behind abstract data types is to describe data structures without unveiling their implementation. Essentially, an abstract data type encapsulates a data structure D together with the operations which manipulate it. Each value in D is expressed by means of these operations only. It then becomes possible to axiomatize D in an algebraic manner.

Let us illustrate the idea with the very simple case of the natural integers. To this end, instead of using directly $\mathbb{N}$, we would introduce an abstract data type called nat, together with names for the regular arithmetical operations. Of course, $\mathbb{N}$, endowed with appropriate operations defined by means of set-theoretic primitives, would provide a model for this abstract data type. Note that the typing discipline itself does not depend on this interpretation. The rule we should conform to would be that only expressions built up from the operations declared in the type are accepted as arithmetical (integer) expressions.

---

[2]In a "high-level" (or abstract) specification stage, a less strict viewpoint is acceptable.

We can proceed in the same manner with even numbers, as suggested at the end of § 10.2.6: introduce a data type called **even_nb**, with appropriate operations, and consider that 2N (endowed with $+_e$ and $*_{e_1}$) is a model for them.

Finally, consider the example of a binary tree. In order to be able to use binary trees, one needs to construct a new tree, to compose a tree from two previously constructed ones, to compare two trees, etc. One has also to know whether information items are stored at the leaves or at the nodes. But implementation choices such as the use of pointers, arithmetical operations in arrays, or whatever, are not relevant here. One has a **concrete type** when the representation of data items, and of the functions for accessing or modifying them, are described, whereas one has an **abstract data type** when only properties of these data items and functions are described. There is an analogy in logic: a concrete type would correspond to the notion of a model, while properties defined in an abstract data type would be represented by formulas.

10.3.1 Sorts, Signatures

To define an abstract data type, one first gives a name, termed a **sort**, to the various kinds of data items to be used. For example, we will need trees, integers and Boolean values, having **tree**, **nat** and **bool**, respectively, as their sort. We also need to designate operations over these objects. For example, for a binary tree whose leaves contain an integer, we can consider the operations **bin** which constructs a tree from its two sub-trees, **leaf** which constructs a one-leaf tree, **lft** (respectively **rgt**) which extracts the left (respectively right) sub-tree, **depth** which yields the depth of a tree (the length of its longest branch), **bal** which indicates whether the tree is balanced, and so on.

The operations we consider are side-effect free; that is, they have no effect other than the production of a **value**. Let us give some examples: the expression **bin(leaf(3),leaf(1))** represents a tree having two leaves containing 3 and 1, respectively; **rgt(bin(leaf(3),leaf(1)))** represents a tree having exactly a leaf containing 1. A value is always designated by means of previously declared operations, without reference to a particular model. In most cases there are operations, called the **constructors**, which play a special role. They allow one to designate all possible values, and only those values. In the case of binary trees, the constructors are **leaf** and **bin**. Using axioms, one should be able to prove that every expression is equal to an expression using only constructors.

The **signature** of an operation consists of the declaration of the sorts of its arguments and of its result. For example **nat → tree** is the signature of a function which takes an integer as input, and returns a tree, while **tree × tree → tree** is the signature of a function which takes two trees as inputs, and returns a tree. The previous operations would then be declared as follows:

```
leaf   :  nat → tree        bin  :  tree × tree → tree
lft    :  tree → tree       rgt  :  tree → tree
depth  :  tree → nat        bal  :  tree → bool .
```

10.3.2 Axioms

We need to have more knowledge about the contents or about the behavior of those operations. In the case of a programming language (Ada, CLU), the semantics of operations is expressed by programs. Their internals are based on a concrete representation of data types; the formal interface is then made up of only the signatures. In the case of B, a set-based model is provided for the operations. This model is quite a high-level one, however, as it uses unbounded choices and general operations over sets and relations. The more concrete representations, described in *refinements* and finally in *implementations*, are then hidden behind an abstract specification. In the framework of an algebraic specification language, one does not provide any model, abstract nor concrete, but rather a number of properties which are expected from the operations. Those properties are expressed by logical formulas, which can be axioms or theorems.

We have an analogous situation in mathematics. As a well-known example, groups can be characterized by three axioms. Similarly, the effect of the operations of an abstract data type can be characterized by appropriate axioms. For example, natural numbers can be seen as an abstract data type, described by the signature:

```
zero :               → nat    succ : nat → nat
plus : nat × nat → nat    mult : nat × nat → nat
eq   : nat × nat → bool   inf  : nat × nat → bool ,
```

and the axioms of Peano.

In the example of binary trees, here is an axiom stating that a tree, made up of two sub-trees is balanced, if these sub-trees are themselves balanced and if the difference between their respective depths does not exceed 1 (we use here a liberal syntax for the arithmetical parts of the formula):

$$\forall a, b \quad (\mathrm{bal}(a) \wedge \mathrm{bal}(b) \wedge \\ |\mathrm{depth}(a) - \mathrm{depth}(b)| \leq 1) \\ \Rightarrow \mathrm{bal}(\mathrm{bin}(a, b)) \ . \tag{10.1}$$

The functions lft, rgt and depth are determined by the following axioms:

$$\forall a, b \quad \mathrm{lft}(\mathrm{bin}(a, b)) = a$$
$$\forall a, b \quad \mathrm{rgt}(\mathrm{bin}(a, b)) = b$$
$$\forall n \quad \mathrm{depth}(\mathrm{leaf}(n)) = 1$$
$$\forall a, b \quad \mathrm{depth}(\mathrm{bin}(a, b)) = 1 + \mathrm{max}(\mathrm{depth}(a), \mathrm{depth}(b)) \ .$$

If the axioms are arbitrary formulas, we have an **axiomatic abstract data type**; the name **algebraic abstract data type** is preferred when the axioms are equations[3] (or, sometimes, formulas of the form $E_1 \wedge \cdots \wedge E_n \Rightarrow E_0$, where E_i are equations). We can state the axiom about bal in the form of an implication between equations in the following manner:

---

[3] Recall that in mathematics, the axioms used to define algebraic structures such as groups, rings, vector spaces, etc. are (universally quantified) equations.

$$\forall a, b \quad (\mathtt{bal}(a) = \mathtt{true} \;\wedge\; \mathtt{bal}(b) = \mathtt{true} \;\wedge$$
$$\mathtt{infeg}(|\mathtt{depth}(a) - \mathtt{depth}(b)|, 1) = \mathtt{true}) \qquad (10.2)$$
$$\Rightarrow \mathtt{bal}(\mathtt{bin}(a, b)) = \mathtt{true} \;.$$

All these formulas are quantified universally for every variable; the general framework is first-order logic, as described in § 5.2.

In our specification, bool is a sort corresponding to an abstract data type in the same way as arb. Its constructors are true and false. It can be endowed with usual Boolean operations (negation, conjunction, disjunction, etc.).

Note that bal is a predicate in (10.1), whereas it is an ordinary function in (10.2). If we consider semantics, in both cases bal is interpreted as a function to $\mathbb{B}$. In a way, explicitly using bool places a formula such as $\mathtt{bal}(a)$ at the level of semantics.

For the sake of completeness, the specification of binary trees should make explicit that any tree constructed with bin is different from any tree constructed with leaf, and that the constructors are injections, and finally, that all trees are constructed with bin and leaf. This is easy to express using first-order axioms, for example:

$$\forall a, b, n \quad \neg(\mathtt{bin}(a, b) = \mathtt{leaf}(n))$$
$$\forall m, n \quad \mathtt{rgt}(\mathtt{bin}(a, b)) = b$$
$$\forall m, n \quad \mathtt{leaf}(m) = \mathtt{leaf}(n) \;\Rightarrow\; m = n \;.$$

In order to simplify the specification, it is agreed that such axioms are implicitly stated. One does not then have the freedom to interpret two expressions as the same object, except if this is a consequence of the axioms. In short: two objects which are not explicitly (or provably) equal must be distinct.

10.3.3 First-order and Beyond

The formulas considered so far in this chapter are first-order. However, higher-order logic may be useful if we want to express generic operations. Let us illustrate the idea on (an abstract view of) the sum of nat and nat × nat — another abstract data type, say, for binary trees, would do just as well. We have essentially two manners of constructing a value of this type — let us call it Snat:

- by means of the constructor i1 (first injection) of type
  nat → Snat;
- by means of the constructor i2 (second injection) of type
  nat × nat → Snat.

If we want to use a value s of type Snat, in order to build up a Boolean for example, we have to consider the two possible sources of s. To this end we would introduce the operation case with three arguments:

- a value of type Snat;
- a function of type nat → bool to be applied in the first case;
- a function of type nat × nat → bool to be applied in the second case.

These operations come with the following axioms:

$$\texttt{case}(\texttt{i1}(n), f, g) = f(n) \ ,$$
$$\texttt{case}(\texttt{i2}(\langle x, y \rangle), f, g) = g(\langle x, y \rangle) \ .$$

Any model of Snat should satisfy these axioms. In particular, this can be checked with a set-theoretic interpretation based on $\{\texttt{false}\} \times \mathbb{N} \cup \{\texttt{true}\} \times (\mathbb{N} \times \mathbb{N})$ together with an appropriate model of i1, i2 and case.

Note that the function case takes functions as arguments. A polymorphic version of this function would be welcome: its behavior is the same whatever the type of the result. Then one would systematically replace bool with a parameter T.

Actually, one would like to introduce such parameters for the sum itself and for the two injections i1 and i2. All this can be done, provided that one goes beyond first-order. Appropriate devices for doing that will be introduced in the next chapter.

10.4 Semantics

The semantics of a specification defined by an abstract data type is given by a model of the axioms (§ 5.6). For algebraic abstract data types, one generally considers multi-sorted logic. Each sort is interpreted by a previously known set. Expressions such as nat × nat → nat are interpreted by a total function from a Cartesian product to a set, for example, the addition is from $\mathbb{N} \times \mathbb{N}$ to $\mathbb{N}$.

The abstract data type itself is interpreted by a mathematical structure, that is, an n-tuple composed of sets and operations over these sets. For example, the abstract data type nat of the previous section could be interpreted by $\langle \mathbb{N}, \mathbb{B}, 0, \texttt{S}, +, \times, =, < \rangle$.

The example of even_nb, given in § 10.2.6, is interesting. One model for it is 2N, of course, but another one is N itself, endowed with regular addition and multiplication, without any changes. What changes in this interpretation is the coercion function from even_nb to nat, which is no longer the identity function, but the function which returns the double of its argument.

There are several options for defining the semantics of an algebraic abstract data type. Let us mention two of them here: *initial semantics* and *loose semantics*.

In the case of initial semantics, a specific structure, referred to as the initial model, plays a central role. For example, we would take $\langle \mathbb{N}, \mathbb{B}, 0, \texttt{S}, +, \times, =, < \rangle$ in the case of arithmetic. This approach is well suited when we have constructors, such as zero and succ (we want them to have a "no junk – no confusion" property).

On the other hand, *loose semantics* considers the class of all possible models. As a mathematical example, this framework would be more appropriate than initial semantics for group theory.

10.5 Example of the Table

10.5.1 Signature of Operations

Given an arbitrary sort U, we want to represent tables of elements of U, considered here as finite sets of elements of U. We will construct them by means of the operations emptytab (the empty table, this operation, intuitively, corresponds to the creation of a new table) and insert (which inserts an element in a table):

    emptytab : table   insert : $U \times$ table $\rightarrow$ table .

Other operations can be designed, for example, removing an element or building the union of two tables:

    remove   : $U \times$ table $\rightarrow$ table
    tabunion : table $\times$ table $\rightarrow$ table .

The search for an item will be specified by the relation in:

    in   : $U \times$ table $\rightarrow$ bool
    search : table $\rightarrow U$.

10.5.2 Axioms

The following axioms express that the order of insertion is not important, and that possible repetitions are not either.

$\forall x, y, t$ insert$(x, $insert$(y, t)) = $insert$(y, $insert$(x, t))$
$\forall x, t$ insert$(x, $insert$(x, t)) = $insert$(x, t)$.

In order to specify the search for an item we need a predicate P over elements of U:

$\forall x$ in$(x, $emptytab$) = $false
$\forall x, t$ in$(x, $insert$(x, t)) = $true
$\forall x, y, t$ $x \neq y \Rightarrow$ in$(x, $insert$(y, t)) = in(x, t)$
$\forall x, t$ search$(t) = x \wedge$ in$(x, t) \Rightarrow P(x)$.

Note that the failure of a search is not completely specified here. We can only say that, in case of failure, the item that is returned is not an element of t. The two axioms given for in are sufficient because we consider that the order of insertion is not relevant.

10.6 Rewriting

It may be tempting to write the — unfortunately flawed — following specification for the search operation:

$$\forall x, t \quad P(x) \Rightarrow \texttt{search}(\texttt{insert}(x, t)) = x .$$

An undesired consequence of this specification is that the table could not contain two distinct items x and y satisfying P! Indeed, a table containing x and y can be put in the two forms $\texttt{insert}(x, t)$ and $\texttt{insert}(y, t')$ — where x and y are members, respectively, of t' and of t. If we also have $P(x)$ and $P(y)$, we get:

$$x = \texttt{search}(\texttt{insert}(x, t)) = \texttt{search}(\texttt{insert}(y, t')) = y .$$

Yet the latter specification of search might seem quite harmless. The lesson we draw from this is that formal statements, alone, are not a panacea. It is crucial to examine their consequences. This is precisely the job performed by deduction tools. In the case of algebraic specifications, they are generally based on rewriting systems, as explained in the previous chapter. Let us illustrate this technique using the example of binary trees.

An example of a property which is quite easy to check is that the left subtree of a tree a (supposed to be in the form $\texttt{bin}(b, c)$), is less deep than the full tree a:

$$a = \texttt{bin}(b, c) \Rightarrow \texttt{inf}(\texttt{depth}(\texttt{lft}(a)), \texttt{depth}(a)) = \texttt{true} .$$

Indeed, after the substitution of $\texttt{bin}(b, c)$ for a in the right-hand side of this implication, we obtain a formula containing $\texttt{lft}(\texttt{bin}(b, c))$ and $\texttt{depth}(\texttt{bin}(b, c))$, which in turn can be rewritten to b and $1 + \texttt{max}(\texttt{depth}(b), \texttt{depth}(c))$, respectively, according to our axioms. The right-hand side can then be written:

$$\texttt{inf}(\texttt{depth}(b), 1 + \texttt{max}(\texttt{depth}(b), \texttt{depth}(c))) = \texttt{true} ,$$

which is easy to solve using arithmetic rules:

$$\begin{aligned} \texttt{inf}(x, 1 + y) &= \texttt{infeg}(x, y) \\ \texttt{infeg}(x, \texttt{max}(x, y)) &= \texttt{true} . \end{aligned}$$

Note that in this example, we always proceeded by replacing the left-hand side with the right-hand side of an equation. In other words, equations were used as rewriting rules, as indicated in § 9.7.

10.7 Notes and Suggestions for Further Reading

The book [BKL$^+$91] provides an overview of algebraic specifications. One of the main approaches to this topic, using so-called *initial* algebras, is developed in

[EM85, EM90], where the reader can find a description of the language ACT1. This language has been reused in the two "formal description techniques" for communication protocols LOTOS — whose control aspects are derived from process algebras in CSP and CCS style — and a (now obsolete) version of SDL.

Another important algebraic specification language is OBJ, described for example in [JKKM92]. Its more recent successors include Maude [CDE$^+$99] and CafeOBJ [DF98].

11. Type Systems and Constructive Logics

This chapter introduces the relationship between typing, logic, and specification. In fact, a type can be viewed as a kind of specification. This analogy can be carried to a fair extent, at least in the framework of the *constructive* approach to logic, already mentioned on page 42. From this perspective, intuitionistic logic turns out to have better features than classical logic.

In passing, we will introduce the λ-calculus, which is both a plain logical tool and an elementary language which is much appreciated for studying fundamental issues in computer science, including questions related to typing. All that will lead us to the topic of the next chapter, devoted to the *calculus of inductive constructions*, a powerful type system implemented in two software systems, Coq and Lego.[1]

11.1 Yet Another Concept of a Type

11.1.1 Formulas as Types

The most general thing we can say about a type is that it is just a non-interpreted formal expression, which can be attached to the concepts of the language we consider (variables, functions, etc.). An object which has a given type is sometimes referred to as an **inhabitant** of this type.

A typing system tells us how to assign a type to an expression of the language, as soon as we know the type of the components of that expression. For example, if f has the type $A \to B$ and if x has the type A, then $f(x)$ has the type B. *A typing system can then be regarded as a formal specification language*, which is more, or less, refined depending on the richness of the typing system. From this perspective, verifying that a program is well typed amounts to proving that it satisfies its specification.

Note that the concept of a type applies not only to programming languages, but to specification languages as well. Thus, in algebraic specification languages, the basic symbols for types are called **sorts**, for example nat, bool, stack. The operators $\times$ and $\to$ allow one to construct compound types, such as nat $\times$ nat $\to$ bool. In this case we have two levels of specification: the typing specifies

---

[1] We also want to mention NuPRL and ALF, which are based on very similar principles.

something about the algebraic specification. In particular, it forces the axioms which come with the declaration of operations to be well typed. The same comment applies to the specification languages based on set-theoretic notations.

Considering that a specification is, in general, a logical formula, we can still go one step further. Types, regarded as formal expressions, become more precisely logical formulas. This was already suggested in the case of dependent types. We get here the first part of the so-called Curry–Howard isomorphism, to be developed below:

type = logical formula .

We will see that, in this framework, × and → are given a simple logical meaning.

11.1.2 Interpretation

As soon as we consider a type as a formula, we can consider interpretations of it, as in model theory. However, model theory does not provide all possible kinds of interpretation. Instead of interpreting the *truth value* of a formula by means of the two values *true* and *false*, one can examine the space of the *proofs* that conclude with this formula. Such objects turn out to be relevant to computer science: they are just computable functions. More exactly, they are *algorithmic (or, intentional) definitions* of computable functions — recall that usually, "function" is used with its extensional meaning, including the phrase "recursive function" in computability theory. A good framework for expressing intentional presentations of recursive functions, and for studying typing systems, is the λ-calculus. We start with the untyped version of this formalism.

11.2 The Lambda-calculus

> *Une fois rien... c'est rien !*
> *Deux fois rien... ce n'est pas beaucoup !*
> *Mais trois fois rien !... Pour trois fois rien, on peut déjà*
> *acheter quelque chose... et pour pas cher !*[2]
>
>   R. DEVOS

The λ-**calculus**, devised by Alonzo Church, formalizes with remarkably restricted means the concept of a computable function. It can be regarded as a programming language, powerful enough for encoding any algorithm, whereas its rudimentary character simplifies the study of a number of fundamental issues, such as computability and typing. The λ-calculus was also used for defining the semantics of programming languages, and it is the archetype of functional languages. Finally, notations of the λ-calculus are very often re-used, including in languages such as Z or B.

---

[2] *One time nothing... this is just nothing! Two times nothing... this is not much! But three times nothing!... with three times nothing, one can buy something... and with little money!*

In the λ-calculus, the notation $\lambda x.\ x+3$ represents the function f such that $f(x) = x+3$. Evaluating $f(5)$, denoted by $((\lambda x.\ x+3)\ 5)$, consists of substituting 5 for x in the body of the function, which yields $5+3$ (and then, 8).

One can get rid of the symbols "+", "3" and "5". The main idea of the λ-calculus is that everything can be represented by one-argument functions, including data structures and control structures. Only one computation mechanism, called the β-reduction, is available. It formalizes what happens when a function is applied to its argument.

11.2.1 Syntax

In the λ-calculus, programs, or functions, are expressed by λ-terms, built up from **variables** denoted by identifiers $x, y, z \dots$ and only three rules:

- a variable is a λ-term;
- λ-**abstraction**: if T is a λ-term and if x is a variable, $\lambda x.\,T$ is a λ-term (intuitively, it represents the function which maps x to T; for example $\lambda x.\ x$ is the identity function);
- **application**: if F and X are λ-terms, the application of F to X is a λ-term denoted by FX.

As usual, parentheses are used for removing ambiguities, for example, for distinguishing $(\lambda x.\ x)y$ from $\lambda x.\ (xy)$. Application has syntactical precedence over abstraction: $\lambda x.\ xy$ is a shorthand for $\lambda x.\ (xy)$.

The concept of a free, or of a bound, variable is similar to the one in predicate calculus, the role of $\forall$ being played by λ. In the same way, the meta-notation $[x := V]\,T$ denotes the substitution of a term V for each free occurrence of x in T. For example, in $x(\lambda x.\ xy)$ the first occurrence of x is free, the second is bound, and y possesses only one occurrence which happens to be free; $[x := (\lambda z.\ z)]\,(x(\lambda x.\ xy))$ represents the term $(\lambda z.\ z)(\lambda x.\ xy)$.

The mechanism of the β-**reduction** is quite natural: when a function F with a parameter x, say $\lambda x.\,T$, is invoked on an argument V (this situation is called a **redex**), V is substituted for all free occurrences of x in the body T of F; more precisely, the λ-term $(\lambda x.\,T)V$ is rewritten as $[x := V]\,T$. For example, if we take $T = x$, $(\lambda x.\ x)V$ is rewritten as V; thus we check that $\lambda x.\ x$ represents the identity function, as one would expect.

We easily see that $\lambda y.\ y$ is also the identity function. More generally, λ-terms are defined up to a renaming of bound variables (this is called α-**conversion**). In practice, such renaming can be performed systematically before every β-reduction in order to avoid confusion. We will always comply with this discipline in the following.

A redex can occur at the top of a term, but also at the top of an arbitrary sub-term. Of course, the scope of substitutions performed by a given β-reduction extends only over the concerned sub-term. A β-reduction step, from T to T', is denoted by $T \xrightarrow{\beta} T'$. For example, we have

$$(\lambda x.\ x)(\lambda y.\ y) \xrightarrow{\beta} \lambda y.\ y \ . \tag{11.1}$$

When we have a (finite) chain of β-reductions going from T to S this is written $T \stackrel{\beta}{\to}^* S$. When a term has no redex, we say that it is **irreducible**, or in **normal form**. Evaluating a λ-term T consists of looking for an irreducible term S such that $T \stackrel{\beta}{\to}^* S$.

11.2.2 The Pure λ-calculus and the λ-calculus with Constants

It turns out to be possible to encode all useful data structures (integers, Booleans, pairs, lists, trees, etc.), as well as the functions which allow one to manipulate them, by means of λ-terms. One can also represent fixed-point operators, and then recursive functions. Thus, this calculus has the maximal expressive power that one can expect.

Nevertheless, it is sometimes convenient to enrich the syntax with additional operations, together with appropriate reduction rules called δ-rules. These operations are called **constants**,[3] the system thus obtained is called the λ-**calculus with constants**. For example, one could introduce the constants $+$, $0...$ 3, 4, $5...$ with rules such as:

$$5+3 \stackrel{\delta}{\to} 8 \ .$$

An important example is the λ-**calculus with pairs**, which introduces three constants: pair formation $\langle\_,\_\rangle$, and the first (respectively the second) projection p_1 (respectively p_2) which, respectively, extract the first and the second elements from a pair. For example, $\lambda x. \langle p_2 x, p_1 x\rangle$ represents the transposition of the elements of a pair. It is necessary to introduce the δ-rules $p_1\langle x, y\rangle \stackrel{\delta}{\to} x$ and $p_2\langle x, y\rangle \stackrel{\delta}{\to} y$. The system without constants is called the **pure λ-calculus**. We consider the latter up to the end of the current section.

11.2.3 Function and Function

We have seen that, as a function, a λ-term takes a function as input and then returns a function. However, the example given in (11.1) illustrates something less common: a function can be applied to itself! Indeed, applying the identity function to the identity function can be written $(\lambda x. x)(\lambda x. x)$, or, preferably $(\lambda x. x)(\lambda y. y)$; after β-reduction we get $\lambda y. y$, which is the identity function again (as expected). Recall that in set theory it is required that *before* defining a function, its domain and its co-domain are defined, and this prevents a function from being applied to itself: a function from A to B is regarded as a member of $\mathcal{P}(A \times B)$, so that it cannot be a member of A. We will see later that, even though a function of the λ-calculus is not interpreted by a set (of pairs), a

---

[3]Admittedly, these constants represent functions, but these functions do not change. One has to distinguish between the *result* of a function, which generally varies when arguments vary, from the function itself. In contrast, letters x, y are variables, which represent (and may be substituted by) arbitrary functions.

type may be assigned to it. The concept of a function in the λ-calculus turns out to be, from this respect, more powerful than the set-theoretic concept of a function.

Actually, functions in the λ-calculus are computation procedures above all. In this respect, they are quite close to the concept of a function used prior to Dedekind and Cantor. We admit that they represent computable functions *par excellence*. Moreover, recall that there are many more set-theoretic functions than computable functions (cf. § 3.3.4).

The difference between set-theoretic functions and computable functions remains at the root of an important issue for formal specification of software. Indeed, the set-theoretic concept of a function is sometimes easier to handle or to understand, than the constructive concept, whereas the latter is the only one available in programming; the set-theoretic concept is then put to the fore. An essential issue in program construction from formal specifications is to exhibit an algorithm computing a function previously presented in an implicit manner, and one hopes that such an algorithm does exist.

11.2.4 Representing Elementary Functions

In order to illustrate more concretely how the λ-calculus can be used, let us show how a number of common programming constructs can be represented. This will shed new light on a mapping between data structures and control structures. Thus, the concept of a pair is associated with the concept of a projection, the concept of a Boolean is associated with the concept of a test, the concept of an integer is associated with the concept of an iteration. In some respect, a data structure is defined by its typical use cases.

11.2.4.1 Preliminary Conventions, Curryfication. We have to agree on a number of notational simplifications. We will also show how to "curryfy" a two arguments function in order to consider it as a one argument function.

Let us consider informally the case of addition, which maps x and y to "$x+y$" . If we fix the first argument to 2 (respectively to u), we get the function "add 2" (respectively "add u") which maps y to "$2+y$" (respectively "$u+y$"); this function is then $\lambda y.\text{"}2+y\text{"}$ (respectively $\lambda y.\text{"}u+y\text{"}$).

Consider that x, y and "$x+y$" have the type nat. The function $\lambda y.\text{"}x+y\text{"}$ then has the type $\text{nat} \to \text{nat}$. The addition is then regarded as the function which maps x, not to an integer, but to the *function* of type $\text{nat} \to \text{nat}$ we have just seen, $\lambda y.\text{"}x+y\text{"}$. In other words, the addition is represented by $\lambda x.(\lambda y.\text{"}x+y\text{"})$ of type $\text{nat} \to (\text{nat} \to \text{nat})$. We agree that this term is also noted $\lambda x.\lambda y.\text{"}x+y\text{"}$ or even more simply, $\lambda xy.\text{"}x+y\text{"}$.

More generally, $\lambda xy.T$ and $\lambda x.\lambda y.T$ represents $\lambda x.(\lambda y.T)$; this convention generalizes to an arbitrary number of arguments. Consistently, the application operation associates *to the left*: gfy does not represent $g(fy)$ as one would expect at a first glance, but $(gf)y$: this expression should be interpreted as the application of the two-arguments (curryfied) function g to f and y. Thus we do have, assuming that the only occurrences of x and y are in T:

$$(\lambda xy. T) U V \overset{\beta}{\to}^* [x, y := U, V] T \ .$$

11.2.4.2 Concept of a Combinator. A **combinator** is a closed λ-term (that is, without free variables). Three examples of combinators are:

$$\mathbf{I} \overset{\text{def}}{=} \lambda x. x \qquad \mathbf{K} \overset{\text{def}}{=} \lambda xy. x \qquad \text{and} \qquad \mathbf{S} \overset{\text{def}}{=} \lambda xyz. (xz)(yz) \ .$$

A theorem states that every combinator can be obtained using only **I**, **K** and **S**. One can even dispense with **I**, because $\mathbf{SKK} \overset{\beta}{\to}^* \mathbf{I}$ (recall that abc is read $(ab)c$).

11.2.4.3 Booleans and Tests. The very purpose of a Boolean b is to choose, from two arguments X and Y, the first if the value of b is "true" and the second otherwise. In a functional language, this would be expressed by:

$$\textbf{if } b \textbf{ then } X \textbf{ else } Y \ . \tag{11.2}$$

Beware: this expression designates the *value* of X or of Y, and not a command. We take for "true" and for "false" two terms denoted by **t** and **f**, respectively, and defined by

$$\mathbf{t} \overset{\text{def}}{=} \lambda xy. x \qquad \text{and} \qquad \mathbf{f} \overset{\text{def}}{=} \lambda xy. y \ .$$

The fact that $\mathbf{t} X Y \overset{\beta}{\to}^* X$ and that $\mathbf{f} X Y \overset{\beta}{\to}^* Y$ allows us to represent the test (11.2) by bXY (which reads: $(bX)Y$).

Now Boolean functions are easy to program. For example, the disjunction is obtained by computing $x \vee y$ by means of **if** x **then t else** y:

$$\textbf{or} \overset{\text{def}}{=} \lambda xy. x \mathbf{t} y \ .$$

11.2.4.4 Integers and Iteration. There are several means of encoding integers with the λ-calculus. The most popular, due to Church, consists of representing the integers n by the function which iterates a function given as an argument n times, that is, intuitively:

$$\lambda f. f^n = \lambda f. \underbrace{f \circ f \cdots \circ f}_{n} = \lambda f x. \underbrace{f(f \cdots (f\, x) \cdots)}_{n} \ .$$

We need to represent the two constructors 0 and S. Let us observe that f^0 is the identity function, while $f^{n+1}(x)$ is $f(f^n(x))$. This idea is implemented in the following combinators:

$$\mathbf{0} \overset{\text{def}}{=} \lambda f. \lambda x. x \qquad \text{and} \qquad \mathbf{S} \overset{\text{def}}{=} \lambda n. \lambda f x. f(n f x) \ .$$

We will also use:

$$\mathbf{1} \overset{\text{def}}{=} \lambda f x. f x \quad , \qquad \mathbf{2} \overset{\text{def}}{=} \lambda f x. f(f x) \quad , \quad \text{etc.}$$

As an illustration of the use of iteration, let us represent arithmetical operations. The addition $m+n$ is obtained by m successive increments of n: "$m+n$" = "$\mathbf{S}^m(n)$" = mSn, that is, formally, **plus** $\stackrel{\text{def}}{=} \lambda mn.\,m\mathbf{S}n$.

In the following we prefer a slightly different definition, which comes directly from $f^{m+n}(x) = f^m(f^n(x))$, that is,

plus $\stackrel{\text{def}}{=} \lambda mn.\,\lambda fx.\,(mf)(nfx)$.

This version is actually shorter than the previous one, because we should expand $\mathbf{S}$ in the latter.

We can get the multiplication $m \times n$ by iterating m times the addition of n to 0, which yields the expression: $\lambda mn.\,m(\lambda x.\,\mathbf{plus}\,n\,x)0$. However we get a shorter definition from $f^{mn} = (f^m)^n$:

mult $\stackrel{\text{def}}{=} \lambda mn.\,\lambda f.\,n(mf)$.

Remarkably, the exponential function is still simpler to represent, since a Church integer is precisely the exponential operation:

exp $\stackrel{\text{def}}{=} \lambda mn.\,nm$.

Finally, comparison to zero is expressed by

zer $\stackrel{\text{def}}{=} \lambda n.\,n(\lambda x.\,\mathbf{f})\mathbf{t}$.

Indeed, $(\lambda x.\,\mathbf{f})^0$ is the identity, which obviously yields $\mathbf{t}$ when applied to $\mathbf{t}$; in contrast, for $n > 0$, $(\lambda x.\,\mathbf{f})^n$ is $\lambda x.\,\mathbf{f}$ which, applied to $\mathbf{t}$ yields $\mathbf{f}$. As an exercise, the reader may calculate **zer(plus 1 x)** and **zer(plus 0 0)**.

11.2.4.5 Pairs and Projections. The combinator for constructing a pair takes two "data" items X and Y as inputs (for example integers, Booleans, but actually arbitrary λ-terms) and it returns $\langle X, Y \rangle$. The first (respectively the second) projection takes a pair as input and it returns X (respectively Y). The abstract type "pair" is actually characterized by three functions `pair`, `pr1` and `pr2` which must verify $\text{pr1}(\text{pair}(x,y)) = x$ and $\text{pr2}(\text{pair}(x,y)) = y$.

Natural definitions of the curryfied projections **pc1** and **pc2** are $\lambda xy.\,x$ and $\lambda xy.\,y$. Let us represent the construction of the pair $\langle x, y \rangle$ by a term taking a curryfied projection p as an argument and applies it to x and y:

$\langle x, y \rangle \stackrel{\text{def}}{=} \lambda p.\,pxy$ $\qquad$ and $\qquad$ **pair** $\stackrel{\text{def}}{=} \lambda xy.\,\lambda p.\,pxy$.

The projections are then in the form $\lambda c.\,c\pi$ where c is a pair and π is a curryfied projection:

pr1 $\stackrel{\text{def}}{=} \lambda c.\,c(\lambda xy.\,x)$ $\qquad$ and $\qquad$ **pr2** $\stackrel{\text{def}}{=} \lambda c.\,c(\lambda xy.\,y)$.

11.2.4.6 Paradoxical and Fixed-point Combinators. The paradoxical combinator is

$$\Omega \overset{\text{def}}{=} (\lambda x. xx)(\lambda x. xx) \ .$$

It has a feature which was not present in the combinators introduced so far: there is a redex inside it. One can even perform an arbitrarily high number of successive β-reductions from Ω, since $\Omega \overset{\beta}{\to} \Omega$. This term represents a "looping" program. On first inspection, one might blame $\lambda x. xx$, because it contains the "self-application" xx; but there are combinators T such that $(\lambda x. xx)T$ converges (terminates). The simplest is **I**:

$$(\lambda x. xx)(\lambda x. x) \overset{\beta}{\to} (\lambda x. x)(\lambda x. x) \overset{\beta}{\to} (\lambda x. x) \ .$$

We can even get an infinite number of such terms, by taking for T the Church encoding of an arbitrary integer n: this yields a representation of n^n which, after successive reductions, reaches the normal form

$$\lambda fx. \underbrace{f(f \cdots (f x) \cdots)}_{n^n} \ .$$

This is an example of a term yielding a very long chain of β-reductions.

A slight modification of Ω provides a **fixed-point combinator**:

$$\mathbf{Y} \overset{\text{def}}{=} \lambda f. \big(\lambda x. f(xx)\big)\big(\lambda x. f(xx)\big) \ ,$$

which, applied to any term F, yields after a β-reduction step a term F' that reduces itself to FF', hence the infinite chain:

$$\mathbf{Y}F \overset{\beta}{\to} F' \overset{\beta}{\to} FF' \overset{\beta}{\to} F(FF') \overset{\beta}{\to} F(F(FF')) \overset{\beta}{\to} \cdots \ . \tag{11.3}$$

This term then yields an infinite loop too, *but not necessarily*! Let us consider a "recursive" definition of the form $g = \lambda x. G$, where G contains free occurrences of g. We know that, from a semantical viewpoint, we should interpret it as "g is the least fixed point of G" (cf. § 3.6). Even though the set-theoretical concept of a function turns out too narrow for developing this idea in λ-calculus, the intuitive idea remains valid: in some sense, $\mathbf{Y}F$ is a fixed point of F. Let us see what happens if we take $F = \lambda gx. G$ and we apply $\mathbf{Y}F$ to a given term t. In the example of the sequence of Fibonacci, G would be a translation of

if $x = 0 \vee x = 1$ then 1 else $g(x - 1) + g(x - 2)$.

When we reach $FF't$, that is, $(\lambda gx. G)F't$, two new redexes appear successively, so that we get $[g := F', x := t]G$. If the function G is programmed as expected, this term contains redexes that, intuitively, check the value of x. Then there are two options:

- "recursive" case, the β-reduction yields a term containing other occurrences of g — but recall that g has been replaced with F'; then a reduction step $F' \xrightarrow{\beta} FF'$ can be applied, and we get one or several sub-terms under the form $FF't_i$, so that we have a situation similar to the previous one, with new values for t ("$t-1$" and "$t-2$" in the above example);
- "base" case, the β-reduction yields a term which contains no occurrence of g ("1" in the above example); we no longer enter in the infinite loop (11.3); all additional reductions that may have been performed on F' turn out to be useless.

To sum up, we observe that suitable applications of the β-reduction mechanism allow us to simulate the evaluation of a "recursive" function.

11.2.5 Functionality of β-reduction

We presented β-reduction as an evaluation mechanism which transforms any given term into an irreducible one. A number of phenomena appear in previous examples:

1. As a term may contain several redexes, it can be reduced in several ways; is it possible to get different irreducible forms from the same term?
2. In a number of cases, such as Ω, a term does not possess a normal form; can this be decided *a priori*?
3. In other situations, a term can be transformed into an irreducible form, along some paths, while successive reductions along other paths do not terminate. A very simple case is $(\lambda x. T)\Omega$, where T is a normal term containing no free occurrence of x. Can we find a strategy for the choice of redexes such that an irreducible form will eventually be reached if there is one?

The first issue is about functionality: we wish the result to depend only on the initial expression, and not on the manner of performing computations. This property is called **confluence**, or the Church–Rosser property. The following result is quite difficult to prove.

Theorem 11.1 (Church–Rosser)
If a term T can be reduced to two different terms U and V, that is, if $T \xrightarrow{\beta}{}^* U$ and $T \xrightarrow{\beta}{}^* V$, then there exists a term W such that $U \xrightarrow{\beta}{}^* W$ and $V \xrightarrow{\beta}{}^* W$.

As a corollary, if the irreducible form exists, it is unique (hence the name **normal form**).

This result can also be regarded as stating a kind of **consistency** of the λ-calculus, in the following sense. Let us consider $\xleftrightarrow{\beta}$, the reflexive and symmetrical closure of $\xrightarrow{\beta}{}^*$. By construction, this is an equivalence relation. Each term is in a unique equivalence class modulo $\xleftrightarrow{\beta}$, and each class represents the "value" denoted by one of its members. For example, we have (**or t f**) $\xleftrightarrow{\beta}$ (**or t t**). It is clear that β-reductions preserve the equivalence class. However, it is important that the values represented by **t** and **f** are different, that all the values of Church integers are different, and similarly for each data structure: otherwise computations would have no interest, one would say that the calculus is inconsistent. The Church–Rosser property guarantees the consistency of the calculus because the normal form forms of **t** and **f** are syntactically different (and similarly for Church integers).

The second and third point concern the termination of computations. The answer to the second can only be negative, because the λ-calculus has the power of Turing machines. In contrast, a strategy exists which reaches the normal form if it exists. This strategy, called the **normal** strategy, consists of reducing the redex whose "λ" is on the left-most position. For example, in $(\lambda x.\,T)\Omega$, one has to choose the reduction $(\lambda x.\,T)\Omega \overset{\beta}{\to} [x := \Omega]T$ (which yields T if T does not contain a free occurrence of x), and not the redex which is inside Ω. These two results are summarized in the following theorem.

Theorem 11.2
The existence of a normal form of an arbitrary λ-term is a semi-decidable problem.

Confluence and termination properties play a pivotal role in the study of rewriting systems. The tools developed in the framework of the λ-calculus are widely used in this theory.

To summarize, we can recall that the λ-calculus is a formalism well suited for representing the concept of a computable function (or of a recursive function, as defined in § 3.7). Indeed, it is a consistent and Turing-complete calculus. In practice, it is present in several important specification languages, and also in functional programming languages, though the notations used there are more user-friendly. In these languages, integers and other data structures are generally represented by common encodings close to the machine, instead of λ-terms, for obvious reasons of efficiency. However, understanding the behavior of Church integers and other combinators is quite useful, because they are a good illustration of fundamental manipulations of functions to be met in the practice of modeling and of programming.

11.3 Intuitionistic Logic and Simple Typing

11.3.1 Constructive Logics

In mathematics, it can sometimes occur that one proves the existence of an object verifying some property without exhibiting this witnessing object. A frequently cited example is the proof that there is an irrational number r such that $r^{\sqrt{2}}$ is a rational — so that we have two irrational numbers r and s such that r^s is rational. Consider $a \overset{\text{def}}{=} \sqrt{3}$ and $b \overset{\text{def}}{=} a^{\sqrt{2}}$. If b is rational, we can take $r = a$; otherwise, we can take $r = b$ because $(a^{\sqrt{2}})^{\sqrt{2}} = \sqrt{3}^2$. By the law of the excluded middle, the existence of r is ensured, without needing to say whether r is a or b.

Such proofs are called non-constructive, because they do not provide an effective manner to obtain the witness possessing the desired property. Constructive proofs are, however, quite common. A simple case is when the witness is explicitly provided, for example 3 in the property $\exists x \; 2x = 6$. Many proofs by induction are constructive, because they implicitly contain a construction process allowing one to compute a witness. For example, in order to prove that every integer is even or odd, the induction step consists of taking an integer n which is already in the form $2k$ or $2l + 1$, and then proving that $n + 1$ is in the form $2k'$ or $2l' + 1$; here one has to consider $k' = l + 1$ and $l' = k$. During this proof, we implicitly described an algorithm which performs the Euclidian division of n by 2.

Most theorems in basic arithmetic are proved constructively, as well as theorems which are involved in program proving. Note that the axiom of choice, in set theory, is essentially non-constructive. Actually, it is scarcely used in computer science, because one is often interested in the algorithmic contents of proofs. Unfortunately, classical logic, that is, the kind of logic that everyone uses regularly, turns out to be inappropriate for the development of constructive proofs.

We need constructive logics, which not only allow us to extract an algorithmic content from proofs, but provide proof spaces themselves with an interesting mathematical structure. The most commonly used constructive logic in computer science is *intuitionistic logic*, which originated at the beginning of the 20th century. In constructive logics, provability and proof structure become more important concepts than truth values.

More recently, a considerable amount of research work has been devoted to a new and promising constructive logic called *linear logic*. We want to also mention that subtle variants of classical logic can be made constructive. This is a recent discovery, related to the interpretation of control structures such as exceptions and `call/cc` [Gir91, Gri90, Mur91].

11.3.2 Intuitionistic Logic

Intuitionistic logic was already presented in Chapter 9. In the following we rely on natural deduction. Let us recall that, in the system NJ, one manipulates deductions made under some hypotheses. These are displayed in a tree whose root contains the conclusion. A finished proof is a deduction where all hypotheses have been discharged. The formation of deductions can be interpreted in the following manner (disjunction and quantifiers will be discussed later):

$\wedge_i$ given a deduction a of A and a deduction b of B, one forms a deduction of $A \wedge B$; the latter is represented by the pair $\langle a, b \rangle$;

$\wedge_e$ given a deduction c of $A \wedge B$, under the form $\langle a, b \rangle$, one forms a deduction of A (respectively of B) which is a (respectively b), obtained from c by a projection;

$\Rightarrow_e$ given a deduction f of $A \Rightarrow B$ and a deduction a of A, one forms a deduction of B; f is then regarded as a function which maps every deduction of A to a deduction of B;

$\Rightarrow_i$ given a deduction of B under the hypothesis A, one forms a deduction of $A \Rightarrow B$; it can be regarded as a function from the space of deductions concluding to A, to the space of deductions concluding to B;

$\bot$ the space of deductions of $\bot$ is empty.

This interpretation is the **interpretation of Heyting**, also called the BHK interpretation (Brouwer–Heyting–Kolmogoroff). Here, the semantics of a proposition P is not a truth value, but the space of proofs concluding to P. The propositions which can be proved with the connectors considered here are essentially implications (for example $P \Rightarrow P$ or $P \wedge Q \Rightarrow Q \wedge P$), since the conjunctions which can be proved are conjunctions of implications (for example $(P \Rightarrow P) \wedge (P \wedge Q \Rightarrow Q \wedge P)$). The main spaces we consider are then essentially sets of functions.

We use the term "space" because the study of the semantics of these objects shows that they live in spaces, in the common mathematical sense, that is, sets endowed with algebraic or topological properties.

11.3.3 The Simply Typed λ-calculus

It will be seen that these functions are actually nothing but λ-terms, more precisely terms of the simply typed λ-calculus with pairs which we introduce now.

First we define types as expressions formed by the means of **type variables** A, B, C, etc. and of binary connectors $\rightarrow$ and $\times$. The terms of the simply typed λ-calculus with pairs are just the terms of the λ-calculus with pairs which are compatible with typing rules. Thus, we are given, for each type τ, typed variables of type τ, for example $x : A$, $y : A$, $z : A \times B$, $t : A \rightarrow B$. Let τ and σ be types:

- if x is a variable of type σ and if t is a λ-term of type τ, $\lambda x{:}\sigma.\, t$ is a λ-term of type $\sigma \rightarrow \tau$; to simplify reading, we also write $\lambda x^\sigma.\, t$;
- if f is a λ-term of type $\sigma \rightarrow \tau$ and if s is a λ-term of type σ, $f s$ is a λ-term of type τ;
- if s and t are λ-terms of types σ and τ, respectively, $\langle s, t \rangle$ is a λ-term of type $\sigma \times \tau$;
- if c is a λ-term of type $\sigma \times \tau$, $p_1 c$ and $p_2 c$ are λ-terms of types respectively σ and τ.

Each typed λ-term can trivially be mapped to an untyped λ-term: just remove typing information.

11.3.3.1 Examples.

- $\lambda x^A.\, x$ has the type $A \rightarrow A$; the underlying untyped λ-term is $\lambda x.\, x$;

- $\lambda x^A.\,\lambda y^B.\,x$ has the type $A \to B \to A$; the underlying untyped λ-term is $\lambda x.\,\lambda y.\,x$;
- $\lambda f^{A \times B \to C}.\,\lambda x^A.\,\lambda y^B.\,f\langle x, y \rangle$ has the type $(A \times B \to C) \to (A \to B \to C)$; the underlying untyped λ-term is $\lambda f.\,\lambda xy.\,f\langle x, y \rangle$ — this is curryfication.

11.3.3.2 Properties. When T is the underlying untyped λ-term of a typed λ-term of type τ, we say that T is **typable** with type τ. As one may expect, β-reduction is compatible with typing: if $T \overset{\beta}{\to} S$ and if T is typable with type τ, then S is typable with type τ. Theorem 11.3 will soon provide a much more interesting property.

11.3.4 Curry–Howard Correspondence

The semantics of Heyting amounts to interpreting deductions by λ-terms of the above system. In this interpretation, $\wedge$ is regarded as a *product*, since proof of $A \wedge B$ boils down to a proof of A and a proof of B. The connector $\Rightarrow$ is still more interesting: $A \Rightarrow B$ allows one to construct a proof of B from any proof of A; implication is then interpreted as the construction of a space of functions. This yields the following systematic translation:

- a proposition P is translated into a type $P^\sharp$, obtained from P by replacing $\Rightarrow$ with $\to$ and $\wedge$ with $\times$;
- a packet of hypotheses P (see page 161) is translated into a variable $x : P^\sharp$ (or $x^{P^\sharp}$);
- if two deductions of A and B are respectively translated into the λ-terms a of type $A^\sharp$ and b of type $B^\sharp$, the deduction of $A \wedge B$ obtained by $\wedge_i$ is translated into $\langle a, b \rangle$ of type $A^\sharp \times B^\sharp$;
- if a deduction of $A \wedge B$ is translated into the λ-term c of type $A^\sharp \times B^\sharp$, the deduction of A (respectively of B) obtained by $\wedge_{e1}$ (respectively $\wedge_{e2}$) is translated into $p_1 c$ (respectively $p_2 c$);
- if a deduction of $A \Rightarrow B$ is translated into the λ-term f of type $A^\sharp \to B^\sharp$, and if a deduction of A is translated into the λ-term a of type $A^\sharp$, the deduction of B obtained by $\Rightarrow_e$ is $f a$ of type $B^\sharp$;
- if a deduction of B, done under a packet of hypotheses A, is translated into the λ-term b of type $B^\sharp$ — the latter must contain a free variable of type $A^\sharp$, say x, which translates the packet of hypotheses — the deduction of B obtained by $\Rightarrow_i$ by discharging this packet of hypotheses is translated into $\lambda x^{A^\sharp}.\,b$ of type $A^\sharp \to B^\sharp$.

Conversely, every λ-term of type τ can be regarded as a deduction concluding to $\tau^\flat$ which is τ after the replacement of $\times$ with $\wedge$ and of $\to$ with $\Rightarrow$. The two reciprocal translations $\sharp$ and $\flat$ constitute the Curry–Howard correspondence (also called the Curry–Howard isomorphism), which can be summarized as follows:

$$\begin{aligned} proposition &= type &, \\ proof &= function &, \end{aligned}$$

where "function" should be understood as "λ-term". This sheds a new light on the typed λ-calculus: it is a concise notation for deductions. Here are some examples:

$$\left.\begin{array}{c} \overset{(x)}{\overbrace{A}} \\ \hline A \Rightarrow A \end{array} \Rightarrow_{i(x)} \right\} \text{ is noted } \lambda x^A.\, x \ ,$$

$$\left.\begin{array}{c} \overset{(x)}{\overbrace{A}} \\ \hline B \Rightarrow A \end{array} \Rightarrow_{i(y)} \atop \dfrac{}{A \Rightarrow B \Rightarrow A} \Rightarrow_{i(x)} \right\} \text{ is noted } \lambda x^A.\, \lambda y^B.\, x \ ,$$

$$\left.\begin{array}{c} \dfrac{\overset{(f)}{\overbrace{A \wedge B \Rightarrow C}} \qquad \dfrac{\overset{(x)}{\overbrace{A}} \quad \overset{(y)}{\overbrace{B}}}{A \wedge B}\wedge_i}{C} \Rightarrow e \end{array} \right\} \text{ is noted } f\langle x, y \rangle \ ,$$

$$\left.\begin{array}{c} \dfrac{\dfrac{\overset{(f)}{\overbrace{A \wedge B \Rightarrow C}} \qquad \dfrac{\overset{(x)}{\overbrace{A}} \quad \overset{(y)}{\overbrace{B}}}{A \wedge B}\wedge_i}{C}\Rightarrow e}{\dfrac{\dfrac{B \Rightarrow C}{A \Rightarrow B \Rightarrow C} \Rightarrow_{i(y)}}{(A \wedge B \Rightarrow C) \Rightarrow (A \Rightarrow B \Rightarrow C)} \Rightarrow_{i(x)}} \Rightarrow_{i(f)} \end{array} \right\} \begin{array}{l} \text{is noted} \\[6pt] \lambda f^{A \wedge B \Rightarrow C}.\, \lambda x^A.\, \lambda y^B.\, f\langle x, y \rangle \ . \end{array}$$

From now on, types will be considered and noted as propositions, and we will take "$\rightarrow$" as the symbol for implication.

It is then natural to ask how to interpret the β-reduction from the perspective of logic. In other words, what is the meaning of the evaluation of a function in the space of proofs? In order to simplify the discussion, let us here limit ourselves to the implicative fragment of NJ, that is, the fragment having $\rightarrow$ as its only connector. The corresponding λ-calculus is the **simply typed λ-calculus**.

Let us write the deduction corresponding to a λ-term containing a redex $(\lambda x^A.\, b)\, a$ where a has the type A and b has the type B:

$$\dfrac{\dfrac{\begin{array}{c} \overset{(x)}{\overbrace{A}} \\ \vdots\, b \\ B \end{array}}{A \rightarrow B} \rightarrow_{i(x)} \qquad \begin{array}{c} \vdots\, a \\ A \end{array}}{B} \rightarrow e \ . \tag{11.4}$$

The deduction corresponding to $[x := a]\, b$ is:

$$
\begin{array}{l}
\vdots\, a \\
A \\
\vdots\, b \\
B
\end{array}
\qquad (11.5)
$$

In (11.4), the deduction a can be seen as a proof of the lemma A, and this lemma is used an arbitrary number of times in b for proving B. In (11.5), each occurrence of the hypothesis A which is a member of the packet denoted by x, is replaced with the deduction a in b. In some sense, the proof (11.5) is more direct than (11.4), because it avoids the passage by $A \to B$, which is not a sub-formula of the conclusion B to be proved. The transformation of (11.4) into (11.5) is called a **normalization step**. This provides the third part of the Curry–Howard correspondence:

$$
normalization \;=\; \beta\text{-}reduction \quad.
$$

This transformation is similar to cut-elimination in the sequent calculus. The deduction (11.4) corresponds to:

$$
\dfrac{\dfrac{\Gamma, A \vdash B}{\Gamma \vdash A \to B}\ {\to}_r \qquad \dfrac{B \vdash B \qquad \Gamma' \vdash A}{\Gamma', A \to B \vdash B}\ {\to}_l}{\Gamma, \Gamma' \vdash B}\ \text{cut} \qquad.
\qquad (11.6)
$$

while (11.5) corresponds to:

$$
\dfrac{\Gamma' \vdash A \qquad \Gamma, A \vdash B}{\Gamma, \Gamma' \vdash B}\ \text{cut} \qquad.
\qquad (11.7)
$$

The transition from (11.6) to (11.7) is the one that was already discussed on page 167. By iterating such steps, one reaches a proof without lemma, also called a **normal** proof, as in the Hauptsatz of Gentzen. We have even a theorem stating that this normal form is reached *whichever reduction strategy is employed*: this property is called **strong normalization**. One says also that the relation $\overset{\beta}{\to}$, restricted to typed terms, is Noetherian (see page 52).

Theorem 11.3
The above procedure for transforming deductions (11.4) $\to$ (11.5) in the implicative fragment of NJ, or, equivalently, the β-reduction of simply typed λ-calculus, has the strong normalization property.

We have seen that there exist λ-terms, such as Ω and $\mathbf{Y}F$, which are not normalizable. There is no contradiction, because these terms are not simply typable. Actually, even $\lambda x.\, xx$ is not.

Let us take stock. The untyped λ-calculus has the power of Turing machines. In general one cannot know in advance whether the evaluation of a λ-term t

does or does not terminate. Moreover, it can terminate for some reduction strategies only, for example, if t encodes a recursive function by the means of a fixed-point combinator. We also learned that the evaluation of t terminates in all cases if t is simply typed. We will see that there are yet other strongly normalizable terms.

11.4 Expressive Power of the Simply Typed λ-calculus

The fact that the simply typed λ-calculus prohibits some terms is, by itself, quite legitimate. The very purpose of a typing discipline is precisely to detect irrelevant combinations, such as the application of a function defined over integers to a Boolean value. But what is preserved from the expressive power of the λ-calculus? Let us consider some of the combinators presented above.

11.4.1 Typing of the Natural Numbers

A possible typing of $\lambda f x. \underbrace{f(f \cdots (f\, x) \cdots)}_{n}$ is

$$\lambda f^{X \to X}. \lambda x^X. \underbrace{f(f \cdots (f\, x) \cdots)}_{n}$$

which has the type $(X \to X) \to (X \to X)$. Any other expression where X is uniformly substituted for a given proposition φ is also suitable; we will abbreviate this formula to N_φ. It is easy to verify that we can give $\mathbf{0}$ the type N_X, $\mathbf{S}$ the type $N_X \to N_X$, **plus** and **mult** the type $N_X \to N_X \to N_X$.

The exponential function $\mathbf{exp} \stackrel{\text{def}}{=} \lambda mn.\, nm$ raises a problem. If we give m and the result the type N_X, we are driven to give n the type $N_X \to N_X$, which yields $\lambda m^{N_X}. \lambda n^{N_X \to N_X}.\, nm$ of type $N_X \to (N_X \to N_X) \to N_X$, while one would expect the same type as for **plus** and **mult**.

Indeed, note that $N_X \to N_X$ is also of the form N_φ: it is $N_{X \to X}$. However, it is not very satisfactory to have to consider different formulas for the type of the natural numbers within the same term.[4]

We have a more serious issue:[5] it is impossible to give a type to a term as simple as $\lambda n.\, (\mathbf{exp}\, n\, n)$ — which can be simplified to $\lambda n.\, nn$ — and, more generally, to any term in which one would use a variable with different instances of the same type.

---

[4]Note that a similar problem would be raised with other definitions of the addition and of the multiplication, for example **plus** $\stackrel{\text{def}}{=} \lambda mn.\, m\mathbf{S}n$.

[5]Warning: we consider here *one* among the possible encodings of the function that maps n to n^n. Other encodings admit a simple type, but they are more complicated.

11.4.2 Typing of Booleans

A typing of $\lambda xy.\,x$ is $\lambda x^X.\,\lambda y^Y.\,x$ of type $X \to Y \to X$, and a typing of $\lambda xy.\,y$ is $\lambda x^X.\,\lambda y^Y.\,y$ of type $X \to Y \to Y$. As we want **t** and **f** to have the same type, we are led to take $X = Y$. The Boolean type is then $X \to X \to X$ (abbreviated to B_X) or any instance of B_X.

Consider a possible expression for the negation: $\lambda b.\,b\mathbf{ft}$. Assume we give **t** and **f** the type B_X, we are led to taking the expression $B_X \to B_X \to B_X$, as the type of b; this expression is of the form B_φ, as desired, with $\varphi = B_X$. But this yields $B_{B_X} \to B_X$ as the type of $\lambda b.\,b\mathbf{ft}$, so we again get a quite unsatisfactory situation, as with the exponential. We can, however, use another expression for the negation, which is $\lambda b.\,\lambda xy.\,byx$, that is, $\lambda b^{X \to X \to X}.\,\lambda x^X.\,\lambda y^X.\,byx$, whose type is $B_X \to B_X$ as expected. We get a similar problem if we encode disjunction by $\lambda bc.\,b\mathbf{t}c$, whereas $\lambda bc.\,\lambda xy.\,bx(cxy)$ has the type $B_X \to B_X \to B_X$.

There is no relationship between the terms **t** or **f** and the logical interpretation of the typing system. The type of Booleans is essentially an enumerated type with two values, which could quite legitimately be named **aa** and **bb** instead of **t** and **f**; the fact that the latter convention is preferred can be regarded as tradition. It is sometimes convenient to introduce an enumerated type with, for example, three values. It would be $C_X \stackrel{\mathrm{def}}{=} X \to X \to X \to X$, it is inhabited by $\lambda x^X.\,\lambda y^X.\,\lambda z^X.\,x$, $\lambda x^X.\,\lambda y^X.\,\lambda z^X.\,y$ and $\lambda x^X.\,\lambda y^X.\,\lambda z^X.\,z$. For this type one would get a "three-cases if" control structure.

Several types can be given to the same λ-term. For example, a possible typing of $\lambda f.\,\lambda x.\,x$ is $\lambda f^{X \to X}.\,\lambda x^X.\,x$, that is, **0**; another is $\lambda f^X.\,\lambda x^X.\,x$, that is, **f**.

11.4.3 Typing of the Identity Function

The aforementioned problems can readily be observed in a very simple example, the identity function: $\lambda x.\,x$ is typable by $X \to X$ and by every proposition under the form $\varphi \to \varphi$. What do we think about the term $(\lambda x.\,x)(\lambda x.\,x)$? Possible typings are of the form $(\lambda x^{\varphi \to \varphi}.\,x)(\lambda x^\varphi.\,x)$, which forces us to consider two different identity functions within the same expression.

11.4.4 Typing of Pairs, Product of Types

It is not difficult to propose a type for the curryfied projections **pc1** and **pc2**, with **pc1** $\stackrel{\mathrm{def}}{=} \lambda xy.\,x$ and **pc2** $\stackrel{\mathrm{def}}{=} \lambda xy.\,y$: just take $X \to Y \to X$ and $X \to Y \to Y$. However, the implicative fragment of NJ turns out to be insufficient for coping with pair formation. The typed version of the combinator **pair** $\stackrel{\mathrm{def}}{=} \lambda xyp.\,pxy$ is of the form:

$$\mathbf{pair} = \lambda x^X.\,\lambda y^Y.\,\lambda p^{X \to Y \to U}.\,pxy \ .$$

The variable p represents here a projection, which means that U must be either X, or Y. However X and Y are *a priori* distinct — for example, if we want to form pairs composed of an integer and of a Boolean.

This drives us to consider a λ-calculus where the formation of pairs and projections are primitive: this is the simply typed λ-calculus with pairs. Its typing system corresponds to the fragment $\{\rightarrow, \wedge\}$ of NJ. Note that, in propositional logic, $\wedge$ cannot be defined using $\rightarrow$ only.

11.4.5 Sum Types

Given two types σ and τ, we can form their sum $\sigma + \tau$. Let y be an inhabitant of $\sigma + \tau$, y comes from an inhabitant u which is either in σ, or in τ.

How can we use y? A function from $\sigma + \tau$ to φ is obtained by providing a function f of type $\sigma \rightarrow \varphi$ and a function g of type $\tau \rightarrow \varphi$. Then one considers a construction **case** yfg, designed in such a way that, if u is of type σ, the result is obtained by applying f to u and if u is of type τ, the result is obtained by applying g to u.

In order to form y, we are given two injections, which are **i1** of type $\sigma \rightarrow \sigma + \tau$, and **i2** of type $\tau \rightarrow \sigma + \tau$. We assume that the abstract type **sum** satisfies **case** $(\mathbf{i1}\, s)fg = fs$ and **case** $(\mathbf{i2}\, t)fg = gt$.

In the untyped λ-calculus, these operations can be represented by

$$\mathbf{i1} \stackrel{\text{def}}{=} \lambda s.\lambda fg.fs \ ,$$
$$\mathbf{i2} \stackrel{\text{def}}{=} \lambda t.\lambda fg.gt \ ,$$
$$\mathbf{case} \stackrel{\text{def}}{=} \lambda xfg.xfg \ .$$

Again, it is not possible to give a satisfactory type to these operations. As for the product, the sum cannot be recovered from $\rightarrow$ only. It cannot be constructed from $\rightarrow$ and $\times$ either; we thus need a further extension.

The strong normalization theorem stated on page 217 can be extended to the whole NJ calculus; that is, to the simply typed λ-calculus with pairs and sums.

From the viewpoint of logic, the sum corresponds exactly to the intuitionistic disjunction: $\vee_{i_1}$ corresponds to **i1**, $\vee_{i_2}$ to **i2**, $\vee_e$ to **case**. An inhabitant of $\sigma + \tau$ is either an inhabitant of σ, or an inhabitant of τ; similarly, a deduction of $S \vee T$ is formed either from a deduction of S, or from a deduction of T. Moreover, we are able to know which is the right case, depending on whether $\vee_{i_1}$ or $\vee_{i_2}$ was applied: this is typical of the intuitionistic disjunction.

There is, however, a subtle point: if $(\mathbf{i1}\, s)$ is of type $\sigma + \tau$, $(\lambda x.\, x)(\mathbf{i1}\, s)$ is also of type $\sigma + \tau$. There are actually an infinite number of terms of type $\sigma + \tau$ which are not of the form $(\mathbf{i1}\, s)$ or $(\mathbf{i2}\, t)$. Then, how can we justify that every inhabitant of $\sigma + \tau$ comes from a term of type σ or of type τ — and that we know which one? Precisely because of the strong normalization property, which entails that every term of type $\sigma + \tau$ reduces to a *normal* term of type $\sigma + \tau$, and that the latter is necessarily of the form $(\mathbf{i1}\, s)$ or $(\mathbf{i2}\, t)$.

Comment. In the set-theoretic interpretation of types, if we denote the interpretation of the type φ by $\|\varphi\|$, $\|\sigma + \tau\|$ is a **disjoint union** $\{1\} \times \|\sigma\| \cup \{2\} \times \|\tau\|$. Recall that the union cannot be used in a naïve manner because, if $\|\sigma\|$ and $\|\tau\|$ share elements, their origin cannot be distinguished in $\|\sigma\| \cup \|\tau\|$.

11.4.6 Paradoxical and Fixed-point Combinators

We already mentioned that there is no simple type for the paradoxical combinator Ω. The same is true for fixed-point combinators such as **Y**. This is more problematic because they provide a very important expressiveness. In particular they are crucial for simulating a Turing machine. We may add that this is precisely why functional languages such as ML include a typed fixed-point constant (syntactically, it is presented in the form of a **let rec** construct).

11.4.7 Summary

The previous examples illustrate the benefits of simple typing, as well as its limitations: the constraints of simple typing turn out to make it reject too many λ-terms. The expressive power left with the calculus is insufficient for the needs of programming, even if primitives for the product and the sum are introduced. Recursion is not allowed, and iteration itself cannot be employed to its full extent — remember $\lambda n. (\exp n\, n)$. On the other hand, the typing system considered above is still far from what is needed in specification languages. For example, all functions mapping an integer to a Boolean are indiscriminately put in the same category, whereas it would make sense to distinguish the functions which, say, return "true" if their argument is an even integer less than 100.

These two issues can be attacked by generalizing the typing system, and this is done in two independent directions. In both cases, this amounts to considering a more powerful constructive logic. A means to greatly increase the number of typable functions is introducing *second-order* quantification, over propositional variables. To allow for richness of expression, we gain polymorphism. In the second direction, introducing first-order variables and related quantifiers provides a system which includes dependent types, which are interesting for specification purposes.

Note that, in compensation for its coarseness, simple typing has a feature of interest to secure prototyping languages: **type inference**. As indicated by its name, this mechanism infers the type of an expression or of a program where minimal or even no typing information is given explicitly.[6] The typing system implemented in functional languages such as ML and Haskell is a kind of simple typing extended to recursive constructs, in a manner such that type inference is still possible.

---

[6]Type inference relies on the use of a unification algorithm to type expressions. Unification was already described in Chapter 9: it is one of the main basic tools of automated proof.

11.5 Second-Order Typing: System F

The **system F** was devised by Girard and independently rediscovered by Reynolds. It is built on a single logical connector, implication, and on second-order quantification. We will see that in the presence of the latter, the other intuitionistic connectors can be defined.

Let us first illustrate some intuitive ideas behind second-order quantification, by starting with the following deduction of $P \to P$:

$$\begin{array}{c} \overbrace{P}^{(x)} \\ \hline P \to P \end{array} \to_{i(x)}$$

We then deduce $\forall^2 P\ P \to P$ (in the current section we distinguish second-order from first-order quantification by using $\forall^2$ in the former case and $\forall$ in the latter). The second-order quantifier can then be eliminated by substituting an arbitrary proposition for P. This yields, for example:

$$(A \to B) \to (A \to B) \ .$$

We can even substitute $\forall^2 P\ P \to P$ for P, and we get

$$(\forall^2 P\ P \to P) \ \to \ (\forall^2 P\ P \to P) \ .$$

Now consider the deduction:

$$\begin{array}{c} \overbrace{Q \to P}^{(x)} \quad Q \\ \hline P \\ \hline (Q \to P) \to P \end{array} \begin{array}{l} \to_e \\ \\ \to_{i(x)} \end{array} \ .$$

We can deduce $\forall^2 P\ (Q \to P) \to P$, without incident because we still have the hypothesis Q. In contrast, it would be manifestly incorrect to deduce $\forall^2 Q\ (Q \to P) \to P$. For example, $[Q := P]\,((Q \to P) \to P)$ does not hold, even under the hypothesis Q, which does not intervene. The most simple case of that kind is the trivial deduction of P under the hypothesis P; we certainly don't want to deduce $\forall^2 P\ P$! This unprovable proposition actually provides a representation of the absurd proposition $\bot$ of NJ. In the light of the preceding remarks, we can give the introduction and elimination rules of $\forall^2$ (Figure 11.1), the first rule being constrained by a proviso: the deduction of φ must make no hypothesis over P.

Let us consider again the theorem $\forall^2 P\ P \to P$. We have seen that the quantifier carries over the space of all propositions, including $\forall^2 P\ P \to P$. We recognize here the impredicativity previously encountered in set theory. We are going to employ techniques similar to the ones used in § 7.3.1, for inductively defining the product, the sum, the natural numbers, trees, etc.; but here we will

$$\frac{\varphi}{\forall^2 P\ \varphi}\ \forall_i^2 \qquad\qquad \frac{\forall^2 P\ \varphi}{[P := \psi]\,\varphi}\ \forall_e^2$$

In $\forall_i^2$, all undischarged hypotheses must contain no free occurrence of P.

Figure 11.1: Rules of $\forall^2$ in system F

not be disturbed by a constraint corresponding to the one governing the axiom of separation. Moreover, the structures constructed here will be polymorphic right away. The latter feature can already be observed in the previous proof of $\forall^2 P\ P \to P$. It is time to provide a functional syntax for the manipulation rules of $\forall^2$. The introduction of $\forall^2$ will be represented by a Λ-abstraction which does not carry over a regular variable, but over a type variable (a propositional variable). From the proof $\lambda x^P.\,x$ of $P \to P$, we then construct $\Lambda P.\,\lambda x^P.\,x$ of type $\forall^2 P\ P \to P$. Define

Idp $\stackrel{\text{def}}{=} \Lambda P.\,\lambda x^P.\,x$,

$\lambda x^P.\,x$ is the identity function over P, while **Idp** is the polymorphic identity function which may be applied to an inhabitant h of any type H... after P has been explicitly instantiated by means of $\forall_e^2$, that is, in a functional syntax, by means of a "second-order application": $(\Lambda P.\,\lambda x^P.\,x)Hh$, which successively reduces to $(\lambda x^H.\,x)h$, then to h. Impredicativity appears when we take for h the polymorphic function identity itself:

   Idp $(\forall^2 P\ P \to P)$ **Idp**

$\stackrel{\text{def}}{=}$

   $(\Lambda P.\,\lambda x^P.\,x)\,(\forall^2 P\ P \to P)$ **Idp**

$\to$

   $(\lambda x^{\forall^2 P\ P \to P}.\,x)$ **Idp**

$\stackrel{\beta}{\to}$

   Idp

$\stackrel{\text{def}}{=}$

   $(\Lambda P.\,\lambda x^P.\,x)$.

The remainder of this section provides some hints on the expressive capacity of system F.

11.5.1 Typing of Regular Structures

The natural numbers, the Booleans, and the other data structures admit a satisfactory typing if we combine the simple typings previously proposed with suitable quantifications. Let us quickly inspect them. The type of integers is:

$$\mathbf{N} \stackrel{\text{def}}{=} \forall^2 X \ (X \to X) \to (X \to X) \ .$$

Any inhabitant of $\mathbf{N}$ can be regarded as a polymorphic iterator which, given a type T, a function f from T to T and an inhabitant x of T, is able to compute $f(\ldots(x)\ldots)$. The exponential function can be typed as follows:

$$\lambda m^{\mathbf{N}}.\,\lambda n^{\mathbf{N}}.\,\Lambda X.\,n(X \to X)(mX) \ : \ \mathbf{N} \to \mathbf{N} \to \mathbf{N} \ .$$

As a consequence, the function which maps n to n^n, that is $\lambda n.\,nn$, gets a suitable typing.

The type of Booleans is:

$$\mathbf{B} \stackrel{\text{def}}{=} \forall^2 X \ X \to X \to X \ .$$

The constants $\mathbf{t}$ and $\mathbf{f}$ are respectively typed in the following manner:

$$\mathbf{t} \stackrel{\text{def}}{=} \Lambda X.\,\lambda x^X.\,\lambda y^X.\,x \qquad \text{and} \qquad \mathbf{f} \stackrel{\text{def}}{=} \Lambda X.\,\lambda x^X.\,\lambda y^X.\,y \qquad .$$

The *two* versions of the negation are typable in F, including $\lambda b.\,\mathbf{bft}$:

$$\lambda b^B.\,b\,(\mathbf{B} \to \mathbf{B} \to \mathbf{B})\,\mathbf{f}\,\mathbf{t} \ .$$

The initial version of **or** can be typed by following a similar approach.

In order to represent pairs, we first examine the problem from a logical perspective: how can we represent the conjunction by means of $\forall^2$ and of $\to$? The intuitive idea can be explained from the impredicative definitions we have seen in § 7.3.1. The intersection of α and β could have been impredicatively defined as the smallest superset of α and β:

$$\{x \mid \forall e \ \underbrace{\overbrace{(\forall y\ y \in \alpha \wedge y \in \beta \ \to \ y \in e)}^{e \text{ contains } \alpha \cap \beta} \to x \in e}_{x \text{ is in any superset of } \alpha \cap \beta}\} \ .$$

But $P \wedge Q \to R$ can also be written without "$\wedge$": $P \to Q \to R$, hence:

$$\alpha \cap \beta \stackrel{\text{def}}{=} \{x \mid \forall e \ (\forall y\ y \in \alpha \to y \in \beta \to y \in e) \to x \in e\} \ ,$$

that is:

$$x \in \alpha \cap \beta \quad \text{iff} \quad \forall e \ (\forall y\ y \in \alpha \to y \in \beta \to y \in e) \to x \in e \ . \tag{11.8}$$

If we consider that $x \in P$ means "x allows us to prove P", or just "I know how to prove P", (11.8) can read: "I know how to prove $A \wedge B$ if and only if I know how to prove every consequence of A and of B", which can be represented in second-order logic as follows:[7]

---

[7]It is not a completely rigorous justification of (11.9). But we can see an analogy between impredicative constructions in set theory and in type theory.

$$A \wedge B \stackrel{\text{def}}{=} \forall^2 X \ (A \rightarrow B \rightarrow X) \rightarrow X \ , \tag{11.9}$$

At the functional level, we then have a guideline for building $\langle a, b \rangle$ from a^A and from b^B: $\langle a, b \rangle$ is of the form $\Lambda X. \lambda f^{A \rightarrow B \rightarrow X}. \mathcal{E}$, where $\mathcal{E}$ is an inhabitant of X built on a, b and f: the only possibility is fab, which yields

$$\langle a, b \rangle \stackrel{\text{def}}{=} \Lambda X. \lambda f^{A \rightarrow B \rightarrow X}. fab \ .$$

The projections are inhabitants of $A \wedge B \rightarrow A$ and of $A \wedge B \rightarrow B$. Let c be a variable of type $A \wedge B$, c can be specialized as a function to A by an application to A, since cA is of type $(A \rightarrow B \rightarrow A) \rightarrow A$. We still have to find a function of type $A \rightarrow B \rightarrow A$, we naturally consider the curryfied projection $\lambda x^A. \lambda y^B. x$. With a similar reasoning about the second projection, we get:

$$\textbf{pr1} \stackrel{\text{def}}{=} \lambda c^{A \wedge B}. cA(\lambda x^A. \lambda y^B. x) \qquad \text{and}$$
$$\textbf{pr2} \stackrel{\text{def}}{=} \lambda c^{A \wedge B}. cB(\lambda x^A. \lambda y^B. y) \ .$$

We still have to check that $\textbf{pr1}\langle a, b \rangle \stackrel{\beta}{\rightarrow} a$ and $\textbf{pr2}\langle a, b \rangle \stackrel{\beta}{\rightarrow} b$, but this was previously done, since removing types provides exactly the definitions of untyped λ-calculus. This example illustrates the help provided by types for designing a program (in λ-calculus here).

The sum of types is designed along the same lines, only the steps are given here :

$$\alpha \cup \beta \stackrel{\text{def}}{=} \{ x \mid \overbrace{\forall e \ (\forall y \ (y \in \alpha \rightarrow y \in e) \wedge (y \in \beta \rightarrow y \in e))}^{e \text{ contains } \alpha \cup \beta} \underbrace{\rightarrow x \in e \}}_{x \text{ is in any superset of } \alpha \cup \beta}$$
$$= \{ x \mid \forall e \ (\forall y \ y \in \alpha \rightarrow y \in e) \rightarrow \\ (\forall y \ y \in \beta \rightarrow y \in e) \rightarrow x \in e \} \ ,$$

which means "I know how to prove $A \vee B$ if and only if I know how to prove any X which is both a consequence of A and a consequence of B":

$$A \vee B \stackrel{\text{def}}{=} \forall^2 X \ (A \rightarrow X) \rightarrow (B \rightarrow X) \rightarrow X \ ,$$
$$\textbf{i1} \ a \stackrel{\text{def}}{=} \Lambda X. \lambda f^{A \rightarrow X}. \lambda g^{B \rightarrow X}. fa \qquad\qquad \text{for } a \ : \ A \ ,$$
$$\textbf{i2} \ b \stackrel{\text{def}}{=} \Lambda X. \lambda f^{A \rightarrow X}. \lambda g^{B \rightarrow X}. gb \qquad\qquad \text{for } b \ : \ B \ ,$$
$$\textbf{case} \stackrel{\text{def}}{=} \Lambda T. \lambda s^{A \vee B}. \lambda f^{A \rightarrow T}. \lambda g^{B \rightarrow T}. sTfg \ .$$

11.5.2 Systematic Construction of Types

The scope of the method explained in the previous subsection goes far beyond propositional connectors: it can be generalized to inductive definitions such as the natural numbers (one recovers the representation of Church), all kind of trees, the (polymorphic) lists, etc. In this way, one can represent all sets of closed terms obtained by the means of a finite set of constructors, which are employed in algebraic specification.[8]

The reciepe consists of starting with the curryfied signature of each constructor, in which the desired type is systematically replaced with the variable X, where X has previously been universally quantified. If this type is made up of n constructors, having $\sigma_1, \ldots \sigma_n$ as their respective signatures, we represent it by $\forall^2 X\ \sigma_1 \to \ldots \sigma_n \to X$. For example, the two constructors of $A \lor B$ are i1 of type $A \to A \lor B$ and i2 of type $B \to A \lor B$, which yields $\sigma_1 = A \to X$ and $\sigma_2 = B \to X$; $A \lor B$ is then represented by $\forall^2 X\ (A \to X) \to (B \to X) \to X$. For a little variety, let us consider binary trees, as defined by

```
leaf : int → tree
bin  : tree × tree → tree
```

Here we represent the integers by $\mathbf{N}$. After curryfication and replacement with X, the constructor signatures become $\mathbf{N} \to X$ and $X \to X \to X$, which yields for tree:

$$\forall^2 X\ (\mathbf{N} \to X) \to (X \to X \to X) \to X \ .$$

This idea goes far beyond regular algebraic data types, because we may introduce constructors having more complex types. For example, here is a type of trees where each node may possess 0, 1, or an infinite number of children:

```
init : arbi
next : arbi → arbi
lim  : (int → arbi) → arbi.
```

It is represented in system F by

$$\forall^2 X\quad X \to (X \to X) \to ((\mathbf{N} \to X) \to X) \to X \ .$$

This structure provides a representation of ordinal numbers, where init is interpreted by 0, next is interpreted by the successor function and lim is interpreted by the formation of a limit ordinal number. This gives some idea of the expressive abilities of system F.

11.5.3 Expressive Power and Consistency of System F

System F includes an extremely rich class of functions. The following theorem states that almost all functions we need in practice can be represented in F.[9]

---

[8] See the concept of an initial algebra on page 88.

[9] However, there is a restriction: system F does not always provide a type for the most efficient algorithm which computes a given function. Here the term "function" takes its set-theoretical meaning — something uncommon in this chapter.

Theorem 11.4

Any total function whose termination can be proved by means of regular mathematics[10] can be represented in system F.

It is then quite remarkable that the strong normalization property of the simply typed λ-calculus still holds.

Theorem 11.5

The β-reduction is strongly normalizing in system F.

This means that the termination of computations is decidable (ensured, in fact) as soon as typing is checked. This property has a good consequence: it guarantees that system F is free of logical paradoxes (such problems may have been caused by the impredicativity of the system). Note that, the expressive power of system F does not come from recursion (in the sense of computer science): fixed-point combinators cannot be represented.

11.6 Dependent Types

11.6.1 Introduction of First-order Variables

The interpretation of Heyting allows one to distinguish the (many) proofs of a given formula. For example, the most simple proofs of $N \to B$ are $\lambda n^N. t$ and $\lambda n^N. f$; but we can find many others:

$\lambda n^N.$ "if $n = 0$ then t else f" ,
$\lambda n^N.$ "if n is even then t else f" ,
etc.

From the viewpoint of specification, only the domain (N) and the co-domain (B) of these functions are specified, but we would like to go further: stating a relation between the result and the argument. To provide an analogy with abstract data types, system F declares only the signatures — however, we have a new feature with relation to algebraic data types: higher-order signatures are allowed here.

In order to tackle this problem, we take predicates instead of propositions. For example, in the case of natural integers, we introduce the symbols S and 0 in the logical language, and N becomes a 1-argument predicate. The idea is that $N(x)$ is provable if x is obtained by successive applications of S to 0. Intuitively, $N(x)$ represents $x \in \mathbb{N}$. Let us write the expected induction schema, where the induction step comes first;[11] here, X is a unary predicate variable:

---

[10] The precise meaning of this phrase should be explained but this is beyond the scope of this chapter. A precise statement can be found in [GLT89].

[11] We could also consider the base case first; the integer n will then be represented by $\lambda x f. \underbrace{f(f \cdots (f x) \cdots)}_{n}$ instead of $\lambda f x. \underbrace{f(f \cdots (f x) \cdots)}_{n}$. Of course, all operations defined over the natural numbers have to be rewritten accordingly.

$$\forall^2 X\ (\forall x\ Xx \to X(Sx)) \to X0 \to \forall n\ \mathbf{N}(n) \to Xn\ .$$

The quantification over n can also be written at the beginning of the formula, which yields:

$$\forall n\ \mathbf{N}(n) \to \forall^2 X\ (\forall x\ Xx \to X(Sx)) \to X0 \to Xn\ .$$

Formally, we will actually *define* **N** by:

$$\mathbf{N}(n) \stackrel{\text{def}}{=} \forall^2 X\ (\forall x\ Xx \to X(Sx)) \to X0 \to Xn\ . \tag{11.10}$$

Note that, if we remove first-order information, we recover the definition given in system F:

$$\mathbf{N} \stackrel{\text{def}}{=} \forall^2 X\ (X \to X) \to X \to X\ .$$

The system we just sketched was introduced by Krivine under the name **second-order functional arithmetic** (AF2) [Kri93]. The class of functions, which can be described in it, is the same as in system F; but typing provides a real specification language.

The previous type expressions may seem somewhat mysterious at first sight, but looking at them as **Prolog** programs may help. From this perspective, one should ignore issues related to the special resolution strategy of **Prolog**, and concentrate on the proof trees that could be constructed using a fair strategy. The **Prolog** program corresponding to the type of the integers in system F would be

```
nat:- nat.
nat.
```

The **Prolog** program corresponding to the type of the integers in AF2 would be

```
nat(S(x)):- nat(x).
nat(0).
```

11.6.2 Sums and Products

11.6.2.1 Products of Sets. Until now, we considered the product $S \times T$ or the sum $S + T$ — also denoted, respectively, by $S \wedge T$ and $S \vee T$, thanks to the Curry–Howard correspondence — of *two* types. In mathematics, these concepts can be generalized to the product, and to the sum, of a family of sets (T_i) indexed by a set I.

For the sake of simplicity, let us first take for I an interval of integers of the form $[1..n]$. The elements of the **product** $\prod_{i \in I} T_i$ are the tuples $\langle x_1, \ldots, x_n \rangle$ where, for all i from I, we have $x_i \in T_i$. In the case where all T_i are identical, we can write $T_i = T$, then we can view the tuples $\langle x_1, \ldots, x_n \rangle$ as functions from $[1..n]$ to T. For example, there is a natural bijection between $T \times T$ and $\{1, 2\} \to T$.

More generally, for an arbitrary I, $\prod_{i \in I} T$ is defined as the set of mappings from I to T, which is commonly denoted by $I \to T$.

In conclusion, $I \to T$ is a kind of product.

In the general case, where T_i are distinct sets, $\prod_{i \in I} T_i$ is termed a **dependent product**. It is seen as a function, whose domain is I, and whose co-domain T_i depends on the element to which it is applied.[12] The following notation is often used:

$$\prod_{i \in I} T(i) \quad .$$

The product $R \times S$ is a simple example of a dependent product, it is:

$$\prod_{i \in \{1,2\}} T(i) \quad \text{with } T(1) = R \text{ and } T(2) = S \quad .$$

Examples:

- In communication protocols, it is not uncommon that, in a message, the type of a field depends on a value, or on a combination of values, which come from a previous field.
- In the example of a calendar, which was mentioned at the beginning of the previous chapter, suppose that we would like to select a day in each month of the year 2002. This will be represented by a 12-tuple, that is, at a first approximation, a member of $[1, 12] \to [1, 31]$. But, for a more accurate specification, we would consider it as a member of the dependent product:

$$\prod_{i \in [1..12]} \text{month}(i), \quad \text{with} \begin{cases} \text{month}(1) = [1, 31], \\ \text{month}(2) = [1, 28], \\ \text{etc.} \end{cases}$$

11.6.2.2 Sum of Sets. In a similar manner, $\sum_{i \in I} T(i)$ is composed of pairs $\langle i, \text{month}(i) \rangle$ whose first element i is taken from $[1, 12]$ and the second element is taken from $\text{month}(i)$. This is a **dependent sum**. In this case, it represents the type of the dates of a non-leap year.

11.6.2.3 Products and Sums of Types. The previous constructs over sets can be translated into constructs over proof spaces. We know that providing a proof of $T_1 \wedge T_2$ amounts to providing a proof of T_1 and a proof of T_2.

By generalizing this remark, providing a proof of $\forall i \; T(i)$ amounts to providing a proof of $T(i)$ for each i: this corresponds to the dependent product $\prod_i T(i)$.

Finally, providing an intuitionistic proof of $\exists i \; T(i)$ amounts to providing an i and a proof of $T(i)$, that is, an inhabitant of $\sum_i T(i)$.

When a type is represented by a formula which contains parameters, it is termed a **dependent type**, because the type of the result of a function, which inhabits such a type, depends on the value of its argument.

---

[12]In order to recover the common concept of a set-theoretic function, one has to build its co-domain: it is the union of all T_i.

11.6.3 Specification Based on Dependent Types

A function to be implemented can be specified using dependent types, in a formula such as:

$$\forall x\!:\!E\ P(x) \to \exists y\!:\!S\ Q(x,y)\ , \tag{11.11}$$

where E is the type of the input argument, S is the type of the output, P is a precondition and Q is a post condition. Intuitively, the above formula tells us that, for all x from E satisfying P, there exists a y from S such that $Q(x,y)$ is satisfied. A constructive proof of (11.11) forces us to make the witness y explicit, or more precisely to make it explicit how y can be computed from x (in classical logic, we could content ourselves with proving that, if all y satisfy $\neg Py$, a contradiction can be derived).

After closer examination, an inhabitant φ of type (11.11) is a function which takes, first, an inhabitant x from E and then, an inhabitant from — a proof of — Px, and which returns a pair $\langle y, q \rangle$ such that y inhabits S and q inhabits $Q(x,y)$. This is then more complicated an object than a function from E to S. Nevertheless, it is possible to extract from φ a function f of type $E \to S$ such that:

$$\forall x\!:\!E\ P(x) \to Q(x, f(x))\ . \tag{11.12}$$

This operation is termed **program extraction** or **program synthesis**. It is implemented in several software tools such as Nuprl and Coq; we will return to this idea in Chapter 12.

11.7 Example: Defining Temporal Logic

In order to illustrate the expressive power of the notions presented in the previous sections, we formalize here the definition of CTL* given in § 8.5.

We assume that we are in an environment which includes a type state for the states and a type nat for the natural integers. We define traj, the type of trajectories, and suff, the function which computes the kth suffix.

$$\text{traj} \stackrel{\text{def}}{=} \text{nat} \to \text{state} \qquad \text{suff} \stackrel{\text{def}}{=} \lambda k^{\text{nat}}.\, \lambda \sigma^{\text{traj}}.\, \lambda n^{\text{nat}}.\, \sigma(k{+}n)$$

For the sake of clarity, we distinguish various kind of predicates by giving them a type: Pstate, for the predicates over states, and Ptraj, for the predicates over trajectories. They are defined from the type of propositions, which is denoted by Prop (as in the next chapter).

$$\text{Pstate} \stackrel{\text{def}}{=} \text{state} \to \text{Prop} \qquad \text{Ptraj} \stackrel{\text{def}}{=} \text{traj} \to \text{Prop}$$

Then we formalize the start operator ∂ and logical connectors. Here, we give only the conjunction andst over state predicates, the conjunction andtr over trajectory predicates, and the universal quantification `forallst` over state predicates.

$$\partial \overset{\text{def}}{=} \Lambda P^{\texttt{Pstate}}.\, \lambda \sigma^{\texttt{traj}}.\, P(\sigma 0)$$

$$\texttt{andst} \overset{\text{def}}{=} \Lambda P^{\texttt{Pstate}}.\, \Lambda Q^{\texttt{Pstate}}.\, \lambda s^{\texttt{state}}.\, Ps \wedge Qs$$

$$\texttt{forallst} \overset{\text{def}}{=} \Lambda P^{A \to \texttt{Pstate}}.\, \lambda s^{\texttt{state}}.\, \forall a^A\, Pas$$

$$\texttt{andtr} \overset{\text{def}}{=} \Lambda \varphi^{\texttt{Ptraj}}.\, \Lambda \psi^{\texttt{Ptraj}}.\, \lambda \sigma^{\texttt{traj}}.\, \varphi \sigma \wedge \psi \sigma$$

Finally, we have the temporal and the branching operators.

$$\mathsf{X} \overset{\text{def}}{=} \Lambda \varphi^{\texttt{Ptraj}}.\, \lambda \sigma^{\texttt{traj}}.\, \varphi(\texttt{suff}\,1\,\sigma)$$

$$\mathsf{F} \overset{\text{def}}{=} \Lambda \varphi^{\texttt{Ptraj}}.\, \lambda \sigma^{\texttt{traj}}.\, \exists n^{\texttt{nat}}\, \varphi(\texttt{suff}\,n\,\sigma)$$

$$\mathsf{G} \overset{\text{def}}{=} \Lambda \varphi^{\texttt{Ptraj}}.\, \lambda \sigma^{\texttt{traj}}.\, \forall n^{\texttt{nat}}\, \varphi(\texttt{suff}\,n\,\sigma)$$

$$\mathsf{W} \overset{\text{def}}{=} \Lambda \varphi^{\texttt{Ptraj}}.\, \Lambda \psi^{\texttt{Ptraj}}.\, \lambda \sigma^{\texttt{traj}}.$$
$$\forall n^{\texttt{nat}}\, (\forall i^{\texttt{nat}}\, i \le n \to \neg \psi(\texttt{suff}\,i\,\sigma)) \to \varphi(\texttt{suff}\,n\,\sigma)$$

$$\mathsf{U} \overset{\text{def}}{=} \Lambda \varphi^{\texttt{Ptraj}}.\, \Lambda \psi^{\texttt{Ptraj}}.\, \lambda \sigma^{\texttt{traj}}.$$
$$\exists n^{\texttt{nat}}\, \psi(\texttt{suff}\,n\,\sigma) \wedge (\forall i^{\texttt{nat}}\, i < n \to \varphi(\texttt{suff}\,i\,\sigma))$$

$$\mathsf{E} \overset{\text{def}}{=} \Lambda \varphi^{\texttt{Ptraj}}.\, \lambda s^{\texttt{state}}.\, \exists \sigma^{\texttt{traj}}\, \sigma(0) = s \wedge \varphi \sigma$$

$$\mathsf{A} \overset{\text{def}}{=} \Lambda \varphi^{\texttt{Ptraj}}.\, \lambda s^{\texttt{state}}.\, \forall \sigma^{\texttt{traj}}\, \sigma(0) = s \to \varphi \sigma$$

11.8 Towards Linear Logic

Recall that, in sequent calculus, intuitionistic logic appears as a restriction of classical logic, where the right-hand side of sequents can be made up of at most one formula. As an important consequence, the use of the contraction rule is prohibited on the right, and the use of the weakening rule is drastically limited. After a deep analysis of this fact, based on semantical considerations, Girard came to consider a logic where a fine-grain control over the space of hypotheses and conclusions, regarded as resources, is specified by special logical operators [Gir87a]. Typically, regular implication is decomposed into a new kind of implication, which is denoted by $\multimap$ and is termed linear implication, and whose inhabitants are functions which "consume" their argument, and a cloning operator for keeping this argument in memory. Two versions of the conjunction and of the disjunction are distinguished: a multiplicative and an additive version. For example, the multiplicative conjunction can be interpreted as the juxtaposition of resources, while the additive conjunction can be interpreted as their superposition. An interesting property of the multiplicative fragment is that, in the corresponding calculus on proofs (according to the Curry–Howard correspondence) transitions can be performed in parallel *without synchronization problems*.

The new constructive logic thus obtained is termed the **linear logic** (not to be confused with the linear temporal logic considered in Chapter 8). In the same vein, let us mention *interaction nets* [Laf90], an elegant paradigm for parallel computations over graphs, which is based on linear logic.

11.9 Notes and Suggestions for Further Reading

Reference works on the λ-calculus are [Bar84] and [HS86]. An algorithmic formulation, interesting for computer scientists and practitioners, is presented by Gérard Huet in [Hue92].

The book [Hue90] edited by Huet contains fundamental chapters on type theory. Chapter 2 of [AGM92b], by Barendregt, presents several type systems for the λ-calculus in a uniform and synthetic manner (see also Chapter 16 in [Hue90]). One may also consider the papers of Mitchell in [vL90b]. The book [Tho91] contains a thorough and progressive introduction to type theory. It is based on a predicative version of type theory, due to Martin-Löf, which is particularly influential [ML84].

The relationship between typing, natural deduction and sequent calculus are handled in [GLT89] and [Gal93]. Interesting hints are also given by Coquand in [Hue92, ch. 17].

Reference books on intuitionistic logic and, more generally, constructive mathematics, are [Dum00] and [TvD88].

12. Using Type Theory

Et, comme la multitude des lois fournit souvent des excuses aux vices, en sorte qu'un État est bien mieux réglé lorsque, n'en ayant que fort peu, elles y sont fort étroitement observées ; ainsi, au lieu de ce grand nombre de préceptes dont la logique est composée, je crus que j'aurais assez des quatre suivants, pourvu que je prisse une ferme et constante résolution de ne manquer pas une seule fois à les observer.[1]

R. DESCARTES, discours de la méthode, II.

In the table example, we would like to consider the search criterion P as a parameter. This is not possible in the framework of a formal method based on first-order logic, at least not in a satisfactory manner:

- P may be encoded in the form of a set, but in the framework of B, for example, only certain finite sets are allowed;
- Z is more flexible, but no straightforward mechanism is provided for deriving a program from the specification;
- the axiom for search, in the algebraic specification of Chapter 10 is actually a *schema* of axioms; we then have to write down an instance of this schema for every property of interest.

Furthermore, the proposed expedients hardly survive if one wants to tackle arbitrary situations, for example if P is an argument to be discovered only at call time, or if P is given by an algorithm instead of a data structure, or else when we consider several-level search processes in complex overlapping tables.

If we take a predicate P as an object which may vary, or be manipulated as an argument of a function or of a predicate, we are working in higher-order logic. The version of higher-order logic we will employ in this chapter is the calculus of inductive constructions. This is a very powerful logic, well-adapted to specifying and reasoning about programs. Interactive and reliable tools, such as Coq and Lego, are available for aiding the development of specifications and proofs.

---

[1] And as a multitude of laws often only hampers justice, so that a state is best governed when, with few laws, these are rigidly administered; in like manner, instead of the great number of precepts of which logic is composed, I believed that the four following would prove perfectly sufficient for me, provided I took the firm and unwavering resolution never in a single instance to fail in observing them.

12.1 The Calculus of Inductive Constructions

We start with a pragmatic presentation of the logic, then we will indicate how it is related to type theory as introduced in the previous chapter.

12.1.1 Basic Concepts

The calculus of constructions includes the ordinary logical operators $\wedge$, $\vee$, $\neg$ and the implication denoted $\rightarrow$. Quantifications are typed. Thus, a property which holds true for every natural integer is expressed by $\forall n : \mathtt{nat}\ P\,n$. (Comment on the notation: as in the λ-calculus, we henceforth omit parentheses for function application whenever possible. For example, $\forall n : \mathtt{nat}\ P\,n$ would be denoted $\forall n : \mathtt{nat}\ P(n)$ in standard mathematical notation.)

The notational confusion between a proposition $P \rightarrow Q$ and function space $P \rightarrow Q$ is intentional: according to the Curry–Howard correspondence (§ 11.3.4), a proof of $P \rightarrow Q$ can be interpreted as a *total* function which computes a proof of Q from a proof of P.

Propositions themselves have a type named $\mathtt{Prop}$. For example, the predicates over natural integers have the type $\mathtt{nat} \rightarrow \mathtt{Prop}$. We can express that, for any given proposition P, P implies P, by the formula $\forall P : \mathtt{Prop}\ P \rightarrow P$. Let us point out that P is quantified here: this would be impossible in first-order logic.

The data types such as $\mathtt{nat}$ themselves have a type named $\mathtt{Set}$. Thus we can build up functions whose type depends on the first argument. The most simple example is the identity function, which is defined (without types) by $\mathtt{Id}\,x = x$. Its behavior is the same, independently from the type of x, which could be an integer, a Boolean, or even a function itself. It is assigned the type $\forall X : \mathtt{Set}\ X \rightarrow X$. The typed version of $\mathtt{Id}$ is then $\mathtt{Id}(X : \mathtt{Set}\,;\ x : X) = x$. For example, $\mathtt{Id}$ could take $\mathtt{nat}$ as its first argument, then 3, and its result is then 3. We can also consider the expression $\mathtt{Id}\,\mathtt{nat}$, and take it as the definition of $\mathtt{Idn}$. $\mathtt{Idn}$ is then the specialization of $\mathtt{Id}$ to natural integers.

Similarly, data structures can be parameterized by a data type. The classical example is lists: given an arbitrary type X, $\mathtt{list}\,X$ is the type of lists of elements from X; $\mathtt{list}$ then has the type $\mathtt{Set} \rightarrow \mathtt{Set}$, its constructors are $\mathtt{nil}$, of type $\forall X : \mathtt{Set}\ \mathtt{list}\,X$, and $\mathtt{cons}$, of type $\forall X : \mathtt{Set}\ X \rightarrow \mathtt{list}\,X \rightarrow \mathtt{list}\,X$.

The expression $A_1 \rightarrow A_2 \rightarrow \ldots A_n \rightarrow B$ denotes the type of a function which has n arguments of types $A_1 \ldots A_n$, respectively, and which returns a result of type B. Similarly, on the side of propositions, we have seen, in the equation (3.11) on page 47, that $P \rightarrow Q \rightarrow R$, which means "if I have P, then if I have Q, then I have R" can replace $P \wedge Q \rightarrow R$.

A type such as $\mathtt{nat} \rightarrow \mathtt{nat}$ is still of type $\mathtt{Set}$. This allows us to form, for example, lists of functions over integers. Thus, we can legitimately apply $\mathtt{Id}$ to $\mathtt{nat} \rightarrow \mathtt{nat}$. For example, $\mathtt{Id}\,(\mathtt{nat} \rightarrow \mathtt{nat})\,\mathtt{Idn}$ returns $\mathtt{Idn}$. We can even apply $\mathtt{Id}$ to itself as follows: $\mathtt{Id}\,(\forall X : \mathtt{Set}\ X \rightarrow X)\,\mathtt{Id}$, and this expression reduces to $\mathtt{Id}$.

12.1.2 Inductive Types

The calculus of inductive constructions also includes a mechanism for defining data types from constructors, as in algebraic data types. The integers, the Booleans, and the lists are defined in this way. However, as we can use higher-order features, we have polymorphic lists from the outset, (also termed "generic" lists, in the terminology of programming languages such as Ada).

The inductive types that we will use in the table example are specializations of very general inductive types, which we present in an informal manner for the moment (we will give the formal definitions in § 12.2.8). The first is $\{x : S \mid P\,x\}$ where S is of type Set and P is of type $S \to$ Prop. As is suggested by the notation, this type plays the role of the set of elements x from S which satisfy $P\,x$.

However, the reader must be aware that $\{x : S \mid P\,x\}$ does not denote exactly the same thing in set theory and in type theory. Here, the inhabitants of $\{x : S \mid P\,x\}$ are the pairs $\langle x, \rho \rangle$ where ρ is a proof of $P\,x$. We will see, in § 12.3.4, how the logical part ρ can be removed.

The sum of two data types is yet another general inductive type. The most common form is:

$$A + B, \qquad \text{with } A, B : \text{Set.} \tag{12.1}$$

The elements of type $A + B$ are elements of type either A, or B, together with a piece of information for indicating their origin.

The following construct uses two propositions:

$$\{P\} + \{Q\}, \qquad \text{with } P, Q : \text{Prop.} \tag{12.2}$$

There are two kinds of inhabitants from this type, the first tells us that P is true and the second tells us that Q is true. As we use a constructive logic here, this means that we can effectively *compute* whether P or Q is satisfied. In the case where Q is the negation of P, this type can also be regarded as an enriched version of bool: an inhabitant of $\{P\} + \{\neg P\}$ yields the truth value of P; providing such an element simply amounts to saying that P *is decidable* (in our example, on page 16, we employed the term P *is defined.*)

The last construct we will use is a kind of mixture of the two previous ones. Its elements are either inhabitants of A which satisfy the predicate P, or an indication that Q is true:

$$\{x : A \mid P\,x\} + \{Q\}, \text{ with } A : \text{Set}, P : A \to \text{Prop et } Q : \text{Prop.} \tag{12.3}$$

This construct is an enriched version of the option type of ML.

12.1.3 The Table Example

12.1.3.1 Specification. We are given an arbitrary universe U of type Set and an arbitrary predicate P over U. The table is represented by its characteristic predicate *Ptable*. We first state the precondition: P is defined for all elements from the table. To this end we write, using (12.2) — the identifiers inside the square brackets U, *Ptable* and P, are simply the parameters of the function def_tbl:

$$\text{def\_tbl}[U:\text{Set};\ Ptable,\ P:U \to \text{Prop}] \overset{\text{def}}{=}$$
$$\forall x{:}U\ Ptable\,x \to \{P\,x\} + \{\neg P\,x\}\ .$$

The expression on the right-hand side can also be interpreted as the type of a total function which, for every x which satisfies *Ptable*, returns the truth value of $P\,x$. This expresses the idea of a "table where every element can be tested". If we consider § 12.3.4, an inhabitant D of type def_tbl could simply be an array of Booleans which represent truth values of P. But a Boolean function defined over an infinite domain would do the job just as well. The specification written above assumes nothing about the future realization of D.

Let us consider the type of the result. It should be either an element from the table verifying P, or an indication that there is no such element. Its type is defined using (12.3):

$$\text{resu\_tbl}[U:\text{Set};\ Ptable,\ P:U \to \text{Prop}] \overset{\text{def}}{=}$$
$$\{x{:}U \mid Ptable\,x \wedge P\,x\} + \{\forall x{:}U\ Ptable\,x \to \neg P\,x\}\ .$$

12.1.3.2 Specialization to an Array. With the aim of developing a program, we consider the case where U is the type of the natural integers and where *Ptable* characterizes an interval of integers. P is left free. The considered interval is defined by its two bounds p and q, which are also considered as parameters, for which we assume that $p \leq q$. This context is concretely declared in the following manner:

Variable P: nat $\to$ Prop .
Variable p, q: nat .
Hypothesis lepq: $p \leq q$.

Now we just apply def_tbl and resu_tbl to nat and to Pinterv, once the latter is defined:

between$[a, b, c{:}\text{nat}] \overset{\text{def}}{=} a \leq b \wedge b < c$.
Pinterv$[x{:}\text{nat}] \overset{\text{def}}{=}$ between $p\,x\,q$.

def_tbl_int $\overset{\text{def}}{=}$ def_tbl nat Pinterv P .
resu_tbl_int $\overset{\text{def}}{=}$ resu_tbl nat Pinterv P .

The definition chosen for between $p\,x\,q$ corresponds to the interval $[p..q[$ that we used in Chapter 2.

12.1.3.3 Specialization to a List. We can also specialize the general specification above to the search for an element in a list. We don't need to specialize U: the table will be represented by a list of elements from U and we will assume that P is defined for all elements of this list. Formally, we first stipulate that a list contains u if, and only if, it is of the form cons $u\,l$, or of the form cons $v\,l$, where u is in l.

```
Inductive contains:list → U → Prop def
      contains_head : ∀l:list ∀u:U contains(cons u l) u
    | contains_queue :
          ∀l:list ∀u,v:U contains l u → contains(cons v l) u.
```

We then consider a given list l, and we write the definition of Ptablist in order to state the desired specification.

```
Variable l: list .
Ptablist[u:U] def contains l u .

def_tbl_lis def def_tbl U Ptablist P .
resu_tbl_lis def resu_tbl U Ptablist P .
```

12.2 More on Type Theory

The **calculus of inductive constructions** is obtained from system F, introduced in Chapter 11, using three independent extensions that we consider in turn:

– introduction of an additional type level on top of propositions;
– introduction of predicates and of dependent types;
– introduction of inductive types.

This system allows one to represent a strict superset of the functions representable in system F, while preserving the strong normalization property.

12.2.1 System Fω

We introduced the symbol Prop for representing the type of propositions. A type quantified using second-order quantification, denoted $\forall^2 X \varphi$ in Chapter 11, is henceforth denoted $\forall X$:Prop φ; similarly, $\Lambda X. \varphi$ becomes λX:Prop. φ . For example, the formula expressing that P implies P, for any proposition P, is $\forall P$:Prop $P \to P$. It is inhabited by the polymorphic identity λP:Prop.λx:$P.x$.

This provides a more uniform syntax, but the main point is that we are now allowed to consider expressions such as Prop $\to$ Prop $\to$ Prop — the type of logical connectors — and even quantifications over connectors.

$$\forall c: (\text{Prop} \to \text{Prop} \to \text{Prop}) \ \varphi$$

Thus, from now on, we can define the logical connectors as functions, using a λ-term. For example, for $\wedge$, we adapt (11.9):

$$\textbf{and} \ \overset{\text{def}}{=} \ \lambda A:\text{Prop}. \ \lambda B:\text{Prop}. \ \forall X:\text{Prop} \ (A \to B \to X) \to X \ . \tag{12.4}$$

In system F, we could only represent $A \wedge B$ for *given* A and B.

○ $P \to Q$ is actually only a simplified notation for $\forall x : P \; Q$, that we use when Q does not depend on P. Indeed, we have seen, in § 11.6.2, that the regular product is a particular case of a dependent product. This still holds if we take Prop instead of P. The only primitive logical operation is then the universal quantification.

As in P:Prop, we can construct other inhabitants of Prop, such as $P \to P$. We have to give a type to expressions such as Prop, Prop $\to$ Prop, etc. This type is named Type. The process continues with a hierarchy of types Type_1, Type_2, and so on. The important point is that polymorphism is not allowed within Type and beyond, because this would leave room for paradoxes.

12.2.2 The Calculus of Pure Constructions

We have seen how to define data types in system F, such as $\mathbf{N}$, the natural integers, or $\mathbf{B}$, the Booleans. We then have three levels: objects from the bottom level, such as $\mathbf{0}$ or $\mathbf{S}$, inhabit objects from the second level, such as $\mathbf{N}$ or $\mathbf{N} \to \mathbf{N}$, which themselves inhabit an object of the third level, Prop.

The calculus of constructions authorizes products such as $\mathbf{N} \to \text{Prop}$, which are simply predicates over the integers. If P is of type $\mathbf{N} \to \text{Prop}$, the formula $\forall n : \mathbf{N} \; Pn$ expresses that this property is verified for every integer.

12.2.3 Inductive Definitions

There is another way of introducing objects such as the natural integers, the Booleans, binary trees and the like: using an **inductive definition**, which consists of an exhaustive enumeration of the constructors of the type to be defined, together with their respective signatures. For example, here is the definition of bool and of nat:

Inductive bool: Set:=   true: bool | false: bool.

Inductive nat: Set:=   0: nat | S: nat $\to$ nat.

Note that bool and nat have the type Set instead of Prop. We can ignore the difference between Set and Prop at the moment. Distinguishing them will become important later, in the context of program extraction, for separating data structures from proofs. In the following example, which defines binary trees, we have a two-argument constructor:

Inductive tree: Set:=
    leaf: nat $\to$ tree | bin: tree $\to$ tree $\to$ tree.

Thanks to inductive definitions, not only does the representation of data structures become clearer, but we gain automatically generated induction principles, which are essential for reasoning about objects or programs. We will come back to them in § 12.2.6. The definitions inspired from system F keep their interest as control structures. For example, an inhabitant of $\mathbf{N}$ is an iterator which applies a function to an argument for a given number of times.

12.2.4 Inductive Dependent Types

In the calculus of inductive constructions, we can also define *predicates* in an inductive manner. For example, here is a definition of the predicate which states that a given natural integer is even:

Inductive even: nat → Prop:=
    p0:   even 0
 | p2:   ∀n:nat even n → even$(n+2)$.

The assertion p0 stipulates that 0 is even, the assertion p2 stipulates that for any integer n, if n is even, then $n+2$ is even; finally, an integer is even only if this can be proved using p0 and p2 only — similarly, the definition of **nat** says that any integer can be constructed with 0 and S only.

A proof of even n, where n is non-zero, can be given in the form p2 k p where p is a proof of even k. For example, the tree

which represents the term p2 2 (p2 0 p0), is a proof of even 4. Let us observe that, in p2 k p, the *type* of the component p depends on the *value* of the previous component k.

A Prolog definition of even would be composed of clauses similar to p0 and p2, but here we can write $n+2$ instead of S(S n). The next definition of even, called even1 below, does not correspond to a Prolog program.

12.2.5 Primitive Recursive Functions

In order to define a function such as the addition, one indicates how to construct the result by means of a case analysis on the possible constructors of nat, which are S and 0. More precisely, one expresses that $m+0$ evaluates to m, and that $m + $ S n evaluates to S$(m+n)$. A possible syntax in Coq is (replacing, following common notation, plus a b with $a+b$):

Fixpoint plus $[m, n : $ nat$]$: nat:=
    Cases n **of**
        0   ⇒ m
     | S n ⇒ S$(m+n)$ **end.**

An expression such as S 0 + S 0 is then an unreduced form of S(S 0). Similarly 2, viewed as a constant function without arguments, is an unreduced form of S(S 0).

We can define in the same way functions over binary trees, by exhausting the possible cases. This is sometimes called structural induction. For example, here is a definition of the sum of the leaves of a tree:

```
Fixpoint sumlf [a : tree]  : nat:=
  Cases a of
    leaf n    ⇒   n
  | bin g d   ⇒   sumlf g + sumlf d end.
```

In the case of integers, we can then define the primitive recursive functions which were introduced in § 3.7.1. The system presented here then includes a generalization of primitive recursive functions to arbitrary inductive types. Furthermore, even in the case of the integers, we actually have much more than primitive recursion: we have a large class of total recursive functions (totality is automatically ensured by the theorem of strong normalization). The large size of this class comes partly from the higher-order features of the calculus. For example, we saw in § 3.7.1 that the Ackermann function is not primitive recursive in the ordinary sense, but after curryfication it becomes so.

12.2.6 Reasoning by Generalized Induction

Here is another definition of the property, for a natural integer, to be even:

```
Inductive even1: nat → Prop:=
  p1:    ∀n:nat  even1(n + n).
```

How can we ensure that the two definitions **even** and **even1** are equivalent? Each inductive definition is automatically associated to an *elimination rule*, which allows one to reason by cases on an object which inhabits an inductive type. In the simple case of an enumerated type, such as **bool**, the rule simply states that, in order to prove Pb for any Boolean b, it is sufficient to prove P true and P false.

In the case of a "recursive" type such as **nat**, the rule states that, in order to prove $P\,n$ for any natural number n, it is sufficient to prove $P\,0$ and to prove that, if $P\,m$, then $P\,(S\,m)$: this is a formalization of reasoning by induction, expressed here by *one* axiom, and not by a schema as we did in § 5.3.2.

$$\forall P:\text{nat}\rightarrow\text{Prop}\quad P\,0 \rightarrow [\forall m:\text{nat}\ \ P\,m \rightarrow P\,(S\,m)] \rightarrow \forall n:\text{nat}\ \ Pn.$$

For example, let us consider a proof of **even1** $n \rightarrow$ **even** n. Reasoning directly by induction over n is not a very good idea, because if n is even, then $S\,n$ is odd. However, the hypothesis **even1** n entails that n has the form $m+m$ (which is formally expressed by an elimination on **even1**). We then reason by induction over m, which amounts to proving that $\text{even}(0+0)$ and $\text{even}(k+k) \rightarrow \text{even}(S\,k + S\,k)$, which is trivial by applying, respectively, p0 and p2, and then using very simple arithmetic facts.

Another complete example of a proof by induction was previously presented in § 9.2.2.2.

12.2.7 Induction Over a Dependent Type

For the same reason as in the previous subsection, it is neither easy, nor natural, to prove the formula **even** $n \to$ **even1** n by induction over n. Intuitively, we would like to count the number of occurrences of p2, m, in a proof of **even** n, to construct p1 m and to verify that the latter is of type **even1** n. Formally, we employ the elimination rule associated with **even**, which amounts to examining the different means of constructing a proof of **even** n. Two cases are possible:

- either this proof is p0 and, in this case, n is 0: we can take p1 0;
- or, the proof is of the form p2 $k\,p$ where p is a proof of **even** k and, in this case, n is $k + 2$; by the induction hypothesis we have a proof of **even1** k, which means that k is of the form $m+m$ (here, an elimination of **even1** is used); we can take p1(S m), since we have $m + m + 2 = (m+1) + (m+1)$.

This reasoning needs some care. In such situations, using a software-based proof assistant turns out very helpful. The main lesson we can draw is that common induction over the natural integers is an elimination rule among many others, and that it is often worthwhile to use an induction principle over types which are more complex than nat, such as **even** in the previous example.

Reasoning by induction proceeds by examining the different means of producing the eliminated object. But one always limits oneself to considering that this object is built using only constructors. For example, in an induction over a natural integer, one only considers the case where it is zero and the case where it is the successor of an integer. However, the integer under examination may well be presented in a different form, using multiplications or one of the many other possibilities. Similarly, a proof of **even** n, when n is non-zero, is not necessarily of the form p2 $k\,p$ from the outset: another possibility is th $n\,p'$, where p' is a proof of **even1** n and th a proof of $\forall x{:}$nat **even1** $x \to$ **even** x.

Confining the exploration to constructors is sufficient, because every expression necessarily reduces to the form of a combination of constructors (when no free variable is left). Termination (normalization) properties of the calculus play an essential role there.

In passing, the above discussion illustrates the importance of the computational contents of proofs — the third part of the Curry-Howard correspondence: the argument given above for proving **even** $n \to$ **even1** n relies on the fact that a proof of **even** n is *eventually* (after a number of computation steps) of the form p2 $k\,p$.

12.2.8 General Purpose Inductive Types

We provide the formal definition of the general-purpose inductive types, that we used in § 12.1.3, for the example of the search for an element in a table.

12.2.8.1 Type of Existence. We assume that a type S is given, together with a property P over the elements of S, and one wants to construct the type sig of pairs composed of an element x from S and of a proof p of $P\,x$ — such a pair can be constructed only if x verifies P. We denote $\text{suchthat}\,x\,p$ such a pair, that is, we name suchthat the corresponding constructor. The type sig is parameterized by S and P, we employ the following notation:

Inductive sig $[S\!:\!\text{Set}; P\!:\!S \to \text{Prop}]$: Set:=
    suchthat: $\forall x\!:\!S\ P\,x \to \text{sig}\,S\,P$.

As we have two parameters S and P, suchthat actually constructs a 4-tuple instead of a pair, which is $\text{suchthat}(S, P, x, p)$ where the types of P and of x depend on the value of S and where the type of p depends on the values of P and of x.

The type sig plays a role similar to a definition by comprehension in set theory. For this reason one uses the notation $\{x\!:\!S \mid P\,x\}$ instead of $\text{sig}\,S\,P$. For example, the type of even integers can be defined by $\{n\!:\!\text{nat} \mid \text{even}\,n\}$. However, one must be aware that in set theory, $\{x\!:\!S \mid P\,x\}$ denotes a subset A of S, while in type theory, the same expression denotes a set of pairs. We recover A by deleting the second element of each pair.

The type sig has another interpretation. Indeed, proving that there exists an x verifying $P\,x$, is the same as exhibiting a witness x and a proof of $P\,x$. The definition of $\exists x\!:\!S\ P\,x$ is identical to the definition of $\{x\!:\!S \mid P\,x\}$, with just one difference: the result is a proposition instead of a data type:

Inductive ex $[S\!:\!\text{Set}; P\!:\!S \to \text{Prop}]$: Prop:=
    ex_intro: $\forall x\!:\!S\ P\,x \to \text{ex}\,S\,P$.

The difference between $\{x\!:\!S \mid P\,x\}$ and $\exists x\!:\!S\ P\,x$ is then tiny; it is important only in the framework of program extraction.

12.2.8.2 Sums and Disjunction. Let A and B be two types, which are themselves of type Set. An inhabitant of their sum is constructed from either an inhabitant of A, or an inhabitant of B. The corresponding inductive definition sum $A\,B$ has two parameters (A and B) and two constructors inl and inr:

Inductive sum $[A, B\!:\!\text{Set}]$: Set:=
      inl: $A \to \text{sum}\,A\,B$
    | inr: $B \to \text{sum}\,A\,B$.

One generally uses the notation $A + B$ instead of sum $A\,B$. The inhabitant of $A + B$ which is constructed from an inhabitant a of A is then $\text{inl}\,A\,B\,a$.

In § 12.1.3, we used the similar construct $\{P\} + \{Q\}$, where P and Q play the role of A and B; here, P and Q have the type Prop, while the result has again the type Set:

Inductive sumbool $[P, Q\!:\!\text{Prop}]$: Set:=
      left: $P \to \text{sumbool}\,P\,Q$
    | right: $Q \to \text{sumbool}\,P\,Q$.

The disjunction of two propositions P and Q is another variant of this inductive type, where the result has the type Prop instead of Set.

Inductive or $[P, Q : \text{Prop}]$: Prop:=
 or_introl: $P \rightarrow \text{or } P Q$
 | or_intror: $Q \rightarrow \text{or } P Q.$

The syntax used is $P \vee Q$ instead of or $P Q$. We get yet another useful variant by summing a data type A with a proposition Q.

Inductive sumor $[A : \text{Set}; Q : \text{Prop}]$: Set:=
 inleft: $A \rightarrow \text{sumor } A Q$
 | inright: $Q \rightarrow \text{sumor } A Q.$

The syntax used is $A + \{Q\}$. When A is itself an existential type, of the form $\{x : A \mid P x\}$, we get $\{x : A \mid (P x)\} + \{Q\}$, which is the last general type we used in § 12.1.3. An equivalent definition is as follows:

Inductive option $[A : \text{Set}; P : A \rightarrow \text{Prop}; Q : \text{Prop}]$: Set:=
 success: $\forall x : A \ P x \rightarrow \text{option}(A, P, Q)$
 | fail: $Q \rightarrow \text{option}(A, P, Q).$

An inhabitant of this type is either an inhabitant x of A together with a proof that x verifies the predicate P, or a proof of Q.

12.3 A Program Correct by Construction

In § 12.1.3, we gave a specification for the search for an element in a table. How can we design an algorithm from this specification? Two approaches can be taken. The first is quite standard. We view

⟨def_tbl U Ptable P, resu_tbl U Ptable P⟩

or its specialization to intervals

⟨def_tbl_int, resu_tbl_int⟩

as the precondition and the postcondition, respectively, of an imperative program. We can, for example, formalize the concept of a predicate transformer in the calculus of inductive constructions. Then we get a framework composed of a formal logic and of tools, which can be used to support the development process presented in Chapter 4. This aid is worthwhile if we want to be sure that nothing has been forgotten in the reasoning of § 4.2.2.

 A more sophisticated variant consists of formalizing in Coq the operational and axiomatic semantics of a programming language. It is then possible to automate the production of the lemmas to be proved. This was previously proposed in [BF95, Ter93]. More recent works include [Fil99].

The second approach is typical of constructive logics. It is based on the aforementioned Curry–Howard correspondence. In this approach, a program and its proof are simultaneously developed. This is reminiscent of the techniques of Dijkstra. The main difference is that here we will get functional programs instead of imperative programs. In general, this means that the efficiency of imperative programs may be lost, but that complex recursive functions can be proven to be correct. However, this cannot be illustrated on the table example: we will get a very simple algorithm, and moreover a tail-recursive one, so that modern compilers of functional languages are able to provide code as efficient as for C programs. But our aim is only to illustrate the technique on our now well-known example.

12.3.1 Programs and Proofs

Recall the Curry–Howard correspondence:

$$
\begin{aligned}
specification &= type, \\
proof &= functional\ program.
\end{aligned}
$$

A specification is, from a logical viewpoint, an implication between a precondition and a postcondition. From the functional viewpoint, it is the type of a function, as given by the type of its arguments (together with logical constraints) and the type of the result (together with logical constraints also). In the table example, for a given U, the specification of the search for an element x verifying P, if there is one, in a table characterized by $Ptable$ is then:

$$\texttt{def\_tbl}\ U\ Ptable\ P \to \texttt{resu\_tbl}\ U\ Ptable\ P \ . \tag{12.5}$$

Instead of directly displaying a function in a functional language, the idea is to prove the formula (12.5), using the rules of logic: introduction of hypotheses, case splitting, reasoning by induction, etc. Figure 12.1 gives the main correspondences between reasoning rules and algorithmic constructs.

conjunction	pair of data
case analysis	**case of, if then else**
implication	function
reasoning by induction	(primitive) recursion

Figure 12.1: Logic and functions

A proof constructed in this way contains an algorithm. Of course, different proofs correspond to different programs. Actually, one may perform the proof with a more or less precise algorithm in mind; in much the same way, one is guided by intuition when one writes down a formal proof.

Efficiency issues are not ignored in this approach, and this may give proofs a somewhat artificial taste. For example, suppose we want to find an inhabitant of the type $T\,n$, where n is a given integer. If $T\,n$ is proven by regular induction, the result will be found after $O(n)$ computation steps.[2] But if we use the following induction principle:

$$\forall n\!:\!\mathtt{nat}\ \ P\,0 \to [\forall k\!:\!\mathtt{nat}\ \ P\,k \to P(2k)]$$
$$\to [\forall k\!:\!\mathtt{nat}\ \ P\,k \to P(2k+1)] \to \forall n\!:\!\mathtt{nat}\ \ P\,n \ ,$$

the number of steps will be $O(\log n)$. This is nothing but the logical translation of well-known design principles for algorithms. In summary, the choice of data types and of induction principles are important design decisions in a development — they are expected to be performed by a human. The support provided by software-based proof assistants is more relevant for the management of technical details.

12.3.2 Example: Searching for an Element in a List

According to the above sections, searching for an element verifying a given property, in a given list, amounts to finding a function specified by the type:

$$\mathtt{def\_tbl\_lis} \to \mathtt{resu\_tbl\_lis} \ . \tag{12.6}$$

Let us expand the definitions of `def_tbl_lis` and of `resu_tbl_lis`:

$$(\forall u\!:\!U\ \mathtt{Ptablist}\,u \to \{P\,u\} + \{\neg P\,u\}) \to$$
$$+\begin{matrix} \{u\!:\!U \mid \mathtt{Ptablist}\,u \wedge P\,u\} \\ \{\forall u\!:\!U\ \mathtt{Ptablist}\,u \to \neg P\,u\} \end{matrix} \ . \tag{12.7}$$

The solution is by no means mysterious: `Ptablist` depends on the given list l, and we just make the desired program check each element until a suitable one is found. From a logical perspective, this corresponds exactly to considering the case where l is empty and the case where l is composed of at least one element. More precisely, we proceed by induction over the structure of l as follows:

– if a property is proved for `nil`,
– furthermore, if we prove the property for $\mathtt{cons}\,u\,l$ with the assumption it holds for l,
– we conclude that this property holds for an arbitrary list.

In practice, such reasoning is elaborated step-by-step and interactively with the aid of a tool such as Coq or Lego. In our example, the Coq script takes less than 10 lines, whereas the underlying detailed and complete reasoning is longer. We write it down for the scrupulous reader.

---

[2]Roughly speaking, $O(n)$ is a proportional function of n.

Let us expand Ptablist in order to make l explicit, and, for the sake of simplifying the presentation, suppose that P can be tested for all inhabitants from U:

$$\begin{aligned}(\forall u{:}U\ \{P\,u\} + \{\neg P\,u\}) \to \\ {}_+\{u{:}U \mid \text{contains}\,l\,u \wedge P\,u\} \\ {}^+\{\forall u{:}U\ \text{contains}\,l\,u \to \neg P\,u\}\ .\end{aligned} \qquad (12.8)$$

We consider that $\{x : A \mid \phi x\} + \{\psi\}$ has two constructors named **success** and **fail**. The result is then either of the form **success** $u\,\rho$, where u inhabits U and ρ is a proof of contains $l\,u \wedge P\,u$, or of the form **fail** σ where σ is a proof of $\forall u{:}U\ \text{contains}\,l\,u \to \neg P\,u$. We construct such an object by induction over the structure of the list l:

- the case where l is **nil** is easy to solve: we have a trivial proof σ of $\forall x{:}U$ contains nil $x \to \neg P\,x$, we then construct **fail** σ_{nil}; intuitively, no member of the empty list verifies P, which prevents us from claiming satisfaction of the first choice (success);
- if $l = \text{cons}\,u\,l'$, we will be allowed to use the induction hypothesis expressed by (12.8) where l' is substituted for l; but let us first test P on u: we get either a proof of $P\,u$, or a proof of $\neg P\,u$ — intuitively: we compute the truth value of $P\,u$;
  - in the first case, we get a proof ρ_u of contains $l\,u \wedge P\,u$, from which we construct **success** $u\,\rho_u$;
  - in the second case, we use the induction hypothesis over l': in the case of success, every member of l' verifying P is also a member of l verifying P; in the case of failure, no member of l' verifies P, and then no member of l verifies P, since we already have $\neg P\,u$.

This proof, viewed as a function, has the following form, where the expressions σ_{nil}, ρ_u, ρ_v, ρ'_v, σ and σ' are not detailed, and where $D\,u$ is of type $\{P\,u\} + \{\neg P\,u\}$:

```
list_search ≝ function
    nil ⟶ fail σ_nil
  | cons u l ⟶ case D u of a proof of
      P u   ⟶ success u ρ_u
      ¬P u  ⟶ case list_search l' of
          success v ρ'_v ⟶ success v ρ_v
        | fail σ'        ⟶ fail σ .
```

12.3.3 Searching in an Interval of Integers

In the case where the table is represented by an interval of integers $[p..q[$, the formula to be proven is:

$$\text{def\_tbl\_int} \to \text{resu\_tbl\_int}\ . \qquad (12.9)$$

The previous proof can be adapted by reasoning over the length of the interval, l. For example, we can consider p as a fixed parameter and l such that $q = p + l$, and then reason by induction over l. We could then paraphrase the previous subsection, but we prefer now to follow the line of the program presented in § 2.4.4, where we use an additional piece of information: if there are several integers satisfying P in $[p..q[$, the result is the least of them.

In this version we express the type of the result as an integer contained between p and q inclusive, by imposing that, if $x = q$, then no integer of the table verifies P. We introduce the auxiliary predicate `ini_seg_empty` x, whose meaning is that no integer from $[p..x[$ verifies P:

```
ini_seg_empty[x:nat] def= ∀i:nat between p i x → ¬P i .
resint def= {x:nat | (Pinterv x ∧ P x) ∨
                      (x = q ∧ ini_seg_empty q)} .
```

In order to allow us to recover `resu_tbl_int` from `resint`, we simply construct a converting function specified by:

$$\texttt{resint} \rightarrow \texttt{resu\_tbl\_int} . \tag{12.10}$$

The proof is by case analysis on the value of x contained in `resint`:

- if $x = q$, we deduce, from the definitions of `resint` and `Pinterv`, that `ini_seg_empty` x is verified, the inhabitant from `resu_tbl_int` to return is then `fail` σ, where σ is the object which formalizes the proof of `ini_seg_empty` x;
- if $x \neq q$, we deduce, from the definition of `resint`, that x is in the interval and verifies P, then we take `success` $x \langle \iota, \rho \rangle$, where ι and ρ are the objects which formalize the proofs of `Pinterv` x and $P x$, respectively.

The function corresponding to this proof is:

$$\textbf{if } x = q \textbf{ then fail } \sigma \textbf{ else success } x \langle \iota, \rho \rangle .$$

We still have to prove `resint`. Intuitively, we will once more examine the elements in the interval $[p..q]$ — characterized by `Pdom` — until a suitable one is found, as in the algorithm explained in § 2.4.4. To this end we consider a stronger specification, named `strg_resint` where no integer from $[p..x[$ verifies P, even when x is smaller than q:

```
Pdom[x:nat] def= Pinterv x ∨ x = q .
strg_resint def= {x:nat | Pdom x ∧ ini_seg_empty x ∧
                           (x < q → P x)} .
```

Proving `strg_resint` → `resint` is quite easy, the underlying function preserves the witnessing integer.

Again, following the reasoning line of Chapter 2, we take $q - x$ as our loop variant. Intuitively, it means that we intend to reason by induction over $q - x$. The base case is $x = q$. The only result we can propose in this case is

q, but to this end, we first need a proof of ini_seg_empty x. It is then better to try to prove step x by induction, where we put a precondition in front of strg_resint:

$$\text{step}[x\text{:nat}] \stackrel{\text{def}}{=} \text{ini\_seg\_empty}\, x \rightarrow \text{strg\_resint} \ .$$

The informal reasoning is as follows:

- if $x = q$, a proof of ini_seg_empty x allows us to deduce that the result is q;
- if $x < q$, suppose once again that we have a proof of ini_seg_empty x at our disposal; the induction hypothesis expresses that we are able to find the result from a proof of ini_seg_empty$(x + 1)$; we reason by case analysis on $P\,x$: if $P\,x$ holds, the result is simply x; in the opposite case, $\neg P\,x$ combined with ini_seg_empty x provides a proof of ini_seg_empty$(x+1)$, so that we are allowed to use the induction hypothesis.

Note that, the hypothesis def_tbl_int is needed for reasoning by case analysis on $P\,x$.

A technique for reasoning by induction over $q - x$ is to explicitly determine an integer l such that $x + l = q$. We prove the theorem loop specified by:

$$\text{def\_tbl\_int} \rightarrow \forall l, x\text{:nat} \quad p \leq x \rightarrow l + x = q \rightarrow \text{step}\, x \ .$$

by induction over l, by formalizing the previous reasoning. A better option is to prove the following specification:

$$\text{def\_tbl\_int} \rightarrow \forall x\text{:nat} \quad p \leq x \rightarrow x \leq q \rightarrow \text{step}\, x \ .$$

using well-founded induction; in this way we avoid using l. The well-founded relation to be used is the one named $R_4(q)$ on page 51.

Finally, giving x the value p in loop, (and l the value $q - p$, if we use the former specification of loop), then providing a — very simple — proof of ini_seg_empty p, we obtain an element from strg_resint.

12.3.4 Program Extraction

The program just obtained manipulates pieces of data, such as x, p, q, and proofs, for example the proof of ini_seg_empty x. If we keep this program as it is, its execution will be composed of computation steps not only on data but also on proofs. Intuitively, this means that assertions on data will be dynamically checked, which is obviously pointless. Clearly, we can compare this with type-checking in the common typed programming languages: for example, compile-time type-checking ensures that arithmetical functions will actually be applied on numbers at run-time; then typing information can be removed from the executable code.

The same strategy can be adopted here. In concrete terms, everything related to Prop can be removed from programs such as the ones that were presented above. Thus, one extracts an untyped program which complies, by construction, with the initial specification. We can first illustrate the idea on the type def_tbl. Its complete definition was:

$$\text{def\_tbl}[U:\text{Set}\,;\,\textit{Ptable},\,P:U \to \text{Prop}] \overset{\text{def}}{=}$$
$$\forall x:U\ \textit{Ptable}\,x \to \{P\,x\} + \{\neg P\,x\}\ .$$

In the extraction process, $\{P\,x\} + \{\neg P\,x\}$ is replaced with bool, which does not depend on x; we are left with:

$$\text{def\_tbl}[U:\text{Set}] \overset{\text{def}}{=} U \to \text{bool}\ .$$

For example, an inhabitant D of type def_tbl nat is a function from nat to bool; this function is not necessarily defined for all integers — it could be implemented by an array, but we are supposed to use it only under the preconditions which are written in the original definition. Here, D is only a parameter; let us consider again the converting function which was developed in § 12.3.3. Its type is resint → resu_tbl_int. Expanding resint and resu_tbl_int, we get:

$$\{x:\text{nat} \mid (\text{Pinterv}\,x \wedge P\,x) \vee (x = q \wedge \text{ini\_seg\_empty}\,q\} \to$$
$$\{x:\text{nat} \mid \text{Pinterv}\,x \wedge P\,x\} + \{\forall x:\text{nat}\ \text{Pinterv}\,x \to \neg P\,x\}.$$

The proposed function was:

$$\text{conversion}[\langle x:\text{nat},\pi:(\text{Pinterv}\,x \wedge P\,x) \vee \ldots\rangle] \overset{\text{def}}{=}$$
if $x = q$ then fail σ else success x , $\langle \iota,\rho\rangle$.

Under its expurgated form, the type resu_tbl_int is inhabited by elements of the form success n, where n is a natural integer, or fail. As for the type resint, it simply boils down to nat. The extracted program is then:

$$\text{conversion}[x:\text{nat}] \overset{\text{def}}{=}$$
if $x = q$ then fail else success x .

If we consider the program for searching in a list, as given on page 246, the extraction process yields the following algorithm:

```
list_search ≝ function
      nil → fail
  |   cons u l → if D u then success u
                  else case list_search l' of
                      success v → success v
                    | fail → fail .
```

This program, although it is correct, is somewhat frustrating, because there is clearly no need to test the result of the recursive call. We would prefer:

```
list_search ≝ function
    nil → fail
  | cons u l → if D u then success u else list_search l' .
```

This can be regarded as an optimization, which could be performed by a good compiler, or at the back-end of the extraction process itself. However, we can sharpen the previous development so that we directly obtain the second program.

The main problem is that, as the type of the result is $\{u : U \mid$ contains $l\,u \wedge P\,u\} + \{\forall u : U$ contains $l\,u \to \neg P\,u\}$, an induction over l forces us to distinguish success $v\,\rho'_v$ from success $v\,\rho_v$ and fail σ' from fail σ: indeed, ρ'_v, for example, is of type contains $l\,u \wedge P\,u$, whereas ρ_v is of type contains$(\text{cons}\,v\,l)\,u \wedge P\,u$.

Once this is understood, the solution consists of putting the goal into an equivalent form Cond $l \to \{\ldots\} + \{\ldots\}$ where Cond l is a purely logical expression, and then will be removed at the extraction stage, and where $\{\ldots\} + \{\ldots\}$ is kept constant in the induction step. In this case Ptablist is just the ticket. We prove:

$$(\forall u{:}U \text{ contains}\,l\,u \leftrightarrow \text{Ptablist}\,u) \to$$
$$+\begin{array}{l}\{u{:}U \mid \text{Ptablist}\,u \wedge P\,u\} \\ \{\forall u{:}U \text{ Ptablist}\,u \to \neg P\,u\}\end{array} \qquad (12.11)$$

by induction over l, following the same reasoning line as before.

In the case where the table is represented by an interval of integers, the search function is the following. We give here the ML program actually extracted by the system Coq from the proof given above. Connoisseurs will note that we get a tail-recursive program, which compiles to a common loop. We then get a program quite close to the imperative algorithm given on page 31.

```
let main p q D =
  let rec loop x =
    match q = x with
      true → q
    | false →    match D x with
                   true → x
                 | false → loop (S x)
  in loop p ;;
```

The program extraction mechanism is based on general results of realizability theory, which ensures that the extracted function conforms to the specification of the complete function.

Program extraction allowed us to point out the deep analogy between program and proof design. In this framework, it remains possible to adopt a more traditional strategy, by proposing the function to be extracted [Par95]; then it is up to the system to infer automatically the corresponding proof obligations.

12.4 On Undefined Expressions

A tricky issue about the relationship between logic and programming was raised in Chapter 2: a logical expression may contain undefined terms. This issue was illustrated on the expression $P\,x$. In this chapter, we introduced a computable function D which determines whether or not $P\,x$ holds. There is a clear distinction between the use of P in the mathematical reasoning, the use of D at the same level, and the use of D in the expressions of the final program. The fact that D is not defined everywhere is represented in its specification by the formula $p \leq x < q \rightarrow \{P\,x\} + \{\neg\,P\,x\}$. By this implication, D takes an additional argument which is a proof δ_x of $p \leq x < q$: the complete expression is actually $D\,x\,\delta_x$. It is always defined — that is, it is defined for all *pairs* $\langle x, \delta_x \rangle$ — and hence it always makes sense to use it in our reasoning. Once the latter is finished, we can consider a program obtained by extraction, where only $D\,x$ is present, and we are ensured that x is in the domain of D.

12.5 Other Proof Systems Based on Higher-order Logic

The main calling of typed higher-order logic is to provide a rigorous and very expressive logical framework: as soon as the systems we want to model are complex at all, we need to rely upon a collection of mathematical results formalized in advance. The richness of expression is an important ingredient for expressing problems, reasonings and hopefully solutions in a natural manner, with an adequate degree of generality.

At the same time, any approach having the goal of verifying realistically sized systems must rely upon automated proof techniques which relieve the user of tasks which are often tedious (arithmetical calculations, propositional reasoning) or complex (*model checking* techniques, for example), or both. Using and combining efficiently the know-how accumulated in the different relevant disciplines is still a research topic. At the same time, the issue of the reliability of the analysis and proof tools becomes important, even more so as the tools become larger and implement more complex algorithms.

PVS (Prototype Verification System) is a proof assistant for a higher-order classical logic, which is quite good at automatically discharging proof obligations, thanks to the implementation of state-of-the-art decision procedures. The specification language of PVS includes dependent types and a predicate-based sub-typing mechanism, which are quite powerful for specification purposes, but make type-checking undecidable: type-checking may generate proof obligations. Fortunately, most of them can be automatically discharged thanks to the automated proof procedures of the system. The latter are indeed very convenient for the user, and they tend to be used extensively, so that the user can concentrate his efforts more on the structure of his developments.

The reliability of the approach relies mainly on the expertise of the designers and implementers of the system. PVS is a good laboratory for experimenting with new ideas in the area. However, to prevent obvious potential problems, only a small number of researchers are authorized to integrate new mechanisms into the official version of the system. Even so, if an undiscovered flaw remains, in particular, a flaw which occurs only in rare configurations or is hardly observable in common situations, the chances that it is — unconsciously — used increase when users more frequently use the automated procedures offered to them. Can we prevent such accidents, or, more modestly, restrict or delimit the risk?

This issue motivated one of the key decisions for the design of the architecture of LCF [GMW79], another proof tool for higher-order logic (without dependent types). The main idea is to have a *small* software kernel, the proof checker, which is very carefully written, with only one objective: checking that only legal deduction rules are used in a formal proof. Such an architecture is open: arbitrary complex proof search procedures may be involved, including, typically, new decision procedures for a specialized area, and this without threatening the logical integrity of the approach, since the kernel eventually checks the correctness of all proof steps.

This approach is made possible when the logic itself is composed of a restricted number of primitive elements. For example, the calculus of inductive constructions is essentially based on one logical quantifier ($\forall$), a very general induction principle and the concept of a reduction.

This idea has also been followed in a number of successors to LCF. It is implemented in two ways in actual systems. One of them consists of defining an abstract type for theorems: the latter are created and derived from each other through an interface, which proposes only the formation of axioms, and the use of deduction rules similar to the ones we have presented in Chapter 9. HOL, for example, is constructed according to this architecture.

Another possibility consists of explicitly handling proof terms. This is particularly suitable to intuitionistic logic, since proof terms are λ-terms: λ-terms are already available, since we are considering a higher-order logic. Actually, as a theorem is nothing but the type of a λ-term, verifying that a formula is proven boils down to performing type checking. The advantage of this approach, over the approach based on an abstract type for theorems, is that it maintains and provides an exhaustive trace of formal reasonings. This leaves room for controlling the latter by an independent system, or for extracting a natural language explanation from a formal proof [Cos96]. The difficulty is to keep proof terms to a reasonable size. Coq, which we described earlier, is a typical example of systems based on this principle.

As the reliability of LCF-technology-based proof assistants relies entirely upon their kernel, much attention is paid to the latter by the development teams concerned. However, since a really powerful logic is available, why not try to formalize and mechanically check the kernel itself? Such a task is far from simple: on the one hand, the manipulated algorithmic structures are complex; on the other, at the specification level, representing the logical rules is not sufficient, it is also necessary to prove a number of metatheorems which govern them. These obstacles were successfully tackled in the case of Coq, by B. Barras [Bar99]. In concrete terms, this opens up the possibility that the kernel of a future version of Coq may be obtained by program extraction (see § 12.3.4).

We have just seen that there are several options for higher-order logic-based proof tools. There are also some differences in the logics considered. For example, PVS and Coq include dependent types, but HOL does not; typing judgements are decidable in Coq and in HOL, but not in PVS; HOL and PVS use a classical logic, whereas the logic of Coq is constructive.[3] Among the three systems considered here, Coq is also the only one where types can themselves be computed (by reduction); this allows one to further exploit the possibilities of dependent types.

To illustrate the idea, here is a small but typical example where the latter feature turns out to be useful. We want to represent names, say a and b, and a specific type for each of them:

```
Inductive name: Set:= a: name   |   b: name.
Definition ty:= [x : name] Cases x of
    a   ⇒  bool
  | b   ⇒  nat  end.
```

We can then construct pairs $\langle x, v \rangle$, where x is of type name and v is of type ty x: $\langle b, 3 \rangle$ is such an object. At the type-checking stage, the proof tool performs the reduction $ty(b) \overset{\delta}{\to} nat$.

The proof tools considered in this chapter have been, and are, successfully used in some industrial applications, for example in areas related to security, smart cards, protocols, etc. There is still work in progress for making them more powerful and more efficient, on the one hand (for example their use in combination with fully automated techniques based on rewriting or on model-checking), and easier to use on the other hand, thanks to syntactical devices, or to graphical interfaces, such as Pcoq based on the idea of "proof-by-pointing" [BKT94].

12.6 Notes and Suggestions for Further Reading

The calculus of inductive constructions is described in the Coq manuals [HKPM02, TP02]. The principles for program extraction implemented in Coq are defined in the thesis of Christine Paulin-Mohring [PM89]. A similar system is Nuprl [CAB+86], which allows one to develop constructive mathematical theories in a system inspired by the type theory of Martin-Löf. Amongst other systems based on a higher-order logic, we have HOL, Isabelle and PVS. All are supplied with user and reference manuals [GM93, Pau94, CAB+86, ORS93]. A

---

[3]However, it should be noted that the excluded middle law can be used in the **Prop** universe.

number of articles are also available, for example [ORS92] on PVS, [Pau90] on Isabelle and [Gor88] on HOL.

Valuable principles for designing and implementing a serious proof assistant are described by Larry Paulson in [Pau92]. Readers may then be tempted to try to write their own software. However, before doing so, it is advisable to read the conclusion of Paulson's article several times.

Bibliography

[Aba90]   M. Abadi. An Axiomatization of Lamport's Temporal Logic of Actions. Technical Report 65, Digital Equipment Corporation, Systems Research Centre, October 1990.

[Abr92]   J-R. Abrial. The B-Technology. In *FORTE'92, 5th Int. Conf. on Formal Description Techniques*, 1992.

[Abr96]   J-R. Abrial. *The B-Book: Assigning Programs to Meanings*. Cambridge University Press, 1996.

[AGM92a] S. Abramsky, D.M. Gabbay, and T.S.E. Maibaum, editors. *Handbook of Logic in Computer Science*, volume 1: Background: Mathematical structures. Oxford Science Publications, 1992.

[AGM92b] S. Abramsky, D.M. Gabbay, and T.S.E. Maibaum, editors. *Handbook of Logic in Computer Science*, volume 2: Background: Computational structures. Oxford Science Publications, 1992.

[AL91]   A. Asperti and G. Longo. *Categories, Types and Structures: an Introduction to Category Theory for the Computer Scientist*. MIT Press, 1991.

[AN82]   A. Arnold and M. Nivat. Comportements de processus. In *Colloque AFCET « Les mathématiques de l'informatique »*, pages 35–68, 1982.

[AN01]   A. Arnold and D. Niwiński. *Rudiments of the mu-calculus*. Elsevier, 2001.

[Arn94]   A. Arnold. *Finite Transition Systems: Semantics of Communicating Systems*. International Series in Computer Science. Prentice Hall, 1994.

[Art91]   R.D. Arthan. On free type definitions in Z. In J.E. Nicholls, editor, *Z User Workshop*, LNCS. Springer-Verlag, 1991.

[Art98]   R.D. Arthan. Recursive definitions in Z. In J.P. Bowen, A. Fett, and M.G. Hinchey, editors, *ZUM'98*, volume 1493 of *LNCS*, pages 154–171. Springer-Verlag, 1998.

[AU79]   A.V. Aho and J. Ullman. Universality of data retrieval languages. In *Principles of Programming Languages*, pages 110–120. ACM, 1979.

[Aug98]   L. Augustsson. Cayenne - a language with dependent types. In *International Conference on Functional Programming*, pages 239–250, 1998.

[B-T91]   Edinburgh Portable Compilers Ltd. *B-Tool Version 1.1 - User Manual/ Tutorial / Reference Manual*, 1991.

[B-T93]   Oxford Science Park, UK. *B-Toolkit Beta-Release Version 1.1 - User Manual, Reference Manual*, 1993.

[Bar77]   J. Barwise, editor. *Handbook of Mathematical Logic*. North Holland, 1977.

[Bar84]   H.P. Barendregt. *The Lambda Calculus, its Syntax and Semantics*, volume 103 of *Studies in Logic*. North Holland, 1984.

[Bar90]   H. Barendregt. Functional programming and lambda calculus. In van Leeuwen [vL90b], chapter 7.

[Bar99]   B. Barras. *Auto-validation d'un systèmes de preuves avec familles inductives*. Ph.D. thesis, Université de Paris 7, 1999.

[Bau91]   F.L. Bauer, editor. *Logic, Algebra and Computation*, volume F79 of *NATO ASI Series*. Springer-Verlag, 1991.

[BBC⁺95]   N. Bjorner, I.A. Browne, E. Chang, M. Colón, A. Kapur, Z. Manna, H.B. Sipma, and T.E. Uribe. STeP: the Stanford Theorem Prover, Users's Manual. Technical report, Stanford University, 1995.

[BBF⁺01]   B. Bérard, M. Bidoit, A. Finkel, A. Petit, L. Petrucci, and Ph. Schnoebelen. *System and Software Verification, Model-Checking Techniques and Tools*. Springer, 2001.

[BBFM99]   P. Behm, P. Benoit, A. Faivre, and J.-M. Meynadier. Météor: a successful application of b in a large project. In Wing et al. [WWD99], pages 369–387.

[BBS93]   F.L. Bauer, W. Brauer, and H. Schwichtenberg, editors. *Logic and Algebra of Specification*, volume F94 of *NATO ASI Series*. Springer-Verlag, 1993.

[Berct]   S. Berezin. The SMV web site, 2000
           http://www.cs.cmu.edu/~modelcheck/smv.html/.

[BF95]   Y. Bertot and R. Fraer. Reasoning with executable specifications. In *Int. Joint Conf. on Theory and Practice of Software Development, TAPSOFT*, volume 915 of *LNCS*. Springer-Verlag, May 1995.

[BG92]   G. Berry and G. Gonthier. The Esterel Synchronous Programming Language: Design, Semantics, Implementation. *Science of Computer Programming*, 19:87–152, 1992.

[Bir95]   R.S. Bird. Functional Algorithm Design. In B. Möller, editor, *Mathematics of Program Construction*, volume 947 of *LNCS*. Springer-Verlag, 1995.

[BK90]   J.C.M. Baeten and J.W. Klop, editors. *CONCUR 90*, Amsterdam, volume 458 of *Lecture Notes in Computer Science*. Springer-Verlag, 1990.

[BKK⁺98]   P. Borovansky, C. Kirchner, H. Kirchner, P.-É. Moreau, and C. Ringeissen. An Overview of ELAN. *Electronic Notes in Theoretical Computer Science*, 15, 1998.
           http://www.elsevier.nl/locate/entcs/volume15.html.

[BKL+91]   M. Bidoit, H-J. Kreowski, P. Lescanne, F. Orejas, and D. Sanella, editors. *Algebraic System Specification and Development, a Survey and Annotated Bibliography*, volume 501 of *LNCS*. Springer-Verlag, 1991.

[BKR92]   A. Bouhoula, E. Kounalis, and M. Rusinowitch. Spike: An automatic theorem prover. In *Proc. 1st Int. Conf. on Logic Programming and Automated Reasoning*, volume 624 of *Lecture Notes in Artificial Intelligence*, St. Petersburg (Russia), July 1992. Springer-Verlag.

[BKT94]   Y. Bertot, G. Kahn, and L. Théry. Proof by pointing. In M. Hagiya and J.C. Mitchell, editors, *Proc. of the Int. Symp. on Theoretical Aspects of Computer Software*, volume 789 of *LNCS*, pages 141–160, Sendai, Japan, April 1994. Springer-Verlag.

[BLJ91]   A. Benveniste, P. Le Guernic, and C. Jacquemot. Synchronous programming with events and relations: the SIGNAL language and its semantics. *Science of Computer and Programming*, 16:103–149, 1991.

[BM79]   R.S. Boyer and J S. Moore. *A Computational Logic*. Academic Press, New York, 1979.

[BN98]   F. Baader and T. Nipkow. *Term Rewriting and All That*. Cambridge University Press, 1998.

[Bou94]   A. Bouhoula. SPIKE: a system for sufficient completeness and parameterized inductive proof. In A. Bundy, editor, *Proc. 12th Int. Conf. on Automated Deduction*, volume 814 of *Lecture Notes in Artificial Intelligence*, pages 836–840, Nancy (France), June 1994. Springer-Verlag.

[BR95]   A. Bouhoula and M. Rusinowitch. Implicit induction in conditional theories. *Journal of Automated Reasoning*, 14(2):189–235, 1995.

[Bra92]   J.C. Bradfield. *Verifying Temporal Properties of Systems*. Progress in Theoretical Computer Science. Birkhäuser, 1992.

[Bro89]   M. Broy, editor. *Constructive Methods in Computing Science*, volume F55 of *NATO ASI Series*. Springer-Verlag, 1989.

[BW88]   R. Bird and P. Wadler. *Introduction to Functional Programming*. Prentice Hall, 1988.

[BW90]   M. Barr and C. Wells. *Category Theory in Computer Science*. Prentice Hall, 1990.

[CAB+86]   R.L. Constable, S.F. Allen, H.M. Bromley, W.R. Cleaveland, J.F. Cremer, R.W. Harper, D.J. Howe, T.B. Knoblock, N.P. Mendler, P. Panangaden, J.T. Sasaki, and S.F. Smith. *Implementing Mathematics with the Nuprl Proof Development System*. Prentice-Hall, 1986.

[CDE+99]   M. Clavel, F. Duràn, S. Eker, P. Lincoln, N. Martí-Oliet, J. Meseguer, and J. Quesada. Maude: Specification and Programming in Rewriting Logic. Technical report, SRI International, January 1999. http://maude.csl.sri.com.

[CES83]   E.M Clarke, E.M. Emerson, and A.P. Sistla. Automatic verification of finite state concurrent systems using temporal logic specifications: a practical approach. In *Proc. 10th ACM Symp. on Principles of Programming Languages*, 1983.

[CGP99]   E.M. Clarke, O. Grumberg, and D.A. Peled. *Model Checking*. MIT Press, 1999.

[CGR93a]   D. Craigen, S. Gerhart, and T. Ralston. An international survey of industrial applications of formal methods, 1 : Purpose, approach, analysis and conclusions. Technical Report 93/626 NIST-GCR, National Institute of Standards and Technology, US Dep. of Commerce, Technology Administration, NIST, Computer Systems Laboratory, Gaithersburg, MD 20899, March 1993.

[CGR93b]   D. Craigen, S. Gerhart, and T. Ralston. An international survey of industrial applications of formal methods, 2 : Case studies. Technical Report 93/626 NISTGCR, National Institute of Standards and Technology, 1993. See [CGR93a].

[CH85]   T. Coquand and G. Huet. A theory of constructions. In Kahn et al. [KMP85].

[CK90]   C.C. Chang and H.J. Keisler. *Model Theory*. North Holland, 3rd edition, 1990.

[CL73]   C.-L. Chang and R. Char-Tung Lee. *Symbolic Logic and Mechanical Theorem Proving*. Computer Science Classics. Academic Press, 1973.

[CL00]   R. Cori and D. Lascar. *Propositional Calculus, Boolean Algebras, Predicate Calculus, Completeness Theorems (Mathematical Logic, a Course with Exercises, part II)*. Oxford University Press, 2000.

[CL01]   R. Cori and D. Lascar. *Recursion Theory, Godel's Theorems, Set Theory, Model Theory (Mathematical Logic, a Course with Exercises, part II)*. Oxford University Press, 2001.

[CM89]   K.M. Chandy and J. Misra. *Parallel Program Design*. Addison-Wesley, Austin, Texas, May 1989.

[CM98]   G. Cousineau and M. Mauny. *The Functional Approach to Programming*. Cambridge University Press, 1998.

[CMP02]   E. Chailloux, P. Manoury, and B. Pagano. *Developing Applications with Objective Caml*. O'Reilly, 2002.

[Coh90]   E. Cohen. *Programming in the 1990s: An Introduction to the Calculation of Programs*. Texts and Monographs in Computer Science. Springer-Verlag, 1990.

[Coq86]   T. Coquand. An analysis of Girard's Paradox. In *Proc. IEEE Symp. on Logic in Computer Science*, pages 227–236. IEEE, 1986.

[Cos96]   Y. Coscoy. A natural language explanation for formal proofs. In C. Retoré, editor, *Proceedings of Int. Conf. on Logical Aspects of Computational Liguistics (LACL), Nancy*, volume 1328 of *LNCS/LNAI*. Springer-Verlag, September 1996.

[Cou91]   B. Courcelle, editor. *Logique et informatique: une introduction.* collection didactique. INRIA, 1991.

[CPHP87]   P. Caspi, D. Pilaud, N. Halbwachs, and J. Plaice. LUSTRE, a Declarative Language for Real-Time Programming. In *Proc. 10th ACM Symp. on Principles of Programming Languages*, 1987.

[CW97]   E.A. Cichon and A. Weiermann. Term rewriting theory for the primitive recursive functions. *Annals of Pure and Applied Logic*, 1997.

[dBdRR91] J.W. de Bakker, W.-P. de Roever, and G. Rozenberg, editors. *Foundations of Object-Oriented Languages*, volume 489 of *LNCS*. Springer-Verlag, 1991.

[Dev93]   K. Devlin. *The Joy of Sets.* Undergraduate Texts in Mathematics. Springer-Verlag, second edition, 1993.

[DF98]   R. Diaconescu and K. Futatsugi. *CafeOBJ Report: The Language, Proof Techniques, and Methodologies for Object-Oriented Algebraic Specification*, volume 6 of *AMAST Series in Computing*. World Scientific, 1998.

[Dij76]   E.W. Dijkstra. *A Discipline of Programming.* Prentice-Hall, Englewood Cliffs, NJ, 1976.

[DJ90]   N. Dershowitz and J.-P. Jouannaud. Rewrite Systems. In van Leeuwen [vL90b], chapter 6, pages 244–320. Also technical report 478, LRI.

[dRE98]   W.-P. de Roever and K. Engelhardt. *Data Refinement: Model-Oriented Proof Methods and their Comparison.* Number 47 in Cambridge Tracts in Theoretical Computer Science. Cambridge University Press, 1998.

[DS90]   E.W. Dijkstra and C.S. Scholten. *Predicate Calculus and Program Semantics.* Texts and Monographs in Computer Science. Springer-Verlag, 1990.

[Dub00]   O. Dubuisson. *ASN.1 Communication between Heterogeneous Systems.* Morgan Kaufmann, 2000.

[Dum00]   M. Dummet. *Elements of Intuitionism.* Clarendon Press, Oxford, 2nd edition, 2000.

[EM85]   H. Ehrig and B. Mahr. *Fundamental of Algebraic Specification 1*, volume 6 of *EATCS Monographs on Theoretical Computer Science*. Springer-Verlag, 1985.

[EM90]   H. Ehrig and B. Mahr. *Fundamental of Algebraic Specification 2*, volume 21 of *EATCS Monographs on Theoretical Computer Science*. Springer-Verlag, 1990.

[Eme90]   E.A. Emerson. Temporal and Modal Logic. In van Leeuwen [vL90b], chapter 16, pages 995–1072.

[End77]   H.B. Enderton. *Elements of Set Theory.* Academic Press, 1977.

[FG84]   R. Forgaard and J.V. Guttag. REVE: A term rewriting system generator with failure-resistant Knuth-Bendix. Technical report, MIT-LCS, 1984.

[Fil99]   J.-C. Filliâtre. *Preuves de programmes impératifs en théorie des types.* Ph.D. thesis, Université de Paris-Sud, 1999. English version available at `http://www.lri.fr/~filliatr`.

[Flo67]   R.W. Floyd. Assigning meanings to programs. *Mathematical Aspects of Computer Sciences*, pages 52–66, 1967.

[Gal86]   J. Gallier. *Logic for Computer Science.* Harper and Row, 1986.

[Gal93]   J. Gallier. Constructive logics part I: a tutorial on proof systems and typed λ-calculi. *Theoretical Computer Science*, 110:249–339, 1993.

[GG89]   S. Garland and J.V. Guttag. An overview of LP, the Larch Prover. In N. Dershowitz, editor, *Proc. 3rd Int. Conf. on Rewriting Techniques and Applications*, volume 355 of *Lecture Notes in Computer Science*, pages 137–151, Chapel Hill (NC, USA), April 1989. Springer-Verlag.

[GG90]   P. Gochet and P. Gribomont. *Logique, méthodes pour l'informatique fondamentale,* volume 1. Hermès, 1990.

[GG91]   S. Garland and J.V. Guttag. A Guide to LP, The Larch Prover. Technical Report 82, Digital Systems Research Center, 130 Lytton Av., Palo Alto, CA 94301, USA, 1991.

[Gir87a]   J.-Y. Girard. Linear logic. *Theoretical Computer Science*, 50:1–102, 1987.

[Gir87b]   J.-Y. Girard. *Proof Theory and Logical Complexity.* Bibliopolis, Napoli, 1987.

[Gir91]   J.-Y. Girard. A new constructive logic: classical logic. *Mathematical Structures in Computer Science*, 1:225–296, 1991.

[GLT89]   J.-Y. Girard, Y. Lafont, and P. Taylor. *Proofs and Types*, volume 7 of *Cambridge Tracts in Theoretical Computer Science*. Cambridge University Press, 1989.

[GM91]   P. Gardiner and C.C. Morgan. Data refinement of predicate transformers. *Theoretical Computer Science*, 87:143–162, 1991.

[GM93]   M.J.C. Gordon and T.F. Melham. *Introduction to HOL: A Theorem Proving Environment for Higher Order Logic.* Cambridge University Press, 1993.

[GM00]   J.A. Goguen and G. Malcolm, editors. *Software Engineering with OBJ: Algebraic Specification in Action.* Kluwer Academic Publishers, Boston, 2000. ISBN: 0-7923-7757-5.

[GMW79]   M.J.C. Gordon, R. Milner, and C.P. Wadsworth. *Edinburgh LCF: A Mechanised Logic of Computation,* volume 78 of *LNCS.* Springer-Verlag, 1979.

[Gor79]   M.J.C. Gordon. *The Denotational Description of Programming Languages.* Springer-Verlag, 1979.

[Gor88]   M.J.C. Gordon. HOL: A Proof Generating System for Higher-Order Logic. In C. Birtwistle and P.A. Subrahmanyam, editors, *VLSI Specification, Verification and Synthesis.* Kluwer Academic Publishers, 1988.

[Gri90]   T. Griffin. A formulae-as-types notion of control. In *Proc. 17th ACM Symp. on Principles of Programming Languages*. ACM, Orlando, 1990.

[Gri91]   S. Grigorieff. Décidabilité et complexité des théories logiques. In Courcelle [Cou91], pages 7–97.

[HA28]   D. Hilbert and W. Ackermann. *Grundzüge der theoretischen Logik*. Springer-Verlag, 1928.

[Hal60]   P.R. Halmos. *Naïve Set Theory*. Van Nostrand, Princeton, NJ, 1960.

[Hal93]   N. Halbwachs. *Synchronous Programming of Reactive Systems*. Kluwer Academic Publishers, 1993.

[HB95]   M.G. Hinchey and J.P. Bowen, editors. *Applications of Formal Methods*. International Series in Computer Science. Prentice-Hall, Hemel Hempstead, 1995.

[HB99]   M.G. Hinchey and J.P. Bowen, editors. *Industrial Strength Formal Methods in Practice*. FACIT Series. Springer-Verlag, London, 1999.

[HBG94]   R. Hänle, B. Beckert, and S. Gerberding. 3TAP, The Many Valued Theorem-Prover. Technical report, University of Karlsruhe, 1994.

[HC96]   B. Heyd and P. Crégut. A modular coding of Unity in Coq. In J. Grundy J. von Wright and J. Harrison, editors, *Theorem Proving in Higher Order Logic*, volume 1125 of *LNCS*, pages 251–266. Springer-Verlag, Turku, Finland, 1996.

[HK91]   I. Houston and S. King. *CICS project report, experiences and results from the use of Z in IBM*, volume 551 of *LNCS*. Springer-Verlag, 1991.

[HKPM02]   G. Huet, G. Kahn, and C. Paulin-Mohring. The Coq Proof Assistant, a Tutorial, V7.3. Technical report, INRIA Rocquencourt and CNRS-ENS Lyon, 1999–2002.

[HM85]   M. Hennessy and R. Milner. Algebraic laws for nondeterminism and concurrency. *Journal of the ACM*, 32:137–161, 1985.

[HO80]   G. Huet and D.C. Oppen. Equations and rewrite rules: A survey. In R. Book, editor, *Formal Language Theory: Perspectives and Open Problems*, pages 349–405. Academic Press, New York, 1980.

[Hoa69]   C.A.R. Hoare. An axiomatic basis for computer programming. *Communications of the ACM*, 12(10):576–580, 1969.

[Hoa85]   C.A.R. Hoare. *Communicating Sequential Processes*. Prentice Hall, 1985.

[Hoa89]   C.A.R. Hoare. Notes on an Approach to Category Theory for computer Scientists. In Broy [Bro89], pages 245–305.

[Hol97]   G.H. Holzmann. The model checker Spin. *IEEE Transactions on Software Engineering*, 23(5), 1997.

[HS86]   J.R. Hindley and J.P. Seldin. *Introduction to Combinators and λ-calculus*. Cambridge University Press, 1986.

[Hue90]   G. Huet, editor. *Logical Foundations of Functional Programming.* University of Texas at Austin Year of Programming Series. Addison Wesley, 1990.

[Hue92]   G. Huet. Constructive Computation Theory. In *École des jeunes chercheurs du GRECO de programmation du CNRS*, University of Bordeaux I, 1992.

[HW73]   C.A.R. Hoare and N. Wirth. An axiomatic definition of the programming language Pascal. *Acta Informatica*, 2(4):335–355, 1973.

[ILL75]   S. Igarishi, R.L. London, and D.C Luckham. Automatic program verification I: a logical basis and its implementation. *Acta Informatica*, 4:142–185, 1975.

[isoa]   International Organization for Standardization, Geneva. *Information Processing Systems - Open Systems Interconnection - A Formal Description technique based on Extended State Transition Model.* ISO/IEC 9074.

[isob]   International Organization for Standardization, Geneva. *Information Processing Systems - Open Systems Interconnection - A Formal Description technique based on the Temporal Ordering of Observationnal Behavior.* ISO/IEC 8807.

[isoc]   International Organization for Standardization, Geneva. *Information Processing Systems - Open Systems Interconnection - Guidelines for the Application of ESTELLE, LOTOS and SDL.* ISO/IEC TR 10167.

[JJLM91]   C.B. Jones, K.D. Jones, P.A. Lindsay, and R. Moore. *MURAL: A Formal Development Support System.* Springer-Verlag, 1991.

[JKKM92]   J-P. Jouannaud, C. Kirchner, H. Kirchner, and A. Mégrelis. Programming with Equalities, Subsorts, Overloading and Parameterization in OBJ. *Journal of Logic Programming*, 12(3):257–279, February 1992.

[Jon90]   C.B. Jones. *Systematic Software Development using VDM.* Prentice Hall, second edition, 1990.

[JRG92]   I. Jacobs and L. Rideau-Gallot. A Centaur Tutorial. RT 140, INRIA, Sophia Antipolis, July 1992.

[JS90]   C.B. Jones and R.C. Shaw. *Case Studies in Systematic Software Development.* Prentice Hall, 1990.

[Kah87]   G. Kahn. Natural Semantics. In *STACS'87*, volume 247 of *LNCS*. Springer-Verlag, March 1987.

[Kal90]   A. Kaldewaij. *Programming: The Derivation of Algorithms.* International Series in Computer Science. Prentice-Hall, 1990.

[KB70]   D.E. Knuth and P.B. Bendix. Simple Word Problems in Universal Algebra. In J. Leech, editor, *Computational Problems in Abstract Algebra*, pages 263–297. Pergamon Press, 1970.

[Kin69]   J.C. King. *A Program Verifier.* PhD thesis, Carnegie-Mellon University, 1969.

[KM01]   N. Klarlund and A. Møller. *MONA Version 1.4 User Manual.*
           BRICS Notes Series NS-01-1, Department of Computer Science,
           University of Aarhus, January 2001.

[KMP85]   G. Kahn, D.B. MacQueen, and G.D. Plotkin, editors. *Semantics
           of Data Types,* volume 173 of *LNCS.* Springer-Verlag, 1985.

[KP82]   L. Kirby and J. Paris. Accessible independence results for Peano
           arithmetic. *Bulletin of London Mathematical Society,* 14:285–293,
           1982.

[Kri93]   J.-L. Krivine. *Lambda-calculus, Types and Models.* Series in Com-
           puters and their Applications. Ellis Horwood, 1993.

[KZ95]   D. Kapur and H. Zhang. An Overview of Rewrite Rule Laboratory
           (RRL). *Journal of Computer and Mathematics with Applications,*
           29(2):91–114, 1995.

[Laf90]   Y. Lafont. Interaction nets. In *Proc. 17th ACM Symp. on Prin-
           ciples of Programming Languages,* pages 95–108, Orlando, 1990.
           ACM.

[Lal93]   R. Lalement. *Computation as Logic.* International Series in Com-
           puter Science. Prentice Hall, 1993.

[Lam94]   L. Lamport. The temporal logic of actions. *ACM Transactions on
           Programming Languages and Systems,* 16(3):872–923, May 1994.

[Lei91]   D. Leivant. A foundational delineation of computational feasibility.
           In *Proc. IEEE Symp. on Logic in Computer Science,* pages 2–11.
           IEEE, 1991.

[Les86]   P. Lescanne. REVE, a Rewrite Rule Laboratory. In J. Siek-
           mann, editor, *Proc. 8th Int. Conf. on Automated Deduction,* Lec-
           ture Notes in Computer Science, pages 696–697, Oxford (UK),
           1986. Springer-Verlag.

[LR98]   X. Leroy and F. Rouaix. Security properties of typed applets.
           In *Conference Record of POPL 98: The 25th ACM SIGPLAN-
           SIGACT Symposium on Principles of Programming Languages,
           San Diego, California,* pages 391–403, New York, NY, 1998.

[LS86]   J. Lambek and P. Scott. *Introduction to Higher Order Categorical
           Logic,* volume 7 of *Cambridge Studies in Advanced Mathematics.*
           Cambridge University Press, 1986.

[Mac71]   S. Mac Lane. *Category Theory for the Working Mathematician,*
           volume 5 of *Graduate Texts in Mathematics.* Springer-Verlag, 1971.

[McC60]   J. McCarthy. Recursive functions of symbolic expressions and their
           computation by machine. *Communications of the ACM,* 3(4):184–
           195, 1960.

[McC94]   W.W. McCune. Otter 3.0 reference manual and guide. Techni-
           cal report, Argonne National Laboratory, 9700 South Cass Avenue
           Argonne, Illinois 60439-4801, January 1994.

[McM93]   K.L. McMillan. *Symbolic Model Checking.* Kluwer Academic Pub-
           lishers, 1993.

[Mey88]   B. Meyer. *Object-oriented Software Construction*. Prentice Hall, 1988.

[Mey92]   B. Meyer. *Eiffel: The Language*. Prentice Hall, 1992.

[Mil87]   R. Milner. A proposal for Standard ML. In *ACM Conf. on Lisp and Functional Programming*, 1987.

[Mil89]   R. Milner. *Communication and Concurrency*. Prentice Hall, 1989.

[ML84]   P. Martin-Löf. *Intuitionistic Type Theory*. Bibliopolis, Napoli, 1984.

[Mor90]   C.C. Morgan. *Programming from Specification*. International Series in Computer Science. Prentice Hall, 1990.

[MT91]   R. Milner and M. Tofte. Co-induction in relational semantics. *Theoretical Computer Science*, 87:209–220, 1991.

[Mur91]   C. Murthy. An evaluation semantics for classical proofs. In *Proc. IEEE Symp. on Logic in Computer Science*. IEEE, 1991.

[NN92]   H.R. Nielson and F. Nielson. *Semantics with Applications. A Formal Introduction*. Wiley, 1992.

[NNGG89]   E. Nagel, J.R. Newman, K. Gödel, and J-Y. Girard. *Le théorème de Gödel*. Seuil, 1989.

[ORS92]   S. Owre, J.M. Rushby, and N. Shankar. PVS: a prototype verification system. In *11th Conf. on Automated Deduction (CADE)*, *LNAI 607*, pages 748–752. Springer-Verlag, 1992.

[ORS93]   S. Owre, J.M. Rushby, and N. Shankar. *The PVS Specification Language (Beta Release)*. Computer Science Laboratory, SRI International, 1993.

[Par95]   C. Parent. Synthesizing proofs from programs in the calculus of inductive constructions. In B. Möller, editor, *Proceedings 3rd Int. Conf. on Mathematics of Program Construction, MPC'95, Kloster Irsee, Germany, 17–21 July 1995*, volume 947, pages 351–379. Springer-Verlag, Berlin, 1995.

[Pau90]   L.C. Paulson. Isabelle: the 700 next theorem provers. In P. Odifreddi, editor, *Logic and Computer Science*, pages 361–386. Academic Press, 1990.

[Pau91]   L.C. Paulson. *ML for the Working Programmer*. Cambridge University Press, 1991.

[Pau92]   L.C. Paulson. Designing a theorem prover. In Abramsky et al. [AGM92b], pages 415–475.

[Pau93]   L.C. Paulson. Introduction to Isabelle. Technical Report 280, University of Cambridge, Computer Laboratory, 1993.

[Pau94]   L.C. Paulson. *Isabelle: A Generic Theorem Prover*, volume 828 of *LNCS*. Springer-Verlag, 1994.

[PH77]   J. Paris and L. Harrington. A mathematical incompleteness in Peano arithmetic. In Barwise [Bar77], chapter D.8.

[PJV01]   Benjamin C. Pierce, Trevor Jim, and Jerome Vouillon. Unison: A portable, cross-platform file synchronizer, 1999–2001. http://www.cis.upenn.edu/~bcpierce/unison.

[Plo81]   G.D. Plotkin. a Structural Approach to Operational Semantics. Technical Report DAIMI-FN-19, University of Aarhus, 1981.

[PM89]   C. Paulin-Mohring. *Extraction de programmes dans le calcul des constructions*. Thesis, Université de Paris VII, 1989.

[Pnu77]   A. Pnueli. The temporal logic of programs. In *Proc. 18th IEEE Symp. on Foundations of Computer Science (FOCS'77)*, pages 46–57, Providence, RI, USA, 1977.

[PST91]   B. Potter, J. Sinclair, and D. Till. *An Introduction to Formal Specification and Z*. International Series in Computer Science. Prentice Hall, 1991.

[QS82]   J.P. Queille and J. Sifakis. Specification and verification of concurrent systems in cesar. In *Proc. Int. Symp.on Programming*, volume 137 of *LNCS*, pages 337–351. Springer-Verlag, 1982.

[Rab77]   M.O. Rabin. Decidable theories. In Barwise [Bar77], chapter C.3.

[Rey85]   J.C. Reynolds. Polymorphism is not set-theoretic. In Kahn et al. [KMP85], pages 145–156.

[Rob65]   J.A. Robinson. A machine oriented logic based on the resolution principle. *Journal of the ACM*, 12(1):23–41, 1965.

[Rus93]   J.M. Rushby. Formal methods and the certification of critical systems. Technical Report CSL-93-7, SRI International, Menlo Park, 1993.

[RW69]   G.A. Robinson and L. Wos. Paramodulation and theorem proving in first order theories with equality. *Machine Intelligence*, 4:135–150, 1969.

[Saa97]   M. Saaltink. The Z/EVES system. In *ZUM '97: Z Formal Specification Notation. 11th International Conference of Z Users. Proceedings*, pages 72–85, Berlin, Germany, 3-4 1997. Springer-Verlag.

[Sch77]   K. Schütte. *Proof Theory*. Springer-Verlag, Berlin, 1977.

[Sch88]   D.A. Schmidt. *Denotational Semantics. A Methodology for Language Development*. Wm.C. Brown Publishers, Dubuque, Iowa, 1988.

[SDM92]   C. Da Silva, B. Dehbonei, and F. Mejia. Formal Specification in the Development of Industrial Applications: the Subway Speed Control Mechanism. In M. Diaz and R. Groz, editors, *FORTE'92*. North Holland, 1992.

[Set89]   R. Sethi. *Programming Languages: Concepts and Constructs*. Addison Wesley, 1989.

[Sho77]   J.R. Shoenfield. Axioms of set theory. In Barwise [Bar77], chapter B.1.

[Sho93]   J.R. Shoenfield. *Recursion Theory*, volume 1 of *Lecture Notes in Logic*. Springer-Verlag, 1993.

[Sif90]   J. Sifakis, editor. *Proc. 1st Int. Workshop on Automatic Verification Methods for Finite State Systems*, volume 407 of *Lecture Notes in Computer Science*. Springer-Verlag, 1990.

[SOR93a]   N. Shankar, S. Owre, and J.M. Rushby. A Tutorial on Specification and Verification Using PVS. In *Tutorial Material of FME'93*, pages 357–406b. IFAD, 1993.

[SOR93b]   N. Shankar, S. Owre, and J.M. Rushby. The PVS Proof Checker: a Reference Manual (Draft). Technical report, SRI, Menlo Park, CA, January 1993.

[Spi88]   J.M. Spivey. *Understanding Z: A Formal Language and its Formal Semantics*, volume 3 of *Cambridge Tracts in Theoretical Computer Science*. Cambridge University Press, 1988.

[Spi89]   J.M. Spivey. *The Z Notation: A Reference Manual*. International Series in Computer Science. Prentice Hall, 1989.

[Sti92]   C. Stirling. Modal and temporal logics. In Abramsky et al. [AGM92b], chapter 5, pages 477–563.

[Sto77]   J.E. Stoy. *Denotational Semantics: The Scott-Strachey Approach to Programming Language Theory*. MIT Press, 1977.

[Tak75]   G. Takeuti. *Proof Theory*, volume 81 of *Studies in Logic*. North Holland, Amsterdam, 1975.

[TBK92]   L. Théry, Y. Bertot, and G. Kahn. Real theorem provers deserve real user-interfaces. RR 1684, INRIA, Sophia-Antipolis, May 1992.

[Ter93]   D. Terrasse. Translation from Typol to Coq. In J. Despeyroux, editor, *Proc. of the Technical Workshop BRA on Proving Properties of Programming Languages*, INRIA, Sophia-Antipolis (France), September 1993.

[Tho91]   S. Thomson. *Type Theory and Functional Programming*. International Computer Science Series. Addison Wesley, 1991.

[TP02]   The Coq Development Team and LogiCal Project. The Coq Proof Assistant Reference Manual, V7.3. Technical report, INRIA, 1999–2002.

[TvD88]   A.S. Troelstra and D. van Dalen. *Constructivism in Mathematics: An Introduction I and II*, volume 121, 123 of *Studies in Logic and the Foundations of Mathematics*. North-Holland, Amsterdam, 1988.

[TVD00]   I. Toyn, S.H. Valentine, and D.A. Duffy. On Mutually Recursive Free Types in Z. In *ZB2000*, LNCS. Springer-Verlag, 2000.

[Var01]   M.Y. Vardi. Branching vs. linear time: Final showdown. In T. Margaria and W. Yi, editors, *Tools and Algorithms for the Construction and Analysis of Systems*, volume 2031 of *LNCS*, pages 1–22. Springer-Verlag, April 2001.

[vG90a]   A.J.M. van Gasteren. *On the Shape of Mathematical Arguments*, volume 445 of *LNCS*. Springer-Verlag, 1990.

[vG90b]   R.J. van Glabbeek. The linear time – branching time spectrum. In Baeten and Klop [BK90], pages 278–297.

[vH67]   J. van Heijenoort, editor. *From Frege to Gödel, a Source Book in Mathematical Logic, 1879-1931*. Harvard University Press, 1967.

[vL90a]   J. van Leeuwen, editor. *Handbook of Theoretical Computer Science*, volume A: Algorithms and Complexity. Elsevier, 1990.

[vL90b]   J. van Leeuwen, editor. *Handbook of Theoretical Computer Science*, volume B: Formal Models and Semantics. Elsevier, 1990.

[Wai91]   S.S. Wainer. Computability - Logical and Recursive Complexity. In Bauer [Bau91], pages 237–264.

[Wai93]   S.S. Wainer. Four Lectures on Primitive Recursion. In Bauer et al. [BBS93], pages 377–410.

[WL88]   J.C.P. Woodcock and M. Loomes. *Software Engineering Mathematics*. Pitman, 1988.

[Wor92]   J.B. Wordsworth. *Software Development with Z*. International Computer Science Series. Addison Wesley, 1992.

[WWD99]   J.M. Wing, J.C.P. Woodcock, and J. Davies, editors. *FM'99 – Formal Methods*, volume 1708-1709 of *LNCS*. Springer-Verlag, 1999.

Index

The numbers in the form p^n refer to footnote n on page p. The bold numbers refer to definitions.

Printed in the United Kingdom
by Lightning Source UK Ltd.
115799UKS00001B/110

9 781852 332471